Kev

KT-140-764

3814

Farming for the future

An introduction to
Low-External-Input and
Sustainable Agriculture

Coen Reijntjes, Bertus Haverkort and
Ann Waters-Bayer

LACKHAM LRC
5 FEB
WITHDRAWN

MACMILLAN

ILEIA

ILEIA, PO Box 64, NL-3830 AB Leusden, Netherlands

LACKHAM
COLLEGE
LIBRARY

Macmillan Education
Between Towns Road, Oxford OX4 3PP
A division of Macmillan Publishers Limited
Companies and representatives throughout the world

www.macmillan-africa.com

ISBN 0 333 57011 1

Text © ETC/ILEIA 1992
Design and illustration © Macmillan Publishers Limited 1992

First published 1992

All rights reserved; no part of this publication may be
reproduced, stored in a retrieval system, transmitted in any
form or by any means, electronic, mechanical, photocopying,
recording, or otherwise, without the prior written permission
of the Publishers.

Printed and bound in Malaysia

2006 2005 2004 2003 2002
13 12 11 10 9 8 7

The production of this book was financed by the
Ministry of Development Cooperation of the
Netherlands.

Cover photograph courtesy of Panos Pictures

Contents

Part III: Linking farmers and scientists in developing LEISA technologies

Appendices   163

List of tables

List of figures

List of boxes

Abbreviations and acronyms

This list contains only those acronyms mentioned in the main text of this book. Further acronyms mentioned in the references and appendices are explained in Appendix C4.

AME	Agriculture, Man and Ecology Programme
ATIP	Agricultural Technology Improvement Programme
B.t.	*Bacillus thuringiensis*
BIG	bio-intensive gardening
CDTF	Community Development Trust Fund
CGIAR	Consultative Group on International Agricultural Research
CIDICCO	International Cover Crop Clearing House
CIMMYT	Centro Internacional de Mejoramiento de Maíz y Trigo (International Maize and Wheat Improvement Center)
CSIRO	Commonwealth Scientific Industrial and Research Organization
CTA	Technical Centre for Agricultural and Rural Co-operation
ELCI	Environment Liaison Centre International
ENDA	Environnement et Développement du Tiers-Monde (Environment and Development Activities in the Third World)
FAO	Food and Agriculture Organization of the United Nations
FASE	Federação de Orgãos para Assistência Social e Educacional
FSR	Farming Systems Research
GRAAP	Groupe de Recherche et d'Appui pour l'Autopromotion Paysanne
HEIA	high-external-input agriculture
HYV	high-yielding variety
IARCs	international agricultural research centres
IBSRAM	International Board of Soil Research and Management
ICLARM	International Centre for Living Aquatic Resources Management
ICRISAT	International Crops Research Institute for the Semi-Arid Tropics
IIRR	International Institute of Rural Reconstruction
IITA	International Institute of Tropical Agriculture
IK	indigenous knowledge
ILCA	International Livestock Centre for Africa
ILEIA	Information Centre for Low-External-Input and Sustainable Agriculture
INADES	Institut Africain Développement Economique
IPM	Integrated Pest Management
IRRI	International Rice Research Institute
KENGO	Kenya Energy and Environment Organization
KWDP	Kenya Woodfuel Development Programme
LEIA	low-external-input agriculture
LEISA	low-external-input and sustainable agriculture
NARSs	national agricultural research services

NCSU	North Carolina State University
NGO	nongovernmental organisation
OTA	Office of Technology Assessment
PAF	Project Agro-Forestier
PRATEC	Proyecto Andino de Technologías Campesinas
PRONAT	Protection Naturelle
PTD	Participatory (or People-centred) Technology Development
R&D	research and development
RRA	Rapid Rural Appraisal
SALT	Sloping Agricultural Land Technology
SATIS	Socially Appropriate Technology International Information Services
sp/spp	species (singular/plural)
TAC	Technical Advisory Committee
ToT	transfer of technology
UNESCAP	United Nations Economic and Social Commission for Asia and the Pacific
UNESCO	United Nations Educational, Scientific and Cultural Organization
UPLB	University of the Philippines at Los Baños
WCED	World Commission on Environment and Development
WN	World Neighbors

Foreword

This is a book about farming. But, more than that, it is a book about farmers, about men and women farmers. It is seldom that farmers – particularly those in the Third World – have an opportunity to make themselves heard. In this book, an effort has been made to give as much room as possible to the knowledge and experience of small-scale farmers in developing countries. In recent years, there has been an enormous increase in studies about small-scale farming and its potential for development, but I have the strong feeling that most of these studies are only meant to be food for discussion among experts.

This is also a book about sustainability. Achieving and maintaining sustainable agriculture has become one of the focal points, not only within Dutch agricultural and environmental policies, but also within those of the international development community. Until now, agricultural policies – whether oriented toward export production or local food production – have focused too narrowly on maximising short-term profits rather than on long-term sustainable management of local resources by farmers. Although this is understandable from the point of view of policy makers confronted with questions of food security, employment, foreign exchange and population growth, it does not take  into account sufficiently the interests of individual farmers and rural communities and does not lead to their empowerment.

The numerous examples in this book show that the aim of production growth should coincide with sustainable resource management, if long-term well-being has the same priority as immediate needs. Moreover, the book quite clearly shows that sustainable agriculture can be realised only through the individual and collective activities of farmers and communities pursuing their own strategies to secure their livelihoods.

The lessons we can learn from the wealth of information in this book are highly valuable, not only for those concerned with Dutch development cooperation, but also for anyone interested in the development of Third World agriculture.

Jan Pronk
Minister of Development Cooperation, the Netherlands.

Acknowledgements

This volume is the result of eight years of work by the Information Centre for Low-External-Input and Sustainable Agriculture (ILEIA) in the Netherlands. Besides the rapidly growing number of official and 'grey' publications on low-external-input and sustainable agriculture (LEISA) and participatory technology development (PTD), the main sources of information and inspiration have been the experiences in developing LEISA gained by the innovative farmers, fieldworkers and scientists among the 4000 members of the ILEIA network. These experiences have been reviewed and analysed by ILEIA staff members.

The book has been produced under the direction of ILEIA as a joint project of the ETC Foundation (the 'mother' organisation of ILEIA) and the Agricultural University of Wageningen, specifically the Departments of General Agronomy and Extension Science. In addition, a number of staff members from other departments of the University and independent professionals have provided us with information about and interpretation of LEISA experiences and the agroecological principles behind them.

The information was compiled and analysed with the support of an editorial board composed of Hans Van Asseldonk and Niels Röling from the Agricultural University of Wageningen and Adriaan Ferf and Hay Sorée from the ETC Foundation. Invited contributors were Ton Groosman, Inge van Halden, Janice Jiggins, Anita Linnemann, Kees Manintveld, Niels Röling, Gerard Oomen, Hans Schiere, Vincent Seewald, Erik van der Werf and Henk de Zeeuw. Critical reviews and additions were made by Wolfgang Bayer, Henk Breman, Arnoud Budelman, Roland Bunch, John Farrington, David Gibbon, Anton Haverkort, Albin Korem, Clive Lightfoot, Ken McKay, David Millar, Meine van Noordwijk, Rudi Rabbinge, Angel Roldan and Freerk Wiersum. Editorial assistance was given by Sibylle Pich, while Annemiek de Haan and Wim Hiemstra assisted in designing the book and obtaining illustrations.

We would like to express our sincere appreciation to:

- the Ministry of Development Cooperation of the Netherlands for its confidence in ILEIA and for funding the production of this book;

- the many contributors, the editorial board and the external reviewers who persevered in the struggle to clarify the structure, format and content of this book over a period of several years; and

- above all, the members of the ILEIA network who have shared their experiences with us: farm women and their male colleagues, researchers, extensionists and other fieldworkers in nongovernmental and governmental organisations.

The authors and publishers would also like to thank Mr M.R. Wake and Professor H.D. Tindall for valuable comments and advice given during the preparation of this book.

ILEIA

A network of persons and organisations joining hands in the search for Low-External-Input and Sustainable Agriculture

ILEIA was established as the Information Centre for Low-External-Input and Sustainable Agriculture by the ETC Foundation in 1982, and has been funded mainly by the Dutch Ministry of Development Cooperation.

ILEIA's long-term objective is to contribute to a situation in which Low-External-Input and Sustainable Agriculture (LEISA) is:

● widely accepted and adopted as a valid approach to agricultural development, particularly in areas not suited to using high levels of external inputs;

● recognised as a means of combining locally available resources and local knowledge with judicious use of external inputs;

● valued as a useful perspective in planning and implementing agricultural research, education and extension;

● developed and consolidated with respect to its stock of knowledge and scientific basis.

ILEIA seeks to reach these objectives by:

● **networking**, i.e. facilitating communication between organisations and individuals about their experiences, problems, questions and information concerning LEISA;

● **regionalisation**, i.e. supporting the establishment of regional LEISA networks in the tropics, and associated small libraries and information/documentation centres;

● **international workshops**, to which a limited number of key persons are invited to exchange experiences about and seek a better understanding of certain aspects of LEISA, and to draw up plans for future action;

● **documentation** of existing experiences with LEISA and Participatory Technology Development (PTD) and of relevant research findings; some 4000 documents in the ILEIA library have been classified according to SATIS, and on-line computer communication linkages give access to further documents;

● **publication** of the quarterly *ILEIA Newsletter* as a means of mobilising and sharing experiences within the ILEIA network, with each issue focused on a special theme, e.g. alternatives to chemical fertilisers, agroforestry, complementary use of external inputs. Further publications include workshop proceedings, readers, bibliographies and guides to sources of information.

Introduction

Millions of smallholders in the tropics are farming under rainfed conditions in diverse and risk-prone environments. In a constant struggle to survive, farm communities have developed innumerable ways of obtaining food and fibre from plants and animals. A wide range of different farming systems have evolved, each adapted to the local ecological conditions and inextricably entwined with the local culture. A closer look at these 'traditional' farming systems reveals that they are not static: they have changed over the generations – and particularly quickly over the last few decades – primarily as a result of the research and development activities of the local people. These activities have not only been in response to external pressures; they are also an expression of local creativity.

However, increasingly rapid changes in economic, technological and demographic conditions demand increasingly rapid changes in smallholder farming systems. New market opportunities, promotion of chemical inputs, and financial constraints may lead or force farmers to seek short-term profits and pay less attention to keeping their agriculture in balance with the ecological conditions.

Conventional science-based research and extension activities in the tropics have focused on 'modern' agriculture with high levels of external inputs, e.g. agrochemicals, hybrid seed, fuel-based mechanisation. Technologies have been developed on research stations and experimental farms in better-endowed areas, and attempts have been made to transfer ready-made technology packages to farmers. The primary aim of these efforts has been to increase production of certain commodities, e.g. rice, maize, wheat. These research and extension activities have contributed to an overall increase in world food production but have brought little benefit to the majority of smallholders, and have sometimes worsened their situation by forcing them onto more marginal land while capital-intensive cropping and ranching expands over the better land.

In recent years, the negative environmental and social impacts of high-external-input agriculture (HEIA) have become increasingly obvious. At the same time, many disadvantaged communities of smallholders are being forced to exploit the resources available to them so intensively that, here too, environmental degradation is setting in. Alarmed development planners and donors are desperately seeking new approaches to agricultural development which will benefit smallholders, halt degradation of natural resources and, if possible, improve these resources. The call for sustainable agriculture is increasing in volume.

WCED laid out the preconditions and broad outlines for achieving sustainable agriculture. The responsibility for elaborating concepts and taking practical steps lies with the national governments international and national institutions, private development organisations and, not least, the farmers themselves. With its 'call for action', WCED summoned all organisations and individuals to take this responsibility.

A call for sustainable agriculture

In 1987 the World Commission on Environment and Development (WCED) called attention to the immense problems and challenges facing world agriculture, if present and future food needs are to be met, and to the need for a new approach to agricultural development:

"In the remaining years of this century, about 1.3 billion people will be added to the human family . . . the global food system must be managed to increase food production by 3 to 4 per cent yearly.

Global food security depends not only on raising global production, but on reducing distortions in the structure of the world food market and on shifting the focus of food production to food-deficit countries, regions, and households . . . This shift in agricultural production will be sustainable only if the resource base is . . . sustained, enhanced and, where it has been diminished or destroyed, restored" (WCED 1987, pp. 128–30).

"The agricultural systems that have been built up over the past few decades have contributed greatly to the alleviation of hunger and the raising of living standards. They have served their purposes up to a point. But they were built for the purposes of a smaller, more fragmented world. New realities reveal their inherent contradictions. These realities require agricultural systems that focus as much attention on people as they do on technology, as much on resources as on production, as much on the long term as on the short term. Only such systems can meet the challenge of the future" (WCED 1987, p. 144).

In this spirit, we – the many people who have contributed to this book – try to illuminate a path towards sustainable agriculture.

Why this book has been written

But how can this call be translated into action? Smallholders in the tropics need appropriate strategies and techniques which will lead to sufficient and reliable yields but will not deplete the resource base upon which they depend. To help smallholders make their farming systems more productive and sustainable, development workers – in their turn – need appropriate strategies and techniques of working with these farmers. **How can development workers assist smallholders in their constant endeavours to adapt their agriculture to changing conditions?** This is the central concern of this book.

Although such activities as basic agroecological research and establishing equitable international trade relations have important roles to play in creating the conditions for sustainable agriculture, strengthening farmers' capacity to develop and manage technology is of paramount importance for the actual creation of sustainable farming systems. We define 'technology' as a specific combination of knowledge, productive resources, inputs and services which are applied systematically to produce desired outputs. The definition encompasses both 'hardware' (tools, equipment, seed, buildings, etc.), i.e. physical forms of technology; and 'software' (methods, practices and strategies, including forms of social organisation). Technology development is a complicated process, involving deliberate activities to generate, transform, combine, test and adjust new hardware and software.

Innovations such as combining chemical and organic fertilisers, appropriate forms of green manuring or integrating new crops could open new doors for farm households. Transferring knowledge about these technical options and combining the forces of farmers, field-workers and scientists in discovering the opportunities and limitations of these options definitely play a role in sustainable agricultural development. However, in many areas, especially where farmers depend mainly on local resources, modern technologies may not be the first option to improve agriculture. In such areas, better use of local resources and natural processes could make farming more effective and create conditions for efficient, profitable and safe use of modern inputs. Improving the insight of farmers and development agents into the ecological principles behind farming and improving their knowledge of the available technical options is an important step in the process of strengthening farmers' capacity to develop and manage technology for sustainable agriculture.

The focus in this book is on farmers who presently operate with low levels of external inputs, either because they are not available or because they are too costly. Our intention is to provide background theory, practical ideas and sources of further information for persons and organisations who are working together with such farmers in trying to solve technical problems and open up potential at the farm level. The solutions to farmers' problems will be as diverse, complex and site-specific as their farming systems, but the principles involved in finding the solutions will be of wider validity.

Our geographical focus is on the tropics, although many of the principles will also hold in other parts of the world. In any case, we hope that agricultural research and development in the industrialised countries will be able to learn from the knowledge and experience gained in the developing countries and will also be able to make their own agriculture more sustainable.

This book is written primarily for middle-level agricultural development staff in extension, research and training in and for the Third World, e.g. district officers, coordinators of field staff, field researchers, trainers of extension workers. It should also be of interest to teachers and students in universities and agricultural colleges, scientists in research institutes, development planners and administrators, and donors of agricultural development projects and programmes. It is meant to help these readers reflect on the way they are presently trying to develop or transfer agricultural technology and to stimulate them to adjust – if necessary – their work so that it contributes more to the emergence of sustainable forms of agriculture.

How this book is structured

The central concepts of this book are Low-External-Input and Sustainable Agriculture (LEISA) and Participatory Technology Development (PTD).

LEISA is agriculture which makes optimal use of locally available natural and human resources (such as soil, water, vegetation, local plants and animals, and human labour, knowledge and skills) and which is economically feasible, ecologically sound, culturally adapted and socially just. The use of external inputs is not excluded but is seen as complementary to the use of local resources and has to meet the above-mentioned criteria. Neither the conventional Western agricultural technology nor any alternative technology is completely embraced or condemned. The attempt is made, rather, to draw lessons from past experiences in agriculture in industrialised and developing countries and to merge them into a process of technology development which leads to LEISA.

PTD is a path to LEISA. It is a process of creative interaction within rural communities, in which indigenous and scientific knowledge are combined in order to find solutions to farmers' problems and to take the fullest possible advantage of local opportunities. It involves collaboration of farmers and development agents in analysing the local agroecological system, defining local problems and priorities, experimenting with various potential solutions, evaluating the results and communicating the findings to other farmers.

The first part of the book provides background information about the need for sustainable agriculture, and draws attention to the central role played by farmers in achieving it.

In Chapter 1, the concept of sustainable agriculture is introduced and compared with the past record of agricultural development. Trends toward two extremes in tropical agriculture are distinguished:

● Excessive use of external inputs, leading to environmental degradation and depletion of nonrenewable resources.

● Erosive forms of low-external-input agriculture, with the result that the natural resources can no longer support the local people.

Limitations of using artificial external inputs and pursuing a Transfer-of-Technology (ToT) approach in smallholder rainfed farming are outlined briefly, to demonstrate that another type of technology and another approach to technology development are necessary. A potential alternative to high-external-input agriculture (HEIA) and ToT is proposed: combining the insights of agroecological science with the knowledge and practices of local farmers.

The focus in Chapter 2 is on the people making the day-to-day decisions about using resources for agriculture: the farmers. The farm family is part of the farm system but also actively manages it. The site-specific characteristics of farm systems – both the biophysical and the human aspects – are discussed. Against this background, we look at decision-making by farm households in their constant efforts to balance the various objectives they seek.

In Chapter 3, a closer look is taken at the dynamics of 'traditional' farming systems and the potential and limitations of farmers' experimentation and innovation in the process of developing sustainable forms of agriculture.

Part II draws from scientific agroecological findings to give the theoretical background of LEISA. In Chapter 4, some basic concepts of agroecology are introduced. Chapter 5 is devoted to important principles upon which productive and site-appropriate forms of low-external-input farming can be based. In Chapter 6, possibilities for developing LEISA systems in smallholder farming in the tropics are explored, with particular attention being given to the relationship between farm characteristics, technology choice and sustainability.

Part III draws from field experiences in developing smallholder agriculture to show how the process of technology development by farmers can be linked with the insights of agroecological science in a participatory approach to development which strengthens farmers' innovative capacity and complements other methods of technology development. Chapter 7 introduces the various actors involved and indicates the potential of PTD for developing sustainable forms of agriculture in rainfed areas. Some ways in which the PTD path to LEISA could be promoted within the 'mainstream' of agricultural research and extension are discussed.

Chapter 8 is practice-oriented. Here, the major types of activities in the PTD process are outlined and examples of methods which have been developed and tried in the field are presented. Given the enormity of the problems facing smallholders in the tropics, the impending and actual degradation of world resources and the promising nature of the PTD approach in developing LEISA systems, we feel it is important that these initial experiences be made as widely known as possible, so that similar action can be taken quickly with and by farmers throughout the world.

The rather extensive appendices are intended to provide some technical information as well as further sources of information, in order to support fieldworkers and farmers in their combined efforts. Appendix A presents a selection of some technical options for LEISA development. These are not successes proven beyond the shadow of a doubt but rather promising methods and technologies which have been developed for specific sites and which may inspire PTD practitioners in other areas to develop or adapt methods and technologies suited to their conditions.

A glossary of key terms used in the discussion of LEISA can be found in Appendix B, and sources of further information are indicated in Appendix C. Publications are listed from which readers can obtain more detailed information about sustainable agriculture, indigenous knowledge and farming systems, and agroecology, as well as about specific experiences or methods of PTD and specific technologies and systems of LEISA. Finally, the addresses are given of organisations concerned with sustainable agriculture, many of which issue relevant periodicals or other publications.

How this book could be used

If you are not familiar with the concepts of LEISA and PTD, then read the case studies given in Section 7.3, before tackling the theoretical Parts I and II.

If you are working in the field, you might find the best point of entry by reading examples of practical approaches and technologies related to the type of work you are presently doing or to the type of farming with which you are concerned. These examples should be understood not as methods or techniques to copy but rather as stimuli for thinking about how the principles could be applied in your situation and how selected methods and techniques could be modified to suit your situation.

The subject index will help you find both theoretical background information and practical examples related to the topics of primary interest to you. These will inevitably be related to several other topics, as LEISA and PTD are not disciplinary in nature. To be able to understand how 'your' topic relates to the others, you are advised to read the entire section in which it appears.

If you would like to use this book as a basis for training, remember that it is meant as a general introduction, not as a training manual. We recommend that you prepare training units which include extracts from this book referring to the principles on which you would like to focus, and illustrate them with examples from your own knowledge and environment. Some of the material in this book could serve as case studies, with the aid of which, general processes can be illustrated and parallels with the local case can be identified. Students could be encouraged to document indigenous farming practices, farmer innovation and communication along the lines of examples given in this book.

In the appendices, names and addresses of organisations or individuals who have applied the described methods and techniques are given, so that you can contact them directly for further information. We encourage you to make contact with organisations working in your area and to explore the possibilities of exchanging ideas and experiences directly with them for your mutual benefit. The appropriate expert for your needs may be just around the corner.

How this book could be continued

Thus far, the PTD approach to developing LEISA has been taken mainly by individuals and small groups of highly motivated researchers, extensionists and other development workers, often in nongovernmental organisations or in projects. If LEISA is to be applied more widely, this approach must be institutionalised within research and extension services of national and international agencies. The examples from field experiences presented in this book give some indication of the widespread potential of this approach, and thus offer support and encouragement to individuals within public agencies who are seeking ways out of the current dilemma of agricultural development. The general PTD approach presented here requires, of course, interpretation and adaptation by readers in the light of specific ecological, economic, political and sociocultural conditions where they live and work.

The ultimate aim of this book is to stimulate further research and development work in the field of LEISA. To this end, it is important that the present gaps in agroecological knowledge and the present weaknesses in the PTD approach – including the gaps and weaknesses in this book about LEISA and PTD – be identified and candidly discussed.

We invite you to send us your comments and to keep us informed about your experiences with LEISA and PTD, about the methods you have tried and how they worked, about problems encountered and how you tried to deal with them. In this book, we can present only first indications of a path to sustainable agriculture and discuss the potentials and problems known to us thus far. Much more experience and validation are needed. After we have received further information and reflections from you, we hope to be able to assemble a book which is further along the path to LEISA.

Please send your comments and information to:

ILEIA Farming for the Future
PO Box 64
NL-3830 AB Leusden
Netherlands.

Low-External-Input and Sustainable Agriculture (LEISA): an emerging option

1 Agriculture and sustainability

1.1 The concept of sustainable agriculture

The word 'sustainability' is now widely used in development circles. But what does it really mean? According to a dictionary definition, 'sustainability' refers to 'keeping an effort going continuously, the ability to last out and keep from falling'. In the context of agriculture, 'sustainability' basically refers to the capacity to remain productive while maintaining the resource base. For example, the Technical Advisory Committee of the Consultative Group on International Agricultural Research (TAC/CGIAR 1988) states: "Sustainable agriculture is the successful management of resources for agriculture to satisfy changing human needs while maintaining or enhancing the quality of the environment and conserving natural resources."

However, many people use a wider definition, judging agriculture to be sustainable if it is (after Gips 1986):

- **Ecologically sound,** which means that the quality of natural resources is maintained and the vitality of the entire agroecosystem – from humans, crops and animals to soil organisms – is enhanced. This is best ensured when the soil is managed and the health of crops, animals and people is maintained through biological processes (self-regulation). Local resources are used in a way that minimises losses of nutrients, biomass and energy, and avoids pollution. Emphasis is on the use of renewable resources.

- **Economically viable,** which means that farmers can produce enough for self-sufficiency and/or income, and gain sufficient returns to warrant the labour and costs involved. Economic viability is measured not only in terms of direct farm produce (yield) but also in terms of functions such as conserving resources and minimising risks.

- **Socially just,** which means that resources and power are distributed in such a way that the basic needs of all members of society are met and their rights to land use, adequate capital, technical assistance and market opportunities are assured. All people have the opportunity to participate in decision-making, in the field and in the society. Social unrest can threaten the entire social system, including its agriculture.

- **Humane,** which means that all forms of life (plant, animal, human) are respected. The fundamental dignity of all human beings is recognised, and relationships and institutions incorporate such basic human values as trust, honesty, self-respect, cooperation and compassion. The cultural and spiritual integrity of the society is preserved and nurtured.

- **Adaptable,** which means that rural communities are capable of adjusting to the constantly changing conditions for farming:

population growth, policies, market demand etc. This involves not only the development of new, appropriate technologies but also innovations in social and cultural terms.

These different criteria of sustainability may conflict and can be seen from different viewpoints: those of the farmer, the community, the nation and the world. There may be conflicts between present and future needs; between satisfying immediate needs and conserving the resource base. The farmer may seek high income through high prices for farm products; the national government may give priority to sufficient food at prices which the urban population can afford. Choices must continually be made in a never-ending search for balance between the conflicting interests. Therefore, well-functioning institutions and well-deliberated policies are needed on all levels – from village to global – in order to ensure sustainable development.

An intensive and highly diversified land-use system near Kisii, Kenya, which may be reaching its upper limits in terms of sustainable production. (Chris Pennarts, Studio 3)

In agricultural development, raising production is often given primary attention. But there is an upper limit to the productivity of ecosystems. If this is exceeded, an ecosystem will degrade and may eventually collapse, and fewer people will be able to survive on the remaining resources than before. This implies that, when the limits on the supply side are reached, something has to be done on the demand side, e.g. other sources of income, emigration, lower consumption level, population control. Production and consumption have to be brought into balance on an ecologically sustainable level. Although sustainability must be seen as a dynamic concept which allows for the changing needs of an increasing global population (TAC/CGIAR 1988), basic ecological principles oblige us to recognise that agricultural productivity has finite limits.

Why has the concept of sustainability gained increasing importance with reference to agricultural development? This becomes evident if we take a look at the present situation of world agriculture.

1.2 World agriculture: the record to date

FAO figures compiled by Alexandratos (1988) about global and national agricultural achievements and problems may not be very exact in their details, but they do indicate some basic trends. They refer primarily to economic and ecological aspects.

Economic aspects

The performance of agriculture can be partially assessed by comparing the production of food, fibre and fuelwood with the need for these products within a given region or country, and by comparing the growth rate of agricultural production with the rate of population growth. According to Alexandratos (1988), from 1961 to 1985:

● Food consumption by the majority of people of a substantially enlarged world population has increased. On a global average, yields of major food crops have risen impressively: by 41% in rice, by 45% in maize and by 70% in wheat. In Asia and Latin America, growth rates of total and per capita food production have been positive (Table 1.1).

Table 1.1 Growth rates in food production 1970 – 1985 (% per annum)

	Total production	Per capita production
Africa (sub-Saharan)	1.7	− 1.3
Near East and North Africa	2.9	0.2
Asia	3.7	3.0
Latin America	3.1	2.7

Source: Alexandratos (1988).

- Although the average self-sufficiency ratio of the developing countries (without China) stayed above 100, there was a decline from 110 in 1961 to 101 in 1985. In 1985 self-sufficiency was below 100% in 48 countries. Only 19 countries managed to raise their self-sufficiency ratios. In nutritional terms, many low-income countries are no better off and some are worse off now than 20 years ago.

- In large parts of Africa, but also in parts of Latin America and Asia, production per unit area of traditional crops, e.g. millet and sorghum, declined, partly on account of soil depletion and degradation but also because of political instability.

- In 1980 an estimated 780 million people in the Third World (without China) were living in absolute poverty. Of these, 90% were rural people wholly or partly dependent on agriculture. An estimated 30 million rural households were landless and 138 million were almost landless.

Ecological aspects

According to FAO, the environmental problems of developing countries are largely due to overexploitation of land, extension of cropping and deforestation (Alexandratos 1988). Some large irrigated areas are seriously affected by salinisation. Increased use of pesticides and artificial fertilisers is also causing environmental problems. Particularly the degradation of soil fertility and the scarcity of fuelwood indicate the graveness of the situation. Referring to the nondesert areas, 43% of Africa, 32% of Asia and 19% of Latin America is at risk of desertification (FAO 1984a). Forty-two developing countries lack sufficient fuelwood in part or all of their territories and can meet fuelwood needs only by depleting tree stocks; 27 countries face such acute scarcity that even overcutting would not supply their needs. In 1980 more than a billion people had a deficit supply of fuelwood and over 110 million suffered an acute scarcity (Alexandratos 1988). There is a great overlap of the areas at risk of desertification and those deficient in fuelwood. These areas also largely coincide with the countries and regions which have major difficulties in feeding their population (FAO 1984a).

According to more recent global data given by the Worldwatch Institute (Brown 1988), per capita consumption of grain increased by nearly 40% between 1950 and 1984 thanks to a 2.6-fold increase in grain production, but from 1984 to 1988 the output per person fell by 14%. In several populous countries, including China, India, Indonesia and Mexico, agricultural production stagnated whereas populations continued to grow. Brown concludes that the growth in world production after grain prices doubled in 1973 was achieved partly by ploughing highly erodible land and partly by drawing down water tables through overpumping for irrigation.

Farmers can overplough and overpump with impressive results in the short run, but the short run is running to a close. As marginal lands brought under the plough during the boom years of the 1970s become exhausted and as irrigated areas shrink because of falling water tables in key food-producing areas, the growth in world food production is slowing down.

Box 1.1
The degradation threat: loss of rainfed cropland

Without conservation measures on rainfed land, soil erosion or loss by wind or water, salinisation or alkalinisation, depletion of plant nutrients and organic matter, deterioration of soil structure and pollution will lead to the loss of 544 million ha of cropland: losses of 10% in South America, 16.5% in Africa, 20% in Southwest Asia, 30% in Central America and 36% in Southeast Asia. Much of the remaining land will also lose fertility due to loss of topsoil. The total loss in productivity on rainfed cropland will amount to a daunting 29% (FAO 1984a).

Box 1.2
Traditional farming systems that could not adapt

Recent research suggests that deforestation, overgrazing, waterlogging and salinisation in Mesopotamia's fertile crescent were as much the immediate cause for the collapse of its Old Testament societies as conquest by outside invaders. The Mesopotamians were able to bring more land into agricultural use, as well as boost the yields per acre, by developing an elaborate system of irrigation canals. But, by failing to provide adequate drainage systems, the build-up of toxic materials and salts in the soil gradually poisoned it for cropping; and by failing to arrest silting in the slow-moving canals, the dependable supply of water was jeopardised.

One of the main centres of Mayan civilisation seems to have failed after centuries of expansion in Guatemala, because the demands of its growing population depleted the soil through erosion. The shifting sands of Libya and Algeria also testify to the failure of Roman agriculturalists to husband the granary which supplied much of the Roman Empire with food (Douglass 1984).

1.3   Trends in tropical agriculture

Such global and national production figures hide differences between regions and between different types of farming systems within a country. A closer look at the situation of tropical agriculture reveals that change has taken two main paths, to be outlined below. But let us start with the point of origin: traditional agriculture.

Originally, agriculture in the tropics depended on local natural resources, knowledge, skills and institutions. Diverse, site-specific farming systems evolved out of a long process of trial and error in which balances were found between the human society and its resource base. In most cases, production was oriented primarily to the subsistence of the family and the community. Modes of cooperation between community members were highly developed.

Traditional farming systems continued to develop in a constant interaction with local culture and local ecology. As conditions for farming changed, e.g. because of population growth or the influence of foreign values, the farming system was also changed. Where adaptation to the new pressures was not fast enough, the natural resource base was eventually destroyed – as was the society depending upon it. Many farming societies disintegrated because a lack of local capacity to manage change led to severe environmental degradation, e.g. in Mesopotamia along the Tigris and Euphrates Rivers, the ancient Mayan culture in Central America and the ancient Mediterranean civilisations of Phoenicia, Palestine, Egypt, Greece and Rome (Lawton & Wilke 1979, Weiskel 1989; see also Box 1.2).

Many traditional farming systems were sustainable for centuries in terms of their ability to maintain a continuing, stable level of production (TAC/CGIAR 1988). However, these systems have had to cope with particularly rapid changes during and since the colonial period: introduction of foreign education and technology in agriculture and

The creeping desert: olive orchards which thrived in northern Tunisia in the time of the ancient Roman Empire are now being conquered by sand. (Wolfgang Bayer)

Box 1.3
***High-external-input
agriculture (HEIA) in India***

In the State of Punjab, the spread of modern varieties, irrigation, fertilisers and agricultural machinery brought dynamic growth in agriculture, but it also created many problems. The increased area under paddy led to a fall in the water table, and farmers must bear extra costs of pumping up underground water. Continuous puddling led to formation of an impervious layer of soil which prevents uptake of water and nutrients from deeper layers. Farmers must therefore apply more fertilisers. This has disturbed the nutrient balance in the soil and led to a deficiency in micronutrients. The predominant wheat-paddy rotation creates conditions congenial for pests and diseases to multiply. Malaria is on the increase again. Because there is now heavy use of pesticides to protect crops, grains are contaminated with their residues and pose health hazards.

With the new technology, production has expanded rapidly, but the market facilities cannot cope with the larger surpluses. Farmers have to wait for days to dispose of their produce. The availability of liberal credit has led to high capital formation despite the small size of holdings. This has resulted in underutilisation of the capital stock. The excess capacity created has inflated the cost of production. The prices of machinery and other inputs are rising steeply as compared to output prices. The terms of trade are becoming unfavourable to agriculture and the gains of the new technology are dwindling over time (Singh Hara 1989).

health care; increased population pressure; changes in social and political relations; and incorporation into an externally-controlled international market system. Originally subsistence-oriented systems have become increasingly market-oriented, and improved communication has increased the demand for consumer goods.

In response to foreign influence and the needs and growing aspirations of increasing numbers of people, farming systems in the tropics tend to change toward one of two extremes:

- Excessive use of external inputs; referred to here as high-external-input agriculture (HEIA).

- Intensified use of local resources with few or no external inputs, to the extent that the natural resources are degraded; referred to here as low-external-input agriculture (LEIA).

Excessive use of external inputs (HEIA)

High-external-input agriculture depends heavily on artificial chemical inputs (e.g. fertilisers, pesticides), hybrid seed, mechanisation based on fossil fuels and often also irrigation. This type of agriculture is consuming non-renewable resources such as oil and phosphates at an alarming rate. It is capital-intensive and highly market-oriented. The cash needed to buy the inputs is often obtained by selling farm products. HEIA is possible only where ecological conditions are relatively uniform and can be easily controlled (e.g. in irrigated areas) and where delivery, extension, marketing·and transport services are good. Increased needs for agricultural products and the development of new varieties of maize, wheat, rice and other commercial crops made the introduction of HEIA technology appear attractive. HEIA is found in the 'resource-rich', 'high-potential' areas of developing countries, and is most widespread in Asia.

However, the excessive and unbalanced use of artificial inputs in HEIA can have serious ecological, economic and sociopolitical repercussions. For example, Box 1.3 describes the problems now being faced in one of the most highly 'developed' agricultural areas of India. The introduction of HEIA under the banner of the 'green revolution' channelled scarce investment resources into capital-intensive agriculture in limited areas, which became dependent on imported machinery, equipment, seeds and other inputs. The built-in bias towards inequality between regions and persons worsened the material situation of the majority of smallholders, who were bypassed by the 'green revolution' (Sachs 1987).

According to Sachs (1987), two major misjudgements were made before the introduction of the 'green revolution':

- The increase of the prices of chemical fertilisers and fuel and the general decrease in international prices as a result of the worldwide overproduction of grains had not been foreseen. These changes led to higher consumer prices for food and lower farm-gate prices; the main beneficiaries have been the suppliers of the fertilisers and fuel.

- The ever-increasing dependency on pesticides and fertilisers had not been foreseen. These have contaminated the streams and water tables, with serious hazards for the population.

Box 1.4
Low-external-input
agriculture (LEIA) in
Ghana

The traditional bush fallow
system practised by the farmers
of Mampong Valley worked
effectively while there was still
enough land to ensure that each
plot left fallow had sufficient
time to regain its fertility. However, increased pressure on the
land as a result of population
growth has led to shortening of
the fallow period and, con-
sequently, severe degradation of
farm sites. The original semi-
deciduous and evergreen forest
has been reduced to bush and
grassland. Soil is being
depleted, fuelwood has become
scarce, animal fodder is lacking
(especially in the dry season)
and important sources of local
medicine from trees shrubs and
other plants have disappeared.
The rainfall pattern has changed
and farmers now find it difficult
to make exact predictions of
planting times. Streams, rivers
and other water bodies are dry
much longer than before.

One old farmer lamented:
"The sun seems to be hotter
now than it was some 20 years
ago; the rains do not come at
the right time and, when they
come, they bring the wind to
destroy our houses and carry
our topsoil away, leaving us
with bare rocks and gullies. It
seems the gods have deserted
us and, if nothing serious is
done about the situation, then
our grandchildren are really
going to suffer. What can be
done?" (Owusu 1990).

Erosive forms of low-external-input agriculture (LEIA)

Low-external-input agriculture, also called low-resource (OTA 1988), resource-poor (WCED 1987) or undervalued-resource agriculture (Chambers et al. 1989), is practised in what Chambers calls complex, diverse and risk-prone areas. Here, the properties of the physical environment and/or the commercial infrastructure (poorly developed rural transportation and input distribution systems, inadequate saving/lending institutions) do not allow widespread use of purchased inputs. Often, only low quantities of, for example, artificial fertilisers and pesticides are sporadically used and then only for a few cash crops and by a small elite group of farmers.

Wolf (1986) estimates that some 1.4 billion people, or about one quarter of the world's population, depend for their livelihood on this form of agriculture: about 1 billion in Asia, 300 million in sub-Saharan Africa and 100 million in Latin America. It is found in the rainfed, undulating hinterlands of developing countries – the drylands, highlands and forest-lands with fragile or problematic soils. In terms of area, LEIA is most widespread in sub-Saharan Africa. The area under LEIA is growing as rural populations in many countries become increasingly impoverished, as external inputs become more expensive, and as many deeply indebted governments of developing countries which do not manufacture HEIA inputs can no longer afford to import them.

In many LEIA areas, production growth lags behind population growth. As new technologies to intensify land use in a sustainable way have not been developed or are not known to the farmers, they are often forced to exploit their land beyond its carrying capacity. This is particularly the case where they have been ousted from or deprived access to better-quality land reserved for 'modern' cropping or ranching. The overutilisation of LEIA smallholdings and their expansion to new, often marginal farming areas lead to deforestation, soil degradation and increased vulnerability to pest attacks, diseases, torrential rains and extended droughts (see Box 1.4). Many tropical land-use systems are in the midst of such a downward spiral of nutrient depletion, loss of vegetative cover, soil erosion, and economic, social and cultural disintegration.

1.4 Implications for agricultural sustainability

In order to understand what implications these trends have for sustainability, we first need to consider the roles of internal and external inputs in agricultural production.

In well-functioning LEIA systems, crops, trees, herbs and animals have not only productive but also ecological functions, such as producing organic matter, nutrient pumping, creating a nutrient reservoir in the soil, natural crop protection, and controlling erosion. These functions contribute to the continuity and stability of farming; they can be seen as producing internal inputs.

Well-functioning LEIA systems can be compared with mature natural ecosystems, in which nearly all biomass produced is reinvested to maintain fertility and biotic stability of the system. However, the reinvestment is more limited, because man extracts part of the production from an agroecosystem. By replacing natural internal inputs

Most land throughout the world is farmed with very low levels of external inputs. Maize sowing in central Nigeria. *(Ann Waters-Bayer)*

such as manure and compost by external inputs such as artificial fertilisers, more products can be extracted from the agroecosystem. By replacing natural processes by human-controlled processes, such as irrigation, variability in production can be reduced.

By selecting and breeding crops and livestock, people enhance their ability to convert inputs into useful products. In this process, other characteristics such as natural resistance or competitive ability are lost. These functions of nature that have been sacrificed must be assumed by man (Stinner & House 1987, Edwards 1987, Swift 1984, Conway 1987).

In HEIA systems, this replacement of ecological functions by man has gone much further than in LEIA systems. Diversity is replaced by uniformity for reasons of technology efficiency and market opportunities. In the short term, the use of external inputs permits great increases in land-use intensity. This made a substantial contribution to the global increase in food production until the mid-1980s. According to FAO, the enormous increase in fertiliser use in developing countries is the most potent single factor in raising productivity, in combination with numerous other external inputs such as modern varieties and breeds, irrigation and relevant information (Alexandratos 1988). However, the recent stagnation in production increase has raised strong doubts whether long-term productivity of such HEIA systems is secure.

Although concerted efforts have been made by development agents to convince farmers that 'modern' inputs will increase production, the majority of farmers have not adopted them. ''Perhaps as much as 80% of agricultural land today is farmed with little or no use of chemicals, machinery or improved seed'' (Dover & Talbot 1987). Some farmers made a conscious choice against a 'green revolution' package because it did not suit their farming conditions; others simply could not adopt it because the inputs were not locally available or were too costly. However, the tendency noted above towards increased extraction of products and decreased reinvestment of internal resources is leading to soil impoverishment in many areas, rendering such LEIA systems nonsustainable.

1.5 Focus of conventional agricultural research and extension

The activities and procedures of conventional research as taught at agricultural universities and practised in official agricultural research and extension organisations has contributed to the nonsustainability of current world agriculture by tending to concentrate on HEIA systems, and neglecting the needs of LEIA farmers.

To some extent, this is understandable. Conventional technology development is expensive. As the investment must be cost-effective from a national viewpoint, it is usually made in areas that can yield surpluses for industry or export or products to feed urban dwellers, e.g. rice, maize, milk, chickens. Also as a result of various social processes, official agricultural research tends to serve resource-rich farmers, who are able to bias technology development in their favour (Röling 1988).

Conventional agricultural research with its bias toward high-potential areas, export crops and better-off farmers has produced results which are out of the reach of most farmers and inappropriate for LEIA areas. This has been, among other reasons, because of its:

- **Focus on single commodities**. The emphasis has been on maximising production of particular commodities, and not total farm production. Plants which compete with the desired crop for water, nutrients and light are regarded as weeds. Technology development has concentrated on reducing this competition. This has hindered the study and enhancement of positive interactions between different plants, animals and man.

- **Primarily market orientation and associated nutrient drain**. Integration of farms into national or international markets results in a net nutrient drain, if extracted nutrients cannot be replenished. Very few technologies have been developed for returning nutrients from consumer to producer areas.

- **Disregard of environmental effects**. Driven by research procedures and political pressures to focus on short-term productivity, agricultural research has tended to externalise longer-term environmental effects to the future or to other sectors. Given the focus on the crop, field and – at best – farm level, the long-term effects on soil fertility, the regenerative capacity of natural vegetation and fauna, human health etc. have not usually been given sufficient consideration.

Box 1.5
Box 1.5
Scientists have neglected LEIA farmers

In West Africa 70 – 80% of the cultivated area is sown to traditional intercrops. Cowpea, one of Africa's most widely grown food staples, is always planted as an intercrop. Yet only 20% of the crop research effort in sub-Saharan Africa focuses on intercropping. In a country like Sudan, most power is provided by animals such as camels and donkeys. In parts of Africa, the sweet potato is an important staple. Yet, there is hardly a scientist in Africa doing research on sweet potatoes, donkeys or camels (T. Odhiambo, interviewed by van den Houdt 1988).

● **Neglect of rainfed areas and local resources**. Until recently, conventional research largely neglected the rainfed and marginal areas where LEIA farmers live and the local crops and animals which provide their livelihood. Research content and design has had little relevance to the major concerns and methods of LEIA farmers (see Box 1.5).

● **Gender bias**. Unlike most farming, agricultural research tends to be the preserve of males. Western trained researchers are usually imbued with Western models of the division of labour between men and women, in which the men dominate the external economic domain and women the household domain. This scenario, inappropriate as it is even for industrial societies, has blinded researchers to the fact that women play a significant role in agriculture. As a result, agricultural research has given little attention to solving the problems of female farmers and, in the design of new technologies, often disregards important questions of women's influence on decision-making and labour allocation.

● **Neglect of local farmers' knowledge**. The conventional top-down approach to technology development within agricultural research institutions gave scientists little opportunity to become well acquainted with the conditions, objectives and knowledge of LEIA farmers. The situation was not improved by the widespread attitude of extensionists and researchers, instilled already in school and university, that the formal system is the ultimate source of innovations and that information can come only from above.

● **Emphasis on station-based research**. The production conditions of research institutes and experiment stations do not resemble those of farmers and cannot possibly represent the highly variable conditions in rainfed agriculture. As a result, technology tested on-station often does not work under farmers' conditions, while the good qualities of local varieties, which are adapted to local conditions, are not recognised under station conditions (Biggs 1984).

● **Extension of incomplete products**. Conventional technology development tends to be organised in terms of disciplines, and not according to the aggregation level of the farm. As a result, the 'products' delivered for extension are often incomplete: they represent merely the answer to a disciplinary technical problem, without taking account of, for example, production aims, labour allocation between various crops, risk spreading, access to and affordability of external inputs, and other aspects of the socioeconomic context.

Not only the research but also the extension systems are male-dominated and have not provided women – the major food producers in many regions – with relevant services. Nonformal education for women is generally related to their reproductive activities in home economics and nutrition. Most rural women have little access to training related to productive and income-generating activities such as cropping and livestock-keeping. Considering the major role of women in crop and livestock care, this indicates a serious flaw in most extension systems.

Much more serious than the lack of sufficient staff, supplies and technical support for extension systems has been the lack of appropriate information and technologies that they could make available to smallholders. In many rainfed farming systems, the available science-based technology simply does not apply. Interventions have often been introduced in ignorance of the realities of existing farming practices and systems. They harbour the danger of upsetting the equilibrium of old methods of land use without leading to new, balanced farming systems.

As conditions in LEIA areas are diverse, farming recommendations and services must be site-specific. The conventional Transfer-of-Technology (ToT) model of widely disseminating a uniform technology package denies that transformation of technology continues even while it spreads among farmers. The ToT model prevents researchers and extensionists from linking into these transformation processes so as to help develop different techniques for different circumstances.

1.6　Using external inputs in LEIA areas: necessity and limits

It would be unjustified or naive to argue that LEIA farmers today are not interested in incorporating certain components of modern agricultural technology into their farming systems, if they have the means and opportunity to do so. Increasing demands (e.g. population growth, greater market integration, desire for more consumer goods) to extract from their farm systems increase their needs to use inputs that make this higher extraction possible, particularly artificial fertilisers. This book focuses on ways of making LEIA systems in the tropics more sustainable, and this will include the judicious use of external inputs, where these are available and can complement farm-produced inputs. It is therefore essential that the potentials and limitations of using external inputs in LEIA areas are understood. The emphasis in this section is on the limitations; the potentials of external inputs to enhance local resources are discussed in Chapter 5.

Artificial fertilisers

Farmers appreciate artificial fertilisers for their fast effect and relative ease of handling. Only after some time do farmers (and scientists) begin to recognise some of the limitations of artificial fertilisers:

● Their efficiency has often proved to be lower than expected. Dryland tropical crops lose up to 40 – 50% of applied nitrogen; in irrigated rice, losses are seldom less than 60 – 70% (Greenwood et al. 1980, Prasad & De Datta 1979, De Datta 1981, FAO 1990). Under unfavourable circumstances, such as high rainfall, extended dry periods, eroded soils and soils with low organic matter content, efficiency may be even lower.

● They may disturb soil life and soil balance. They increase decomposition of organic matter, leading to degradation of soil structure, higher vulnerability to drought and lower effectiveness in producing yields. Unbalanced application of acidifying mineral N fertilisers may also decrease soil pH and lower the availability of phosphorus to plants.

Small-scale farmers are interested in applying artificial fertiliser, but need to find the most efficient way of using the small quantities they can afford. *(Ann Waters-Bayer)*

- Continuous use of only artificial NPK fertilisers leads to depletion of micronutrients – zinc, iron, copper, manganese, magnesium, molybdenum, boron – which may influence plant, animal and human health; as these micronutrients are not replaced by NPK fertilisers, production eventually declines and the occurrence of pests and diseases increases (Sharma 1985, Tandon 1990).

- In addition to the agronomic limits to using artificial fertilisers, the limits in terms of supply of resources (particularly phosphates) to produce them has become increasingly apparent. At the farm level, this will mean rising prices for fertilisers or – if the country has insufficient foreign exchange to continue importing artificial fertilisers or the raw materials to produce them – complete lack of an input to which some farmers may already have adjusted their enterprise.

Fertiliser use in developed and developing countries contributes to global risks arising from the release of nitrous oxide (N_2O) to the atmosphere and above. In the stratosphere, N_2O depletes the ozone layer and, by absorbing certain wavelengths of infra-red light, increases global temperatures ('greenhouse effect') and destabilises climates. This could lead to changes in patterns, levels and risks of agricultural production. A rise in sea level would have grave consequences for low-lying delta and estuarine regions. In view of these dangers, worldwide restriction on fertiliser use cannot be excluded in the future. Therefore, greater effort is needed to promote more efficient and less polluting use of N fertilisers and the use of alternative sources of N, e.g. crop wastes, animal and green manures, legumes in rotations and as tree crops, and blue-green algae and nitrogen-fixing bacteria in rice paddies (Conway & Pretty 1988).

Box 1.6
Chemical pest control and the small-scale farmer

Especially for smallholders in developing countries, chemical control does not offer an acceptable way to improve agricultural production substantially. In subsistence farming, production gains made in food crops are normally too low to compensate for the cost of applying insecticides over a whole field. Certain applications could probably be very useful under these circumstances, e.g. seed dressings or spot treatment of local infestations, but knowledge and technical advice is generally lacking for the effective use of insecticides in such a manner.

Pesticide use in the tropics has primarily been of benefit to large commercial farmers who can bear the expense of these inputs, but does not provide appropriate means for smallholders. On the contrary, it might even have a negative effect by disrupting useful traditional practices and well-established ecological balances. The drive for increased production has often caused the dislocation of well-adapted traditional farming systems and their sometimes overhasty replacement by systems and technologies that are not compatible with either ecological or socioeconomic conditions (Brader 1982).

Pesticides

Pesticides are chemical or natural substances that control pest populations mainly by killing the pest organisms, be they insects, diseases, weeds or animals. In 1985 roughly 2300 million kg of chemical pesticides were used worldwide. About 15% of this, including 30% of all insecticides, is used in the Third World. Pesticide use is increasing particularly quickly in developing countries, where it is regarded as an easy way to raise production and is often actively promoted and subsidised.

However, some disadvantages and dangers of using pesticides are gradually becoming clearer:

● Yearly, thousands of people are poisoned by pesticides, about half of them in the Third World. For example, in 1983 a total of about 2 million people suffered from pesticide poisoning, and 40 000 of the cases were fatal (Schoubroeck et al. 1990). Because of their toxicity, many types of pesticides, e.g. DDT, have been banned in industrialised countries. However, they are still being used in many developing countries.

● Over time, pests build up resistance to pesticides, which must then be used in ever higher doses to have effect. Eventually, new pesticides must be developed – a very expensive process. Pest resistance builds up more rapidly in tropical than in temperate climates, as biological processes are more rapid at higher temperatures. In 1984, resistance to pesticides was known for 447 insects and mites, 100 plant pathogens, 55 kinds of weeds, 2 kinds of nematodes and 5 kinds of rodents (Gips 1987).

● Pesticides kill not only organisms that cause damage to crops but also useful organisms, such as natural enemies of pests. The incidence of pest attacks and secondary pest attacks may increase after pesticides have killed the natural enemies (resurgence).

● Only a small proportion of the pesticides applied in fields reaches the organisms that are supposed to be controlled. The major part reaches the air, soil or water, where it has a damaging effect on living organisms. Aquatic organisms are particularly sensitive to pesticides.

● Pesticides that do not break down easily are absorbed in the food chain and cause considerable damage to insects, insect-consuming animals, prey birds and, ultimately, human beings.

'Improved' seed

In the past three decades, the area under modern 'high-yielding' varieties of certain basic food crops has increased considerably. The focus of improvement has been on wheat, rice and maize. In developing countries, new wheat and rice varieties cover approximately 24 and 45 million ha, respectively, or 50% and almost 60% of the total planted area of each crop (CGIAR 1985). In addition, the use of hybrid maize varieties has rapidly expanded in several developing countries and covered about 25% of the area planted with maize in Africa by 1986 (CIMMYT 1988). The spread of modern varieties of pulses, oilseeds,

Box 1.7
Hybrid maize replacing local crops in Zimbabwe

The hybrid maize varieties are developed for regions with high rainfall. In wet years, yields are high even in less humid regions but, in dry years, there is a high rate of crop failure. After the dry years of 1982 – 83 and 1986 – 87, innumerable small farmers were not able to regain their investments in inputs and had to depend on food aid. Hybrid maize has replaced traditional varieties of maize, sorghum and millet which are more resistant to drought (Mushita, interviewed by Donkers & Hoebink 1989).

Farmers and local artisans can play a leading role in designing and managing small-scale irrigation technology. A bamboo waterwheel in West Java, Indonesia, which can raise water 6 m above the river. *(VIDOC, Royal Tropical Institute, Amsterdam)*

vegetables and other basic food crops such as sorghum, millets and root crops has been very limited.

Modern varieties are essentially high-response varieties, bred to respond to high doses of chemical fertilisers. If they are sown under conditions of high nutrient and water supply and adequate pest control, modern varieties and hybrids can indeed be high-yielding. If these conditions cannot be guaranteed, risks of yield losses may be higher than with local varieties. When only low levels of external inputs are used, local varieties may outyield the modern ones.

Together with other factors, the promotion of modern varieties has led to the disappearance of many indigenous varieties (genetic erosion). This spells disaster for farmers who have to produce their crops with low external inputs under highly variable and risk-prone conditions – and for all farmers who, for both economic and ecological reasons, will have to produce with less chemical inputs in the future.

Although modern wheat and rice varieties cover considerable areas, their influence on local supply systems has been relatively small. Annual seed replacement for these crops is – with a few exceptions – less than 10%. In other words, farmers who start using the modern varieties do not replace them after 3 – 4 years when the varieties may have lost their varietal purity. This may be due to a lack of capacity of national or private seed supply agencies to produce adequate quantities of improved seed, or to lack of financial means of the farmers to pay for improved seed and the necessary fertilisers and pesticides. Another reason for the lack of widespread adoption is that the limited range of modern varieties supplied does not meet the LEIA farmers' need for a choice of varieties to guarantee stable rather than maximum yields.

Irrigation

Lack of reliable water supply is a major constraint in tropical agriculture. The area of land under irrigation has increased considerably during the last decades. About 17% of the world's arable land is now irrigated. However, the increase is stagnating, partly because the water reserves in many countries are used almost to the maximum, and partly because problems have arisen with respect to rising costs of maintaining the infrastructure, salinisation (about 25% of the total irrigated area), falling water tables and other ecological problems. Construction of big dams has led to severe social problems when many people had to be resettled (World Bank 1984, Brown 1989).

Immense sums have been spent by development projects on large-scale irrigation schemes, in which the water is often used inefficiently. Small-scale alternatives (lift irrigation, small dams, water-harvesting tanks etc.) are much cheaper and more flexible, use water more efficiently and give farmers the possibility to participate in layout and management. For LEIA farmers in dry areas where irrigation is important, these small-scale alternatives could be of interest. But for the majority, improvement of rainfed farming by water conservation and organic matter management is of greater importance, as LEIA farmers' capacity for investment is very limited.

Mechanisation based on fossil fuels

The use of tractors and other machines for land preparation, planting, cultivation, harvesting and processing depends, with very rare exceptions, on nonrenewable fossil fuels. Mechanisation can improve yields through better land preparation, more timely and precise placement of seed and fertiliser, and more efficient harvesting, and thus reinforces the effects of the rest of the 'green revolution' package. Machines and fossil fuels are inputs that come not only from outside the farm but often also from outside the country, i.e. they have to be imported and paid for with foreign exchange, a very scarce resource in the case of most Third World countries.

In LEIA, constraints to this kind of mechanisation include limited availability of the equipment, fuel, capital, skills, maintenance facilities and spare parts; and difficult ecological conditions causing high wear and high risk of breakdown (see Box 1.8). Particularly the use of tractors is increasing the risk of environmental damage by soil erosion, soil compaction, deforestation and pest hazards.

Moreover, the burning of fossil energy for running machines and for other purposes is one of the main sources of carbon dioxide released into the atmosphere. Carbon dioxide is the major greenhouse gas which is contributing to global warming (Leggett 1990).

Depending on the local economic and ecological situation, fossil fuel-based mechanisation can be more or less attractive to the individual farmer. Although there may be a great need for mechanisation in general to improve labour productivity and working conditions, alternatives such as animal traction, improved hand implements and less energy-demanding techniques, especially tillage techniques, can greatly decrease the need for fossil fuel-based mechanisation in farming (Stout 1990).

Box 1.8
Tractors versus oxen in Zimbabwe

Many cooperatives in Zimbabwe see the tractor as a means of lifting agriculture from primitive techniques to a system with more output and thus better income. But mechanisation replaces labour with the scarce and expensive resources, capital and management. The justification is that the tractor prepares the land better and more quickly, and an early-sown crop in a good seed bed is more likely to produce a good harvest. But then the tractors are almost sure to be underutilised for the rest of the year.

In Europe 1000 hours per year is taken as the economic minimum for a tractor. This is likely to be higher in Zimbabwe as machinery is more expensive, maintenance is poorer, spare parts are scarce and expensive, the conditions give rougher wear and more breakdowns, and the life of a tractor is shorter. Finally, as crop prices are generally lower in Zimbabwe than in Europe, gross margins per hectare are lower.

The Institute of Agricultural Engineering in Zimbabwe recommends that tractors be used for large-scale cultivation but, to save on costs, oxen be used for inter-row weeding and on-farm transport. Oxen do not need to be imported using valuable foreign currency and their fuel is home-grown (although the area needed to produce the fodder can be quite extensive). There is virtually no depreciation, as the oxen can be fattened for beef. With large arable areas on each farm (100 ha on average), rather than obtaining a substantial loan to purchase a tractor and machinery, the ideal would be to hire a tillage team (provided in each area by the government) to complete land preparation and planting, and then use oxen for weeding and on-farm transport.

The cooperatives must be trained to appraise the economics of mechanisation themselves and discover any benefits in comparison with using animal draught and their own labour (Cole 1987).

External inputs and LEIA farmers: conclusions

Some of the reasons why LEIA farmers have been loath or unable to use the above-mentioned external inputs are:

● they are not available or their availability is unreliable on account of poor commercial infrastructure and services;

● if they are available, they are costly;

● they are risky and may be inefficient under variable and vulnerable ecological conditions (e.g. erratic rainfall, sloping land);

● they are not very profitable under the above circumstances;

● communication between research and development personnel and farmers is poor, which often leads to serious (and sometimes dangerous) comprehension gaps and to incompatibility of innovations and technology recommendations with local ecological conditions and cultural values.

Dangers involved in promoting the introduction of such inputs into LEIA areas include:

● loss of diversity in the farming system, rendering them unstable and more vulnerable to ecological and economic risks;

● irretrievable loss of local genetic resources and traditional knowledge about ecologically-oriented husbandry and local alternatives to purchased inputs;

● social and cultural disintegration, and marginalisation of poorer farmers, particularly women;

● environmental damage, particularly as a result of excessive use of agrochemicals.

Farmers who are given easy access to credit may be lured into high capital investment and production methods which demand that high levels of external inputs be maintained or increased. At the same time as prices for petroleum-based fertilisers, herbicides, pesticides and fuels are increasing, prices of agricultural products are often being kept artificially low by national governments or by the flooding of local markets with subsidised imports of agricultural products from industrialised countries. Third World farmers are then in great danger of being trapped in a debt spiral.

Where purchased inputs are subsidised by the government or a development project, their use is feasible only for a limited term. As soon as the subsidies are withdrawn, most farmers are forced to abandon the inputs. If they have, in the meantime, adjusted other aspects of their farming systems (e.g. reduced diversity of crop and livestock species, or increased nutritional dependence on crops like maize which require high fertiliser inputs), then they will be in very serious trouble.

It would thus appear that the 'green revolution' package could worsen rather than improve the situation of LEIA farmers in complex, diverse and risk-prone areas. But what alternatives are available?

1.7   Sources of additional options for LEIA

To help LEIA farmers develop productive and sustainable forms of agriculture, alternatives to 'green revolution' technology are being sought – and found – from various sources: agroecological science, indigenous knowledge and farming practices, new directions in conventional agricultural science and the practical experiences of experimenting farmers and fieldworkers.

Agroecology

Agroecosystems are communities of plants and animals and their physical and chemical environments that have been modified by people to produce food, fibre, fuel and other products for human consumption and processing. Agroecology is the holistic study of agroecosystems, including all environmental and human elements. It focuses on the form, dynamics and functions of their interrelationships and the processes in which they are involved. An area used for agricultural production, e.g. a field, is seen as a complex system in which ecological processes found under natural conditions also occur, e.g. nutrient cycling, predator/prey interactions, competition, symbiosis and successional changes. Implicit in adapted agroecological work is the idea that, by understanding these ecological relationships and processes, agroeco-systems can be manipulated to improve production and to produce more sustainably, with fewer negative environmental or social impacts and fewer external inputs (Altieri 1987).

Agroecologists are now recognising that intercropping, agroforestry and other traditional farming methods mimic natural ecological processes, and that the sustainability of many local practices lies in the ecological models they follow. By designing farming systems that mimic nature, optimal use can be made of sunlight, soil nutrients and rainfall.

Indigenous knowledge and farming practices

Most of the indigenous agricultural practices that proved to be non-sustainable have not survived. Other indigenous practices that sustained human populations for centuries became obsolete as conditions changed. This has been the case with several forms of shifting cultivation under increased population pressure. Nevertheless, there are still innumerable land-use systems developed by traditional farming communities that exemplify careful management of soil, water and nutrients, precisely the type of methods needed to make farming sustainable (see Box 1.9).

Traditional farmers have found ways of improving soil structure, water-holding capacity and nutrient and water availability without the use of artificial inputs. In many cases, their farming systems are (or were) sophisticated forms of ecological agriculture fine-tuned to the specific environmental conditions. Evaluation of indigenous farming techniques and systems presents options for improvement of LEIA. Not all LEIA systems have reached a point of causing ecological damage, and those that are in a process of decline often include techniques that are still less destructive than indiscriminately adopted modern technology.

Box 1.9
Starting with indigenous knowledge

Reading current literature, one would be inclined to conclude that agroforestry started only 5 – 6 years ago. But agroforestry has already existed for several hundred years. African farmers used to combine the cultivation of food crops with long-term crops such as trees. At the beginning of this century, however, colonial powers forbade these practices, regarding them as backward. The European did not understand the African. Now we have to go back to the basics to see what the traditional farmer does and why it is done that way (T. Odhiambo, interviewed by van den Houdt 1988).

Rice terraces near Cililin in West Java, Indonesia: a sophisticated farming system adapted to the local environmental conditions. *(VIDOC, Royal Tropical Institute, Amsterdam)*

The major strength of indigenous farming systems lies in their functional integration of different resources and farming techniques. By integrating various land-use functions (e.g. producing food, wood, etc; conserving soil and water; protecting crops; maintaining soil fertility) and the use of different biological components (large stock, small stock, food crops, fodder crops, natural pasture plants, trees, herbs, green manures etc.), the stability and productivity of the farm system as a whole can be increased and the natural resource base can be conserved.

Indigenous knowledge is an important source of information about the local farming system (including traditional practices which have fallen into disuse), experiences, institutions, culture etc. Above all, farmers' knowledge and skills in adapting new ideas to their local conditions and needs form the basis for change within the farming community.

New directions in conventional agricultural science

Although conventional agricultural science has been severely limited by its disciplinary and reductionistic approach, it has undeniably made valuable contributions to agricultural development. It would be a mistake to suggest that we could do without modern agricultural technologies and insights. Rather, it is the way in which they have been applied – in isolation and without sufficient concern for ecological effects – which has rendered them debatable. Within a holistic framework which integrates the various scientific disciplines and is ecologically oriented, many of the conventional agricultural technologies could contribute to LEISA.

In recent years, thinking in terms of systems has been increasingly accepted within agricultural sciences, the Farming Systems Research and Development/Extension approach being the most obvious example. Study programmes and courses have been drawn up at the university and college level, which treat agriculture within a systems context (for an example, see Box 7.4).

New developments in biotechnology can also contribute to LEISA. The criticisms of the dangers of biotechnology are justified: e.g. possible depletion of genetic diversity; limited access to genetic material through patenting; control by multinational companies; substitution of tropical products by synthesised ones. However, biotechnology could also bring benefits to Third World farmers, if research were to be done with a view to their needs. Tissue culture makes it easier to compile germplasm collections and to supply disease-free material and, through genetic engineering, varieties with desired traits can be produced. For example, pest-resistant potato, tomato and tobacco varieties will probably be commercially available within a few years. New products coming out of biotechnological research will be of value to livestock-keepers in the tropics. Research is underway to help trace the sources of *peste des petits ruminants* (PPR), and vaccines are being developed to prevent neonatal bacterial diarrhoea and foot-and-mouth disease. Developments in biotechnology which increase the efficiency of conventional breeding processes could allow breeders to give more attention to producing plants and animals with specific traits best suited to the varied agro-ecological conditions of Third World farmers (Greeley & Farrington 1989).

Also the conventional reductionistic trials and laboratory studies conducted by biologists, nutritionists, agronomists, soil scientists etc. can help explain the effects of components within agroecosystems. If this disciplinary work is focused on questions of relevance to LEISA, it can provide a useful source of information for farmers and fieldworkers.

Practical experience of experimenting farmers and fieldworkers

LEISA's major source of information thus far has been the experiences of farmers and fieldworkers who had the courage and creativity to go their own way and develop technologies which were overlooked by mainstream researchers and which the existing scientific journals and databases did not consider relevant. Nongovernmental organisations (NGOs) have been particularly active in this development.

Also farmers in industrialised countries and a like-minded minority of supportive scientists have been experimenting in nonconventional forms of agriculture, e.g. organic farming, biodynamic farming, permaculture. The ecological movements are mainly based in richer countries with temperate climates, but a rapidly increasing number of experiments in 'ecofarming' are being carried out in the tropics. For example, Garcia-Padilla (1990) found that some 120 NGOs in the Philippines are actively engaged in experimenting with techniques of ecological agriculture. The insights thus gained into the principles behind these techniques – if not the specific techniques themselves – could be of value to farmers trying to develop LEISA in other areas.

A wealth of 'grey' literature (unpublished reports and articles in newsletters, circulars and project reports) reveals the experiences of present-day innovators, both in developing and industrialised countries. In fact, our Information Centre for Low-External-Input and Sustainable Agriculture (ILEIA) started by documenting grey literature and disseminating the information and experiences found within these reports and articles, which now number almost 4000. However, most of the experiences of innovative farmers and fieldworkers throughout the world have not been documented, although much has certainly been spread locally by word of mouth.

1.8   Towards Low-External-Input and Sustainable Agriculture (LEISA)

In view of the limited access of most farmers to artificial external inputs, the limited value of these inputs under LEIA conditions, the ecological and social threats of 'green revolution' technology and the dangers of basing production on nonrenewable energy sources, the strong emphasis on HEIA in agricultural development must be questioned. However, it is also open to question whether it will be possible to raise world food production sufficiently without the use of such external inputs. Besides, natural as opposed to artificial inputs can also have detrimental environmental effects (see Box 1.10).

LEISA is an option which is feasible for a large number of farmers and which can complement other forms of agricultural production. As most farmers are not in a position to use artificial inputs or can use them only in small quantities, it is necessary to concentrate on technologies that make efficient use of local resources. Also, those farmers who now practise HEIA could reduce contamination and costs and increase the efficiency of the external inputs by applying some LEISA techniques. It is important that the agroecological knowledge of both scientists and farmers be applied, so that internal and external inputs can be combined in such a way that the natural resources are conserved and enhanced, productivity and security are increased and negative environmental effects are avoided.

LEISA refers to those forms of agriculture that:

● seek to optimise the use of locally available resources by combining the different components of the farm system, i.e. plants, animals, soil, water, climate and people, so that they complement each other and have the greatest possible synergetic effects;

● seek ways of using external inputs only to the extent that they are needed to provide elements that are deficient in the ecosystem and to enhance available biological, physical and human resources. In using external inputs, attention is given mainly to maximum recycling and minimum detrimental impact on the environment.

LEISA does not aim at maximum production of short duration but rather at a stable and adequate production level over the long term. LEISA seeks to maintain and, where possible, enhance the natural resources and make maximum use of natural processes. Where part of the production is marketed, opportunities are sought to regain the nutrients brought to the market.

Box 1.10
Using internal inputs is no guarantee for sustainability

No single agricultural method has a corner on sustainability. Any farming system, whether 'chemical-intensive' or 'natural', can be in some aspects resource-conserving and in other aspects wasteful, environmentally unsound or polluting. Obviously, serious questions surround how long such external energy and external supply of nutrients, fossil fuels, petrochemicals and mineral fertilisers can be maintained. But simply substituting nonchemical alternatives may not necessarily make agriculture more sustainable, e.g. applying animal manures unwisely can pollute ground and surface water as badly as overuse of chemical fertilisers can and plant-derived pesticides can be as dangerous as chemical pesticides (Dover & Talbot 1987).

Numerous developing countries are now implementing so-called structural adjustment programmes that involve policies such as devaluation of exchange rates, reduction of government spending and intervention, reduction of subsidies and removal of price controls. In this way, the demand for imports is to be curtailed and the purchase of local goods stimulated, so as to reduce the balance-of-payment and government deficits and to promote national economic growth. LEISA appears to fit within this context, as it is less demanding on imports and credits than the conventional approach to agricultural development.

At farm, regional and national level, LEISA implies the need for closely monitoring and carefully managing flows of nutrients, water and energy in order to achieve a balance at a high level of production. Management principles include harvesting water and nutrients from the watershed, recycling nutrients within the farm, managing nutrient flow from farm to consumers and back again, using aquifer water judiciously, and using renewable sources of energy. As these flows are not confined by farm boundaries, LEISA requires management not only at farm level but also at district, regional, national and even international levels. At each level, technologies are sought to make the flow cycle as short as possible and to balance the flows. In this book, the focus is on practices that can be applied at farm level. Questions related to techniques and systems at village level and above are equally important, but should be addressed in a separate study.

LEISA incorporates the best components of indigenous farmers' knowledge and practices, ecologically-sound agriculture developed elsewhere, conventional science and new approaches in science (e.g. systems approach, agroecology, biotechnology). Thus far, conventional science has served mainly HEIA, but the contributions it could make to LEIA should be explored to the full.

PTD aims at strengthening the experimental capacity of farmers. Scientists and farmer in Machakos District, Kenya, examining root nodules on a bean plant to see whether they are fixing nitrogen. *(Ann Waters-Bayer)*

LEISA practices must be developed within each ecological and socioeconomic system. The specific strategies and techniques will vary accordingly and will be innumerable. The experience thus far of developing LEISA systems cannot provide universal, ready-made answers for the problems of farmers in other areas, but can provide some indications of principles and promising possibilities.

The process of combining local farmers' knowledge and skills with those of external agents to develop site-specific and socioeconomically adapted farming techniques has been given the name 'Participatory Technology Development' (PTD). Farmers work together with professionals from outside their community (e.g. extensionists, researchers) in identifying, generating, testing and applying new techniques. PTD seeks to strengthen the existing experimental capacity of farmers and to encourage continuation of the innovation process under local control (Haverkort et al. 1988). PTD is an essential element in the development of sustainable farming systems and a central concept in this book.

The experience of combining indigenous and scientific knowledge through a process of PTD (see Part III) indicates strongly that it is indeed possible to transform LEIA to LEISA: Low-External-Input and **Sustainable** Agriculture. This approach to agricultural development appears to be better adapted to the needs and opportunities of LEIA farmers and to fit better into their cultural context than the conventional approach.

In the following chapters, some of these experiences are presented, so as to enlarge the 'basket' of strategies, methods and techniques from which other development workers and farmers can chose in order to develop sustainable farming systems together. The major agroecological principles behind LEISA and some general considerations about how they can be applied are illuminated to give some guidance in making the right choices and mixes of technologies for the site-specific ecological and socioeconomic conditions.

LEISA cannot be presented as the solution to the world's pressing agricultural and environmental problems, but it could make a valuable contribution to solving some of them. It is, above all, an approach to agricultural development which addresses the situation in areas of rainfed agriculture which have been neglected by conventional approaches.

This book concentrates on the biophysical principles of LEISA and farmer-centred processes of developing LEISA systems. At the same time, however, institutional changes will be needed to ensure that LEISA can be actively sought and found (cf. Chambers 1983). Particularly the political prerequisites are essential: for example, where rural people do not have secure land-use rights, or where wars ravage the land, LEISA has little chance. Without a political will for social justice and peace on all levels from the village to the globe, the pursuit of sustainable development is quixotic.

2 Sustainability and farmers: making decisions at the farm level

Thus far, we have been looking at agriculture and sustainability on a global scale. However, the people making the decisions about day-to-day resource use for agriculture are not the compilers of global statistics and the writers of global reports but rather the farmers. Outsiders can influence the economic and political setting within which farmers make their decisions and can provide them with information, guidance and encouragement, but the ultimate decisions will still be made by the farmers. To be able to communicate with farmers and support them in making their livelihoods sustainable, outsiders must first understand how farms function and how decisions about resource use are made within rural families and communities.

2.1 The farm as a system

In conventional scientific analysis, farming is divided into disciplines and regarded from the professional viewpoint of the agronomist, livestock nutritionist, economist etc. By contrast, farmers are not specialists; they regard farming as a whole, and this whole is more than the sum of the parts seen by the specialists. If we want to understand how farms function and how farming decisions are made, we have to look at farming in a holistic way. As stated by CGIAR (1978): "Farming is not simply a collection of crops and animals to which one can apply this input or that, and expect immediate results. Rather, it is a complicated interwoven mesh of soils, plants, animals, implements, workers, other inputs and environmental influences with the strands held and manipulated by a person called the farmer who, given his (or her) preferences and aspirations, attempts to produce output from the inputs and technology available" (gender differentiation added by the authors).

Farming is inextricably linked with culture and history. Geographical and ecological opportunities and constraints (location, climate, soil, local plants and animals) are reflected in the local culture. This is reflected, in turn, in the local agriculture, which is the result of a continuous historical process of interactions between humans and the local resources. The rural society's values, knowledge, skills, technologies and institutions greatly influence the type of agrarian culture (agriculture) that evolved – and continues to evolve.

The term **'farming system'** refers to a particular arrangement of farming enterprises (e.g. cropping, livestock-keeping, processing farm products) that are managed in response to the physical, biological and socioeconomic environment and in accordance with the farmers' goals, preferences and resources (Shaner et al. 1982). Individual farms with enterprises arranged in a similar way are said to practise that particular farming system. Farming is used here in a wide sense to include not only crops and livestock but also the other natural resources available to the farm households, including resources held in common with others. Hunting, fishing and harvesting honey and other nonwoody products from wooded areas, and also the extensive grazing of livestock on natural pasture can all form part of a farming system.

Over time, a great variety of farming systems have developed throughout the world: e.g. nomadism, shifting cultivation, irrigated cropping, ley farming, horticulture and combinations of these. The orientation of these systems ranges from predominantly subsistence to predominantly commercial. The farm household may depend primarily on local resources and indigenous knowledge, or on 'foreign' external inputs, chemical fertilisers, pesticides, mechanisation and formal scientific knowledge. Several farming systems can exist simultaneously in one area. Many of these systems have been described and compared (e.g. Duckham & Masefield 1970, Grigg 1974, Klee 1980, Ruthenberg 1980). They vary greatly in terms of productivity and efficiency of using land, labour and capital, and in their effects on the environment.

Although farms within a given farming system resemble each other, each individual farm has different physical, biological and human resources. Therefore, each is a unique **'farm system'**.

Within a farm system, physical resources such as soil, water and air interact to create unique conditions of temperature, wind, rainfall etc. These conditions affect the functioning of the biological resources (e.g. crops, livestock, birds, insects, weeds, and micro-organisms). These

---

Box 2.1
A farm as a system

A farm is a unique agroeco-system: a combination of physical and biological resources such as land forms, soil, water, plants (wild plants, trees, crops) and animals (wild and domestic). By influencing the components of this agroeco-system and their interactions, the farm household obtains outputs or products such as crops, wood and animals.

To keep the production process going, the household needs inputs, e.g. seeds, energy, nutrients, water (see Figure 2.1). Internal inputs are those harvested on the farm, e.g. solar energy, rainwater, sediments, nitrogen fixed from the air; or produced on the farm, e.g. animal traction, wood, manure, crop residues, green manure, fodder, family labour

and learning experiences. External inputs are those obtained from outside the farm, e.g. information, hired labour, fossil fuel, mineral fertilisers, chemical biocides, improved seeds and breeds, irrigation water, tools, machinery, and services.

The outputs can be used as internal inputs, consumed by the farm household (reproducing farm labour) or sold, exchanged

or given away. During the production process, some losses occur as a result of, for example, leaching or volatilising of nutrients or soil erosion. The sales provide cash which can be used to buy different goods or services (e.g. food, clothes, education, transport), to pay taxes and/or to obtain inputs. Inputs can also be obtained in direct exchange for outputs.

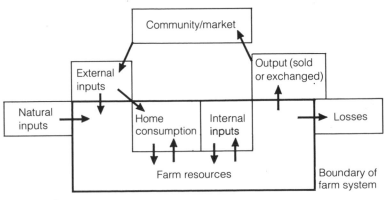

Figure 2.1 Flow of goods and services (indicated by arrows) in a simplified farm system

living organisms, with their life cycles of birth, growth, reproduction and death, interact with each other in processes of, for example, competition, succession, symbiosis and allelopathy. These physical and biological resources and processes are consciously manipulated by the human resources within the system: the farm family with its knowledge, skills, experience and energy.

Farm systems are open systems: they receive inputs (nutrients, water, information etc.) from outside the farm, and some of the outputs leave the farm, e.g. are sold. The very simplified model presented in Box 2.1 helps to explain the concept of inputs and outputs.

LEIA farm systems are usually very complex. Members of a single family may use natural resources in many different ways – cropping, gardening, herding, hunting, gathering wild plants etc. – in order to serve their many different needs. Besides producing food, fibre, wood and various 'minor' products such as medicines, thatching material and ornaments, these activities have other functions, including spreading risks and making sure that production can continue.

Moreover, to gain their livelihood, many LEIA households do not depend solely on farming. Their other income-generating activities may compete with agricultural activities for the limited labour force of the household. It would be more realistic to look at the 'livelihood system' (Chambers et al. 1989) of a rural household, but this would make the model even more complex. We will therefore focus here on the farm system, without losing sight of the fact that this is only part of the entire livelihood system of a rural household.

2.2   Site-specific characteristics of a farm system

The biophysical setting

The genetic resources, techniques and strategies from which farmers can choose in creating, maintaining and developing their farm systems are largely determined by the specific characteristics of the ecosystem, such as those of mountain ecosystems described in Box 2.2.

Farmers have normally investigated local options to the greatest possible extent and exploit them well. In this way, finely tuned farming systems have developed which have survived for generations. Grillo Fernandez and Rengifo Vasques (1988) and Rhoades (1988) give examples of such long-sustained farming systems in mountain areas. However, adaptations to increasing population pressure and changing economic conditions have led, in many cases, to new practices such as expansion of cultivation to more fragile slopes and sole cropping induced by the promotion of modern varieties. These have often failed to match well with the constraints and potentials of the mountain ecosystems (Sanwal 1989).

Also in other ecosystems, factors such as soil characteristics (e.g. acid, alkaline, saline, shallow), typhoons, droughts and prevailing pests may limit farming options. If the scope for further development of LEISA systems in fragile, remote, diverse and marginal areas is to be widened, development agents must understand the implications of the given biophysical specificities for agricultural sustainability.

Box 2.2
Mountain specificities

Marginality: On account of slope, altitude and periodic seasonal hazards (e.g. landslides, snowstorms), mountain areas tend to be poorly accessible and relatively isolated. This is manifested in poor communications and limited mobility and renders such areas marginal to the wider economy.

Diversity or heterogeneity: In mountain areas one finds immense variation among and within ecozones, even at short distances, as a function of the interaction of elevation, altitude, soil condition, steepness and orientation of slope, wind, precipitation and relief of terrain. The biological adaptions and socio-economic responses to this diversity add to the heterogeneity.

Niche or comparative advantage: Owing to their specific environmental and resource-related features, mountains provide a 'niche' for specific activities or products, in which they have comparative advantages over plains, e.g. as a source of unique products (some fruits, flowers etc.) or as a source of hydropower.

The land-use systems evolved by mountain communities are based on adaption of these mountain specificities to suit production requirements, and adaptation of their requirements to suit the mountain conditions. The communities have designed various measures to intensify the use of and protect the land resources, multiply and diversify options, and generate and exchange surpluses despite the constraints of mountain areas. Evaluation of these human adaptation mechanisms could help in identifying elements that can be used in strategies for sustainable development of mountain agriculture (Jodha 1990).

The human setting

Apart from its biophysical setting, a farm system is also determined by its socioeconomic, cultural and political characteristics, foremost those of the farm household. Each household is a unique combination of men and women, adults and children, who provide the management, knowledge, labour, capital and land for farming, and who consume at least part of the produce. The farm household is thus a centre of resource allocation, production and consumption.

The household may consist of more or less autonomous subsystems, such as wives with their subhouseholds and/or farms. Through its external relations, the household functions in the context of the wider economic, social, cultural and political systems but also influences these systems. External relations, such as via the market or mass media will also influence the household and, thus, the farm system.

Ties with the community can be strong. For example, family ties, friendships, common history and culture, and common control over territory can interconnect the individual farm systems. Community members often use land in common and give each other support by sharing or exchanging labour, animals, fields or farm products. These interactions serve to hedge risks and are part of the survival strategies of families and individuals. Different family members may be involved in different networks within the community and possibly also with other communities.

Also in the human setting, certain factors may limit farming options, e.g. availability of land, labour or capital; market demand; transport facilities; human skills. Land and tree tenure is often a major constraint. When farmers are not sure of their rights to use the land cropped or the trees planted, incentives to invest in resource-conserving practices such as erosion control may be weak. Especially women, but also many men, are often obliged to farm under such insecure conditions.

As individuals and as a unit, the farm family decides about use of resources for agricultural production and about use of the products. *(Ann Waters-Bayer)*

Many different socioeconomic and cultural processes influence farm systems, making it necessary to adapt them in order to ensure sustainability. Some of these processes and influences are:

- increasing contacts with industrial/urban society, leading to higher cash needs to purchase industrial products and to pay for education;

- greater exposure via radio, television and other mass media to other life styles, leading to changes in felt needs;

- stronger integration into a commercial market system, requiring changes in kind and quality of produce and leading to dependency on external input supply, market demand, transport, credit and services; under highly variable conditions, this decrease in farmers' self-reliance can threaten the security of farming;

- erosion of knowledge of the local agroecosystem and indigenous farming techniques, strategies and genetic resources, due to lowered status of traditional practices and of farming as a profession; this is a change induced largely by the formal education system and contributes to adoption of 'green revolution' technology or rural exodus;

- population increase, which may lead to reduction in farm size on account of farm splitting and/or make it necessary to extend farming to more marginal areas, or to overexploit resources, or to look for income sources outside the farm;

- labour migration, which may lead to a shortage of young persons and men for farmwork in LEIA areas and which may slow down decision-making processes in *de facto* but not *de jure* female-headed households;

- degradation of the farm system on account of overintensification and the far-reaching negative environmental and socioeconomic effects of 'green revolution' technology (see Chapter 1);

- structural adjustment policies and increases in the price of fossil fuel, which may lead to decreased imports of agricultural inputs and/or increased prices of these inputs, necessitating more efficient use of external and locally available resources and inputs.

But there are also processes which enhance sustainability, such as increasing awareness of the impact of ecological degradation and poverty; shift of attention and action to sustainable development by governments, NGOs and private persons; and initiatives to develop policies and technologies that promote agricultural sustainability and to make the appropriate inputs available. Also some of the processes mentioned above can have positive effects. For example, remittances from labour migration can offer possibilities of investment in farming and resource conservation. Better schooling can increase the self-confidence of rural people. Better communication can also improve exposure to issues of sustainable agricultural development.

Rural women's access to mass media and contacts with urban women, as well as their stronger economic position as (at least, acting) heads of households and as producers and processors of foods for domestic markets, is leading to a change in the position of women in many rural areas. Greater influence of women in decision-making could draw more attention within local development initiatives to food cropping and to sustainable forms of land use.

Not only changes in the wider socioeconomic conditions but also gradual or sudden changes within a household (e.g. increase or decrease in labour capacity because of family growth, marriage, illness or death; increased wealth or indebtedness) may also evoke change in the farm system. Thus, the human setting is very dynamic. In order to understand farm systems, it is necessary to know what has changed and is changing: the historical development and its causes and effects.

2.3   Decision-making within farm households

One of the crucial variables of farm systems is decision-making within the farm household about objectives and how to reach them with the available resources, i.e. the type and quantities of plants grown and animals kept and the type of techniques and strategies applied. The way a farm household makes its management decisions depends on the characteristics of the household, e.g. number of men, women and children; their age, state of health, abilities, desires, needs, farming experience, knowledge and skills; and the relations between household members.

Members of a farm household may function more or less independently and have different needs, preferences and objectives. There may be sub-units within the household, each under the management of an adult (often a woman) responsible for producing and finding food for the sub-unit, but also with responsibilities to the household as a whole. As women are commonly the collectors of water, fuel, medicinal herbs etc., they are directly affected by environmental degradation and, in their decision-making, may place higher value on taking care of the environment than the male family members do (Dankelman & Davidson 1988, Shiva 1988).

The decision-making process is influenced by the culture of the community to which the household belongs. For example, in patriarchal societies, decisions are taken by the household head, a man. In matriarchal societies, this may be true only to a lesser extent and only

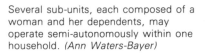

Several sub-units, each composed of a woman and her dependents, may operate semi-autonomously within one household. *(Ann Waters-Bayer)*

with respect to certain distinct responsibilities. However, also the personalities of the different household members will affect decision-making. The man may be the formal decision maker, but – in reality – it may be the women in the household who determine what is done.

Complex factors are involved in decision-making within the farm household, including the biophysical characteristics of the farm, the availability and quality of external inputs and services, and the socio-economic and cultural processes within the community. Household members may interpret these factors in a different way than outsiders perceive them. For example, risk management may be highly valued by farmers but underestimated by outsiders, who may therefore make unacceptable technical proposals. To help farmers develop farm systems that suit the local biophysical and human setting, outsiders must achieve some understanding of how decisions are reached by the household and the logic behind them.

In addition, as the ecological, socioeconomic and cultural environment changes, the farm system must be adjusted accordingly. Thus, farming involves an endless process of making decisions for the short, medium and long term – and the very process by which these decisions are made will also change over time.

2.4   Objectives of farm households

Central to decision-making but also an object of decision-making are the aims of a household with respect to the process and outcome of farming. Each household and each individual within it has specific felt needs and desires, but – judging from smallholders' actions and statements, both those reported in the literature and those known to us personally – farm households appear to have in common various objectives, which can be classified as follows: productivity, security, continuity and identity.

The above objectives may not be classified as such by farmers and are – to a greater or lesser degree – overlapping and mutually reinforcing. However, we make this classification here to give development agents (and the farm communities with whom they are working) a framework for assessing the status and development paths of farm systems in terms of sustainability. By assessing the extent to which these objectives are being attained, it is possible to identify current problems of farm families and their needs for support in developing their farm systems.

The different techniques applied by farmers serve to meet one or several of these objectives. A farmer combines the techniques in such a way that, in the farmer's perception, the household's objectives are best attained, given the limitations of the specific farm.

Productivity

Productivity is the output per unit of land, labour, capital (e.g. livestock, money), time or other input (e.g. cash, energy, water, nutrients). Outsiders tend to measure farm productivity according to total biomass yield, yield of certain components (e.g. grain, straw, protein content), economic yield or profitability, often with a view to maximisation of output per unit of land. Farm families and individuals

Smallholders are not likely to measure productivity only in terms of market values. Preparing millet for home storage and consumption. *(Ann Waters-Bayer)*

within them have their own ways of defining and assessing productivity, measured perhaps per unit of labour expended at planting or weeding time or per unit of irrigation water used etc. It is important that outsiders become aware of these indicators, for these are ultimately the decisive ones for the farmers.

Productivity is the primary objective of farming, also for LEIA households, but they are not likely to measure it solely in terms of market values. The household has a range of needs for consumption, health, housing, education, security, social links etc. Their decisions about what to produce are based not only on market demand but also on what can or cannot be obtained on the market. Specialisation may not be advantageous if the products that then have to be bought (e.g. food, wood) are expensive, of poor quality or not available regularly. For home consumption and for face-to-face local marketing, not only quantity but also quality of the products (e.g. taste, storability, nutritional value, cooking quality, pest resistance) will be important. The household members also value the nonfood products of food crops, e.g. straw, fodder, wood, manure. If modern varieties produce higher yields than traditional ones, farmers may prefer the modern varieties for selling but the traditional ones for eating at home, as these may taste better and suffer fewer losses in storage and processing (ICRISAT 1986).

Normally, smallholder households also attach high value to an even distribution of production over time in order to secure year-round needs and make effective use of the available labour resources.

Security

Seeking security means minimising the risks of production or income losses resulting from variations in ecological, economic or social processes. These variations may involve 'minor' fluctuations in, for example, weather, pests, market demand, resource assessability, labour availability; or 'major' disturbances caused by stress (e.g. nutrient depletion, erosion, salinity, toxicity, indebtedness) or shock (e.g. drought, flood, a new pest or disease, a sharp rise in input prices or fall in output prices). Scientists often express the level of security in terms of variability of production, based on statistical risks of, for example, drought. Farmers may assess the security of their farm systems according to food security, or to degree of self-reliance in obtaining inputs or in marketing products (Conway 1987).

For smallholder farmers, security in production of subsistence goods or income is vital: their very survival is at stake. For this, they need secure access to resources such as land, water and trees. The quest for security affects the choice of techniques and strategies, e.g. in drought-prone areas, the best survival strategy is to keep drought-resistant animals, although they have lower production potential than other animal species or breeds. To be able to apply techniques that raise productivity but also risk (e.g. use of artificial fertiliser in drought-prone areas), it may be necessary to apply, at the same time, techniques that reduce risk (e.g. organic fertiliser). Securing rights of access to resources and externalising risks (e.g. crop insurance, community cereal banks) may make it possible for farmers to use more productive resources and techniques.

The farmers' objectives of productivity and security correspond to the criterion for sustainable agriculture defined in Chapter 1 as 'economic viability'.

Continuity

Farmers who wish that they and their children may continue their way of life have a vested interest in maintaining the potential of the farm system to yield products, i.e. in maintaining the resources which represent the productive 'capital' of the farm. This capital can be lost through, for example, erosion, loss of organic matter in the soil, nutrient depletion, animal mortality, deforestation, pollution, loss of indigenous knowledge, or deterioration of farm implements.

The extent to which the natural resources are conserved so that farming can be continued can be measured in terms of soil condition (e.g. rate of erosion, soil life, soil structure, soil fertility), water reserves (e.g. groundwater level), or nutrient/energy reserves (e.g. stored water, livestock holdings, perennial plants) etc.

In long-established LEIA systems, farmers have developed ways of conserving natural resources for farming, e.g. in shifting cultivation systems, by long fallow periods; in pastoral systems, by intensive animal care; in permanent farming systems, by nutrient harvesting and recycling. In ecologically vulnerable and risk-prone areas, care of the natural resource base is an important traditional management objective, which has often been institutionalised in local regulations, customs or religious rites. However, where greater integration into a market economy has taken place and short-term private consumption needs prevail over long-term communal survival needs, or where poverty has become so extreme that only day-to-day survival can be sought, traditions of deliberate care for the environment may have been lost (Rhoades 1988).

Besides the biophysical capital such as soil, water, trees and animals, the capital of a farm household also includes management capacity, health, links within the community, farm infrastructure (terraces, buildings, implements), financial capital, commercial and farm service infrastructure, and political influence (e.g. representation or lobbying power in local government as well as at higher levels).

To ensure the continuity of their way of life, farmers must also be able to adjust to change. The capacity to adapt to changing conditions ultimately determines the sustainability of agriculture. Vital to such adaptability at the farm level is the capacity to manage farm development: to choose appropriate combinations of genetic resources and inputs; to develop new techniques; to fit innovations into the farm system. In this context, resource-enhancing techniques are particularly important, as they can be used not only to rehabilitate degraded land but also to create new opportunities as new needs arise.

The farmers' objective of continuity embraces the criteria of sustainable agriculture defined as 'ecological soundness and adaptability' in Chapter 1.

Identity

Identity is defined here as the extent to which the farm system and individual farming techniques harmonise with the local culture and the people's vision of their place within nature. It involves aspects such as personal preference (e.g. to keep horses), social status (e.g. to possess many animals that display wealth or can be lent to the needy), cultural traditions (e.g. to perform ceremonies), social norms (e.g. men's and women's roles) and spiritual satisfaction (e.g. being at one with Nature and God). An important aspect of a person's or a community's identity is self-respect.

The farmers' objective of maintaining identity embraces the criteria for sustainable agriculture defined as 'social justness and humaneness' in Chapter 1. It is also related to conservation and transformation of natural resources, as the structure and form of the agrarian landscape is part of the identity of the people who live within it and exercise their influence upon it.

Farmers usually have a strong need to identify with the local culture. History and tradition play an important role in their lives and in their ways of farming. Changes that are incompatible with their social, cultural and spiritual values can cause great stress and counterforces. Being able to gain a decent living befitting the local culture gives an individual or a family self-respect. Self-respect may also be derived from acting in solidarity and striving toward equality of all members of the community. A farm family's or community's feeling of identity is maintained by technologies that permit them to be self-reliant and to control decision-making about use of local resources and products. Within the farm household, the opportunity to influence decision-making, contribute to family welfare and share in the benefits gives self-respect to all members, both male and female.

In assessing the degree to which identity is maintained, outsiders tend to project their own personal and cultural values. We, too, would like to think that criteria such as equal access to productive resources and benefits of production, participation of both men and women in decision-making, and fair division of labour all reflect universal human values, but we are aware that different cultures lay different emphases upon these criteria. Whatever vision the community may have of itself, the fact remains that – if personal and cultural identity is lost or suppressed – the agroecosystem is likely to collapse as a result of social instability, e.g. resignation or revolution.

2.5   Finding the balance

Thus, to be sustainable, a farm system must generate a level of production that satisfies the material (productivity) and social (identity) needs of the farm household, within certain margins of security and without long-term resource depletion. As the objectives of security, continuity and identity usually compete with immediate productivity, an optimal instead of a maximum level of productivity has to be sought in order to ensure sustainability of the farm system. The relative effort needed to attain each of these objectives will be site-specific. For example, in highly risk-prone areas, more effort has to be put into minimising risk and conserving resources and relatively less time and energy of the farm household can be devoted to increasing production.

As the conditions of farming and the needs of the farm household change over time, farm systems are dynamic. Farm households that survive are in a process of constant adjustment to change. As the multiple objectives of farming may also compete to different degrees at different points in time, decision-making by farm households involves constantly seeking a new balance between these objectives. The same applies at the community level.

The first step in seeking a new balance is careful evaluation of the viability of the present way of farming. The existing farming techniques must be assessed in terms of their economic, ecological and sociopolitical sustainability, and available alternatives must be assessed in the same way. In this process, the farm family or farm community – depending on the level of decision-making concerned – will begin to recognise the extent to which their specific objectives and ways of achieving them can be matched with current technical opportunities and limitations.

Also, when new opportunities (e.g. availability of consumer goods, formal education) lead to changes in value systems, or when new crops or new techniques are introduced, farmers must be aware of how pursuit of these values and adoption of these innovations could affect the sustainability of their farm systems. Attractive techniques that quickly increase productivity with little effort, e.g. artificial fertilisers or pesticides, may have long-term negative effects on the natural resources and, hence, on sustainability. These must be openly and critically discussed within the farm community. Development agents have a particularly important role to play in stimulating such discussions and helping farmers obtain the information they need to be able to make sound decisions.

Traditionally, 'keeping the farm system in balance' and, in a broader sense, natural resource management have been the responsibility – consciously or unconsciously – of farmers, individually or as a community. Especially in areas of high environmental risk, such as mountainous or dry areas, the community played an important role in maintaining the land resources. Land use was regulated to ensure that the survival of the community was not endangered by individual mismanagement.

In view of the many site-specific factors involved in finding a balance between the objectives of local resource users, it must be clear that farmers will remain the prime actors in adapting their farm systems to the needs for survival and well-being. Cooperation at the community level may enhance the process in some cases, or may be absolutely necessary in others, e.g. where common land is involved. Experience thus far has shown that natural resources can be managed in a sustainable way only if all men and women who depend on the resources participate in planning and action – including the landless poor who use common lands (Kerkhof 1990, de Leener & Perier 1989, Agarwal & Narain 1989).

Outsiders can support farmers in the process of finding the balance by strengthening their capacity to assess their situation, to develop new technologies and to adapt their farm systems. Outsiders can create favourable conditions for LEISA development, such as favourable prices and secure land or tree tenure. In addition, they can do basic research on particular aspects of agroecology that must be better understood, if farmers are to be supported in their search for sustainability. In Part III, the complementary functions and responsibilities of farmers and outsiders in developing technologies for sustainable agriculture are further elaborated.

3 Technology development by farmers

Research is often seen as a monopoly of scientists who are steadily pushing forward the frontiers of knowledge and developing new technologies. International research institutes are seen as the central source of innovation. Their ideas are passed on to national and regional scientists – also in commercial companies – doing applied and adaptive research, and the products of this research are meant to be spread via extension agents to farmers.

However, the case presented in Box 3.1 illustrates that formal institutions of agricultural research and extension are not the sole agents of innovation and dissemination of new technologies. Empirical evidence from all parts of the world shows that the 'central source of innovation model' does not conform with reality (Röling 1988, Biggs 1989, Chambers et al. 1989). Most agricultural technologies in use in the world today were developed by farmers, not by formally educated scientists. Farming systems based on these technologies provide the food for the majority of the world's population. Innovations are developed and diffused by farmers through processes of which many outsiders are completely unaware. Before discussing the contributions of scientists and other outsiders to developing technologies for sustainable agriculture, we therefore start with the farmers and their achievements and capacities in this field.

Box 3.1
The case of daddawa in central Nigeria

Daddawa (also known as *dawadawa, soumbala* and local *maggi*) is a widely used condiment in West African cooking. It is traditionally made from locust beans from the *Parkia biglobosa* tree. Probably because *daddawa* is an indigenous food, sold at village markets and traded via traditional rather than modern commercial channels, it has received next to no attention in official statistics. Similarly, the locust-bean tree has been neglected in agricultural/forestry research and extension. Nevertheless, amazing changes are occurring in the *daddawa* business in central Nigeria.

Within the last few years, *daddawa* is increasingly made from soybeans. This innovation can be credited to the ingenuity of Nigerian farm women who developed a soybean processing technique which requires much less time and labour than locust-bean processing. The greatest saving is the reduced cooking time (6 instead of 24 hours) and thus the reduced need for cooking water and fuel. Furthermore, the women can now grow the soybeans on their own fields instead of having to buy locust beans or tree-harvesting rights. The trees are usually owned by men.

Information about soybean growing and processing has spread from woman to woman, not only within the indigenous Kaje ethnic group. Also Fulani families who have settled in the area have benefited from the Kaje women's extension activities. Kaje women gave Fulani women some seed to sow in their kitchen gardens, advised them about cropping techniques and showed them how to ferment the beans to make *daddawa* for home consumption.

These activities remain largely unseen by the outsiders who consider themselves to be the major (or sole) catalysts of development (Waters-Bayer 1988).

Much of the *daddawa* sold in Nigerian markets is now made of soybean, a result of technology developed by local women. *(Ann Waters-Bayer)*

3.1 'Traditional' farming and innovation

It has always been basic to rural people's struggle for survival to produce enough food for the family and to maintain the productive capacity of the land, so they can continue producing food for the family and for future generations. In order to succeed in this struggle, technology development through experimentation and integration of new knowledge has always been a necessary part of farming.

Farming systems are in constant change, as experience is accumulated, populations increase or decrease, new opportunities and aspirations arise, and the natural resource base deteriorates or improves. The attempt is continuously made to adapt to the new conditions. As we have seen in Chapter 1, these adaptations have not always been adequate and entire cultures have disintegrated as a result. However, there are also countless farming communities which managed to survive and, in some cases, to thrive by exploiting natural resource bases which their forebears have used for generations. Through a process of innovation and adaptation, indigenous farmers have developed numerous different farming systems, each of which is finely tuned to its ecological, economic, sociocultural and political environment.

According to Richards (1988), although earlier observers admitted that farming practices of pre-industrial societies were well-adapted to local conditions, these traditional practices were often regarded as static – as if reached by happy accident at some point in the evolutionary process and then copied without further thought, generation after generation. This assumption is contradicted by:

● The facts of agrarian history, which reveal long-continued dynamism; for example, how else could two New World crops, maize and cassava, establish themselves as major staple foods throughout tropical Africa in the 450 years since first introduced by the Portuguese?

● The innovativeness of farmers to the present day, which has been well documented by, for example, Johnson (1972), Biggs & Clay (1981), Budelman (1983), Reij et al. (1986), Richards (1986), Altieri (1987), Lightfoot (1987), Millington (1987), McCorkle et al. (1988).

In newer literature, the innovative farmer is now accepted as the norm, not the exception and, in recent years, there has been a growing scientific interest in locally-developed farming systems and technologies. These are seen as a source of sound ideas, locally-adapted cultivars and practices which could lead to sustainable use of local resources.

The local or indigenous knowledge (IK) of a farming population living in a specific area is derived from the local people's past farming experience, both that handed down from previous generations and that of the present generation. When a technology developed elsewhere has been incorporated by local farmers as an integral part of their agriculture, it is as much a part of their indigenous knowledge as self-developed technologies. Farmers' practical knowledge about the local ecosystem – about the natural resources and how they interact – is reflected in their farming techniques and in their skill in using the natural resources to gain their livelihood.

However, IK is far more than merely what is reflected in technical methods. It also entails many insights, perceptions and intuitions related

A prime example of innovation by farmers is this wooden sowing machine developed in China, with which seeds of different sizes can be sown three rows at a time. *(Lex Roeleveld)*

Box 3.2
Consultant admires Indian agriculture in 1889

In 1889, Dr J.A. Voelcker, Consulting Chemist to the Royal Agricultural Society of England, was sent by the British government to study Indian agriculture. Voelcker toured the country extensively for over a year. In his report published in 1893 he stated: ''. . . nowhere would one find better instances of keeping land scrupulously clean from weeds, of ingenuity in device of water-raising appliances, of knowledge of soils and their capabilities as well as of the exact time to sow and to reap, as one would in Indian agriculture . . . It is wonderful, too, how much is known of rotation, the system of mixed crops and of fallowing. Certain it is that I, at least, have never seen a more perfect picture of careful cultivation, combined with hard labour, perseverence and fertility of resources, than I have seen at many of the halting places on my tour'' (Dogra 1983).

to the environment, often including lunar or solar cycles, astrology, and meteorological and geological conditions. This 'folk wisdom' is usually integrated with belief systems and cultural norms, and expressed in traditions and myths. Also traditional methods of communication, e.g. through songs or proverbs, and traditional structures for social organisation and cooperation form part of the local knowledge system. Such knowledge systems are not always easily understood by people trained in Western science (Thrupp 1987).

IK is not static. New techniques developed by a member of the community or introduced from outside, if locally beneficial, spread by word of mouth, imitation or informal education in village meetings, initiation rites etc. and become part of IK. As new experiences are gained, others lose their relevance because of changing circumstances and needs. The capacity of farmers to manage change is also part of their IK system. Thus, IK can be seen as a dynamic and ever-changing accumulation of the collective experience of generations.

3.2   Indigenous farming systems, practices and knowledge: some examples

Already in early colonial times, perceptive observers commended the intricate and careful cultivation methods of 'native' inhabitants (see Box 3.2). Classic studies of Asian and African agriculture were made in the 1940s and 1950s, e.g. de Schlippe (1956), Conklin (1957), Allan (1965). A growing number of publications are now appearing about indigenous knowledge systems and the farming systems based upon them (e.g. Brokensha et al. 1980, Biggs & Clay 1981, Rhoades 1984, Richards 1985, Marten 1986, Wilken 1987, Warren et al. 1989, Dupré 1990), which reveal their complexity and sophistication in dealing with environmental hazards.

The following examples of indigenous practices illustrate how well farmers in the tropics learned to manipulate and derive advantage from local resources and natural processes, applying the principles of agroecology without knowing that this term exists. The principles of agroecology as discerned by scientists will be presented in Part II of this book, but first let us take a look at some of the practical applications evolved by farmers through a process of informal research and development.

Examples of indigenous land-use systems

Forest gardens. In many parts of the humid tropics, indigenous systems of forest gardening (silvihorticulture) have been developed. For example, village agroforests have existed in Java since at least the 10th century and comprise today 15 – 50% of the total cultivated village land. They represent permanent types of land use which provide a wide range of products with a high food value (e.g. fruits, vegetables, meat, eggs) and other products, such as firewood, timber and medicines. In their small plots, often less than 0.1 ha, Javanese peasants mix a large number of different plant species. Within one village, up to 250 different species of diverse biological types may be grown: annual herbs, perennial herbaceous plants, climbing vines, creeping plants, shrubs and trees ranging from 10 to 35 m in height.

Livestock form an important component of this agroforestry system – particularly poultry, but also sheep freely grazing or fenced in sheds and fed with forage gathered from the vegetation. The animals have an important role in nutrient recycling. Also fish ponds are common and the fish are fed with animal and human wastes.

Natural processes of cycling water and organic matter are maintained; dead leaves and twigs are left to decompose, keeping a continual litter layer and humus through which nutrients are recycled. Compost, fish-pond mud and green manures are commonly used on cropland. These forms of recycling are sufficient to maintain soil fertility without the use of chemical fertilisers. Villagers regulate or modify the functioning and dynamics of each plant and animal within the system (Michon et al. 1983).

Shifting cultivation. All over the world, shifting cultivation, also called swidden agriculture, has been and still is practised to manage soil fertility. Shifting cultivation involves an alternation between crops and long-term forest fallow. In a typical sequence, forest is cut and burnt to clear the land and provide ash as 'fertiliser' or 'lime' for the soil. Crop yields are typically high for the first few years but then fall on account of declining soil fertility or invasion of weeds or pests. The fields are then abandoned and the farmer clears another piece of forest. The abandoned field is left to fallow for several years or decades and thus has a chance to rebuild fertility before the farmer returns to it to start the process again.

Shifting cultivation is often characterised by a season-to-season progression of different crops which differ in soil nutrient requirements and susceptibility to weeds and pests. For example, the Hanunóo in the Philippines plant rice and maize the first year after clearing, then root crops such as sweet potatoes, yams and cassava, and finally bananas, abaca *(Musa textilis)*, bamboo and fruits (Conklin 1957).

Shifting cultivation practices throughout the world vary immensely, but there are basically two types of systems:

● **Partial systems**, which evolve out of predominantly economic interests of the producers, e.g. in some kind of cash crop, resettlement and squatter agriculture.

- ● **Integral systems**, which stem from a more traditional, year-round, community-wide and largely self-contained way of life.

Provided that the population pressure does not exceed the carrying capacity of the area at that level of technology, integral systems of shifting cultivation present a good equilibrium between humans and their environment.

Transhumant pastoralism. Where livestock are kept in regions with large seasonal differences in precipitation and temperature, a rational low-external-input management form is to move the livestock with the season. American ranchers use winter and summer pastures; shepherds in European mountain areas use alpine and valley pastures; African pastoralists use wet-season and dry-season pastures. Traditionally, pastoral peoples, such as the Fulani in West Africa, keep their livestock in more arid areas during the wet season, where forage quality is relatively high (Breman & de Wit 1983). In the dry season, when water becomes scarce in the north, they move their animals further south to more humid areas, where the livestock can graze the crop residues in harvested fields and the still-green grass in low-lying areas along streams and rivers. These herds are important sources of manure for arable farming. However, this system of resource use was disturbed by the drawing of national boundaries, the setting up of wildlife reserves and commercial ranches (usually in the best grazing areas), and the expansion of cash cropping as well as subsistence cropping to support rapidly growing populations. Especially, the cultivation of low-lying areas with crops, such as rice, is depriving transhumant pastoralists of vital dry-season grazing areas for their herds.

Integrated agriculture – aquaculture. Particularly in Asia, the productive use of land and water resources has been integrated in traditional farming systems. Farmers have transformed wetlands into ponds separated by cultivable ridges. An outstanding example is the dike-pond system which has existed for centuries in South China. To produce or

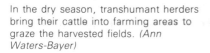

In the dry season, transhumant herders bring their cattle into farming areas to graze the harvested fields. *(Ann Waters-Bayer)*

maintain the ponds, soil is dug out and used to build or repair the dikes around it. Before being filled with river water and rainwater, the pond is prepared for fish rearing by clearing, sanitising and fertilising with local inputs of quicklime, tea-seed cake and organic manure. The fish stocked in the pond include various types of carp, which are harvested for home consumption and sale. Mulberry is planted on the dikes, fertilised with pond mud and irrigated by hand with nutrient-rich pond water. Mulberry leaves are fed to silkworms; the branches are used as stakes to support climbing vegetables and as fuelwood. In sheds, silkworms are reared for yarn production. Their excrements, mixed with the remains of mulberry leaves, are used as fish feed. Sugarcane plants on the dikes provide sugar, young leaves are used to feed to fish and pigs, and old leaves to shade crops, for roofing thatch and for fuel; the roots are also used as fuel. Grass and vegetables are also grown on the dikes to provide food for the fish and the family. Pigs are raised mainly to provide manure but also for meat. They are fed sugarcane tops, byproducts from sugar refining, aquatic plants and other vegetable wastes. Their faeces and urine, as well as human excrement and household wastes, form the principle organic inputs into the fish pond (Ruddle & Zhong 1988).

Soil fertility management practices

Indigenous farmers have developed various techniques to improve or maintain soil fertility. For example, farmers in Southern Sudan and Zaire noticed that the sites of termite mounds are particularly good for growing sorghum and cowpea (de Schlippe 1956). Farmers in Zaachilla, Mexico, use ant refuse to fertilise high-value crops such as tomatoes, chili and onions (Wilken 1987).

In Senegal, the indigenous agrosilvopastoral system takes advantage of the multiple benefits provided by *Faidherbia* (formerly *Acacia*) *albida*. The tree sheds its leaves at the onset of the wet season, permitting enough light to penetrate for the growth of sorghum and millet, yet still providing enough shade to reduce the effects of intense heat. In the dry season, the tree's long tap roots draw nutrients from beyond the reach of other plants; the nutrients are stored in the fruits and leaves. The tree also fixes nitrogen from the air, thus enriching the soil and improving crop yields (see Table 3.1). In the wet season, the fallen leaves provide mulch that enriches the topsoil, as well as highly nutritious

TABLE 3.1   Millet yield under and near *Acacia (Faidherbia) albida* in Senegal (600 mm annual rainfall)

Yield		Near trunk	Location Edge of tree canopy	Beyond tree canopy
Grain	kg/ha	1669	983	600
	%	253	149	100
Protein	kg/ha	179.9	84.2	52.2
	%	345	161	100

Source: Charreau & Vidal (1965), quoted in Kotschi (1990).

forage. The soil is also enriched by the dung of livestock which feed on the *F. albida* leaves and the residues of the cereal crops. These benefits are extremely important in places where few alternatives exist for improving soil fertility, crop yields and animal nutrition (OTA 1988).

Pest management practices

Traditional practices of biological pest control have recently been the subject of increasing scientific interest, and some interesting examples have been documented. For example, a century-old practice among citrus growers in China is to place nests of the predacious ant *(Oecophylla smaragdini* F.) in orange trees to reduce insect damage. The citrus growers even install interconnecting bamboo rods as bridges for the ants to move from tree to tree (Doutt 1964). Ducks, fish, frogs and snakes are traditionally used to control insects in paddy rice cultivation. Traditional crop selection, planting times and cultivation practices often reflect efforts to minimise insect damage (Altieri 1987, Thurston 1990).

In innumerable traditional systems, living and hiding places for natural enemies of crop pests are maintained by conserving part of the natural environment. In Sri Lanka, large trees and wooded upland were traditionally left standing around the paddy tract and threshing floors to provide nesting and resting places for birds, which the farmers regard as the main agents of insect control. When pests appeared, certain rituals were performed. For example, when caterpillars invaded the paddy, an offering of food and light was placed at sunset on an unstable plantain disk fitted to a stake. The light attracted birds. When the birds attempted to perch, the food fell. When the birds went after the fallen food, they saw the caterpillars and ate them (Upawansa 1989).

Weed management practices

Farmers in the Usambara Mountains in Tanzania developed a multi-storey farming system in which they practised fallowing, intercropping and selective weeding. Young crops do not provide ground cover. The farmers understood that, if weeds are left to grow, they cover the soil, prevent it from heating up or drying out excessively, induce a positive competition which stimulates crop growth, and reduce erosion during rainfall. Later in the season, when the farmers regarded weed competition as negative for crop growth, they did superficial hoeing. They left the weeds on the soil surface as protective mulch, to recycle nutrients and to allow nitrogen assimilation through the bacteria decomposing the plants. The crops could then develop fully. A second generation of weeds was allowed to cover the field completely and produce seed, so as to ensure their reproduction in future seasons. When the dry season started, the field was covered with high weeds. The soil remained moist, soft and rich in humus and was thus in good condition for the next growing season. However, the introduction of the principle of weed-free fields led to the collapse of this system of weed-tolerant cropping, so that fertiliser became necessary to replace the green-manuring effect of selective weeding (Egger 1987).

Genetic resource management

Traditional agriculture is characterised by its great diversity of genetic resources. Many LEIA farmers are highly skilled in managing this diversity so as to ensure sustainable farming systems. For example, farmers in East Java, Indonesia, deliberately make use of different soybean varieties to ensure a supply of fresh seed.

About 70% of soybean production in East Java comes from dry-season cropping on wetland after rice, while the remaining 30% is produced on dryland during the wet season (Soegito & Siemonsma 1985). Most farmers use local soybean varieties which they generally call 'local 29', referring to variety No. 29, which was introduced from Taiwan to Indonesia in 1924. This variety was maintained at Indonesian research institutes but was not multiplied and distributed after its initial introduction at farm level. The farmers' local varieties have small, green-yellow seeds and mature in about 90 – 100 days, like No. 29. However, the variation found among farmers' varieties in terms of time to reach maturity and yield levels indicates that 60 years of intensive cultivation has led to the development of many distinct local varieties.

The farmers have difficulties in storing soybean seed so as to maintain its viability for more than about 6 weeks. To obtain good germination and establishment of soybean after wet-season rice, they need access to fresh seed. To achieve this they developed a system called JABAL (Jalinan Arus Benih Antar Lapang), which literally means 'seed flow between fields' (Figure 3.1). Certain villages have specialised in soybean growing on dryland during the wet season. Yields are lower than those of dry-season soybean, but farmers can get a 50% higher price for their wet-season crop.

Not only the local crop varieties but also the numerous local breeds of livestock testify to the skills of traditional livestock-keepers to manage genetic resources. Local breeds are partially a result of natural selection,

Figure 3.1
Flow of soybean seed in relation to annual rainfall distribution and soybean cropping patterns. (*Source:* Soegito et al. 1986)

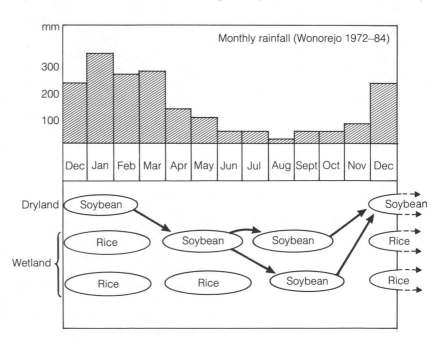

Local breeds are a result partly of natural selection and partly of deliberate selection for specific traits. Trypano-tolerant dwarf zebu cattle in southern Sudan. *(Wolfgang Bayer)*

but they are also a result of deliberate selection for specific traits, above all, for the type of animal that can survive and produce under LEIA conditions. The supposedly 'irrational' marketing behaviour of many livestock-keepers reflects their selection strategies. Animals that are diseased, are weak or have poor mothering qualities are sold; those with proven disease and drought resistance are retained. The animals are also selected to fit into the farming system. For example, in pastoral systems, animals not amenable to herding are culled. Transhumant pastoralists select for animals that can walk long distances. An older animal that knows the route well and keeps the herd going steadily on its way will be kept. Generations of natural and deliberate selection have resulted in local breeds with a high degree of disease resistance or tolerance and capable of subsisting on seasonally scarce and low-quality feed resources (Bayer 1989).

Microclimate management practices

Local climate plays a dominant role in the lives and fortunes of farmers everywhere. Farmers in the tropics have developed several ways of influencing microclimate so as to improve the conditions under which crops and animals can grow. The effects of frost (in tropical highlands), hail, strong wind, extremely dry air and daily peak temperatures on plants and animals can be very great, and buffering these may make the difference between a yield and a complete loss.

Farmers influence microclimate by retaining and planting trees, which reduce temperature, wind velocity, evaporation and direct exposure to sunlight, and intercept hail and rain. They apply mulches of ground-covering plants or straw to reduce radiation and heat levels on newly planted surfaces, inhibit moisture losses and absorb the kinetic energy of falling rain and hail (see Box 3.3). When night frost is expected, some farmers burn straw or other waste materials to generate heat and

---

Box 3.3
Indigenous management of microclimate in Tanzania

In 1980, the indigenous knowledge on microclimate management in Tanzania was tapped via a newspaper contest. Of the many replies, only examples of mulching practices are given here.

Materials used as mulch include tree leaves, dried or green banana leaves, grass, straw, chopped maize stalks, intercrop residues, prunings, weeds, ash, animal dung and household rubbish. Fallen leaves, creeping plants and short intercropped plants were mentioned as having natural mulch effects. Creeping cover crops are also believed to trap dew under certain conditions. In some places, rock mulches are used because of their potential to absorb daytime heat and release it at night.

In rainfed production of rice, tomato and maize in dry areas, a soil mulch layer is created by tillage. The topsoil becomes drier but remains in good condition to receive seeds, and moisture in deeper layers is conserved. In some areas, only the places where seeds are sown are deeply hoed, in the belief that accumulated loss of the topsoil moisture will be slower.

Flood water is used to suppress weeds in rice growing, for soil protection in sugarcane fields and over artificially hardened soil in some coffee-growing areas with heavy rainfall.

In wetter and colder areas or seasons, ridges promote good drainage and root growth. Dark, rotten weed residues are often placed on top of the ridges to absorb heat and thus increase soil temperature (Stigter 1987a).

Box 3.4
Microclimate manipulation in betel vine gardens

Betel vine plants *(Piper betle)* require a cool climate and high humidity during their life span of 2 – 3 years. If the plants are exposed to extreme heat, the leaves become dark green and brittle. If a cool climate and shade are created in the garden, the leaves will be light green and feathery and will fetch a good price in the market.

Farmers in Southern India therefore manipulate the climate in the garden to provide the necessary coolness. They dig long trenches 2.5 feet deep, 2 feet wide and 3 feet apart, on the edges of which they plant *agathi (Sesbania grandiflora)*. After the *agathi* plants have grown 6 feet high, betel vine cuttings are planted beside them. As the *agathi* plants grow taller, they form a canopy which diffuses sunlight. The trenches are impounded with water 2 feet deep. By means of splash irrigation from the trenches, the soil for the betel vine is always kept moist. The garden borders are completely closed by dried banana leaves or plaited coconut leaves. The hot air outside is thus prevented from entering the garden and the dense canopy of the full-grown *agathi* plants provides a cool phytoclimate. The water in the trenches increases the humidity within the garden. All in all, the garden acts like a cool, air-conditioned room and the betel vine grows luxuriantly with broad, light green, feathery leaves (Balasubramainam 1987).

---

produce smog, which traps outgoing radiation. The raised planting beds, mounds and ridges often found in traditional systems serve to control soil temperatures and to reduce waterlogging by improving drainage. Also natural dew is manipulated and exploited (Wilken 1987, Stigter 1987a). An ingenious system of microclimate manipulation by Indian horticulturists is described in Box 3.4.

Local classifications of soil and land use

Most indigenous farmers can quickly identify major soil types and properties according to characteristics such as colour and texture. Farmers' assessment of soil properties often goes beyond the inherent fertility to include an assessment of workability and response to amendments. Also economic and geological factors, e.g. distance to the village, slope, water-holding capacity, presence of rocks and irrigation water, may be taken into account. Examples of such sophisticated classification systems in Mexico and Guatemala are given by Wilken (1987).

Eger (1989) describes a system of land-use classification in Burkina Faso based on local farmers' knowledge. He compared the effectiveness of land-use classification on the basis of aerial surveys and laboratory analysis of soil samples with a classification on the basis of local knowledge, and concluded that farmers' knowledge is far superior to the outsiders' assessment of soil qualities for certain crops.

Farmers often know the soil properties in the wider area, and may deliberately use these differences in soil properties to make optimal use of the available resources and to spread risks (see Box 3.5).

Many other examples of effective indigenous farming practices have been described, e.g. related to risk minimisation strategies (Eldin & Milleville 1989), slope management (Wilken 1987, Mountjoy & Gliessman 1988, Rhoades 1988), water management (Pacey & Cullis 1986, Reij 1990, Ubels 1990) and pastoral resource management and animal health care (Mathias-Mundy & McCorkle 1989, Niamir 1990).

3.3   Common traits of indigenous farming systems

Where farming communities have managed to survive for generations by using only locally available resources, their farming systems tend to have certain principles and processes in common:

- **Holistic world view.** The farming communities commonly believe that nature is given by a superior power to be handled with care. Numerous rituals accompany farming activities, and maintaining the quality of the natural resources is considered vital. The farmers see

Smallholder farming systems are often community-based. Communal rice harvesting in the Philippines. *(Kees Metselaar, Hollandse Hoogte)*

Box 3.5
Using different soils in a catenary sequence

In the seasonal tropics, distinct soils are often found in a regularly occurring sequence or 'soil catena' from valley floor to hill crest. Smallholders are adept at designing cultivation and planting strategies to get the best out of the different soils in a catena. This example comes from Mogbuama, a Mende village in central Sierra Leone, where the people grow at least 49 varieties of rice, mostly Asian (Oryza sativa) but also 3 African rices (O. glaberrima), with up to 20 intercrops including cassava, sorghum, beans, maize, cotton and various vegetables.

The farmers can choose from two types of catenary sequence. On the escarpment to the east, the sequence runs from free-draining gravelly upland soils (kotu and ngongoyo in Mende), to sandy lower slope soils (nganya) and seasonally waterlogged swamp soils (kpete) in valley bottoms. West of the village is a complex series of river terraces and riverine flood plains at the foot of the escarpment. Here, the sequence is of silty river-terrace soils (tumu) running down to seasonally flooded riverine grasslands (bati).

Early-ripening varieties. Most households run short of rice in the weeks before the main harvest. A late start to the rains may delay harvest. Scarce resources of seed and labour may be wasted when there is an unexpected gap in the rains. Early rains make it difficult to burn and clear the farms. A poor burn means poor fertility and too much weed growth,

leading to excessive labour requirements. To beat the problem of preharvest hunger, all households are keen to grow some early-ripening rice. They regard 8 of their rices as fast-growing enough for this. One, pende, ripens in less than 90 days.

Typically, planting begins on the lower part of the catena (on the finer, less free-draining nganya and tumu soils) and proceeds upslope toward the more free-draining soils as the rains become established. On the lowest-lying tumu soils, planting is sometimes commenced even before the first rains.

On flood plain (bati) soils, a typical strategy is to interplant a quick rice with a flood-tolerant one. By the time the river overflows its banks at the peak of the wet season, the short-season rice will have been harvested, leaving the farm to the flood-tolerant one, to be harvested as the water retreats.

Upslope varieties. The second main group of rices (about 25 varieties) comprises the higher-yielding medium-season ones. They are planted upslope, especially on better-draining soils, with the onset of the regular rains. The farmers choose varieties to suit specific conditions, e.g. for very fertile soil or poorly burnt land, to compete strongly where weed growth is likely to be vigorous. Specific intercrop combinations are appropriate to different points on the soil catena. For example, cotton grows well only on gravelly soils. Intercrops belong to the women, who decide what mixtures should be planted where.

Flood-tolerant varieties. A third group of rices comprises

the flood-tolerant and 'floating' varieties (yaka) suitable for wetlands (bati and kpete soils). The original African yaka rices have been replaced by Asian varieties introduced during the last 70 years. Each has acquired a local name. Yaka cultivation requires little labour: grass is uprooted from a swamp cleared in previous years, and seed is broadcast and can be left to fend for itself on the rising flood.

Yaka rices, although heavy yielders, are long-season varieties (150 – 170 days). They are considered inferior in taste and nutritional properties to the rice grown further up the soil catena. Since they ripen after the main harvest when rice is abundant, most yaka rices are sold, often at poor prices. Few farmers give priority to this type of cultivation. Women and young men often cultivate small areas of yaka rice to derive a small independent cash income, fitting the work in among the more important tasks associated with growing medium-season varieties further upslope.

Spreading risks. Such an integrated set of varieties allows for great flexibility in adapting to rainfall irregularities and seasonal labour shortages. Most households find it difficult, even in good years, to surmount labour bottlenecks during planting and weeding. When the rains begin early or late, labour problems are magnified. They minimise these difficulties by planting a catenary farm. This allows them to spread their labour requirements and, by varying the exact proportions of early, medium-season and wetland rice planted, they can adjust to climatic contingencies (Richards 1986).

themselves as fitting into a larger whole – as part of nature, rather than its master. They do not separate animal husbandry from cropping, short- from long-term, economics from ecology etc. Farming is not merely "production"; it is a way of life (see Box 3.6).

● **Community-based farming**. In most farming systems with long and lasting traditions, the community plays an important role. It upholds the local culture and knowledge, organises communal labour, designs and controls land use, and manages change. Often land is communally, not privately, managed or owned. The laborious activities that are often necessary to maintain structures or conditions for farming, e.g. terracing, pond construction, and maintaining irrigation works, are communally organised. A council of elders or the village chief takes decisions about annual division of land, the time for sowing and burning, etc.

● **Optimal use of local resources**. Without external production inputs or technical support, the farming community managed to support itself from natural resources by gaining a detailed understanding of the environment, domesticating local plants and animals, experimenting with different ways of managing and using them, and developing site-specific and often very complex technologies to make optimal use of the local resources. In this process, the community members acquired a wealth of knowledge about the conditions under which local plants and animals thrive best.

● **Reliance on genetic and physical diversity**. A wide variety of genetic resources (crops, livestock, trees etc.) is used, primarily in ways that suit the ecological conditions (rather than trying to make major changes in the environment to permit the growth of otherwise unadapted plants and animals). The farming communities commonly have an ideal of self-sufficiency, and the different genetic resources have complex functions for the family or community: to produce foods, medicines, fuel, building materials, fodder, or to restore soil fertility, create a reserve for poor seasons or years etc. Production for the market usually plays only a minor role. Crop mixtures enhance growth and total yield. Advantage is taken of complementarity between crops, e.g. by combining legumes and cereals in order to improve nutrient availability, control diseases etc. Animals are also used in a complementary way: they are fed on wastes and by-products, provide manure and protein-rich food, and serve as a buffer in case of crop failure.

● **Soil protection and recycling natural nutrients**. Various methods of soil and water conservation are practised; much emphasis is given to fallowing and to recycling plant and animal waste, e.g. by organic manuring through mulching, green manuring and composting. Trees play important roles in protecting the soil and as a source of fertility.

● **Risk minimisation**. Greater importance is attached to reducing or spreading risk than to maximising production. Special strategies have been developed to minimise risks, e.g. selecting crops with built-in resistance to extreme climatic conditions or pests, creating diversity by using mixtures of varieties, crops, animals etc.

Box 3.6
Agrocentric culture in Peru

Some 20 000 years ago, groups of Asian people arrived in what is now Peru. In creating their agriculture, different ethnic groups strengthened their organisation and gained independence by domesticating a great repertoire of plants and some animals, creating varieties and races, and developing techniques of farming, processing and storage.

Andean culture perceives nature as if it were a living and highly sensitive animal, capable of responding positively when handled well and therefore capable of being domesticated, but also capable of responding furiously when mistreated.

The Andean man and woman see the flora, fauna, soil and water as parts of a whole, of which they and their children are a part: "We are part of the earth". This relationship does not imply immobility but rather continuous transformation and domestication of the environment, not for the unilateral benefit of man but for the reciprocal benefit of nature and society.

Andean culture is agrocentric since the prime concern of the society is to assure adequate and sufficient food and to produce raw materials for processing. Agrocentrism means that the social organisation, science, art, philosophy, religion, perceptual framework, language and technology are all functions of the farming activities.

The Andean society seeks an integral interrelationship with its medium, as reflected in the careful organisation of space and the eagerness to create beauty that benefits nature and society. For example, the construction of irrigation systems benefits the society, as it allows an increase in production but, at the same time, benefits nature in the sense that it allows an increase in the total biomass production, i.e. a greater quantity of life in the environment.

In a given place, on the basis of the resources provided (soil, water, flora, fauna, climate, landscape), the farmers create the type of agriculture that is possible at that site. This is given a *chacra*, a name that identifies a particular type of agriculture. The *chacra* of the *campesino* can be seen as the 'concentrate' and the centre of the culture. For a technician a plot is no more than a medium for production; for the *campesino* it is the expression of its agrocentric culture which provides food, is a meeting place and a sacred place where rituals are carried out (Grillo Fernandez & Rengifo Vasquez 1988).

● **Site-specific techniques.** Each farming community has developed different farming techniques to suit the specific local conditions. In Peru, for example, in an area of less than 100 ha, 27 different farming systems have been identified (Grillo Fernandez & Rengifo Vasquez 1988). The principles behind the various low-external-input farming techniques can be understood and generalised, but the specific techniques are usually applicable only at the site for which they were developed or at very similar sites.

3.4   Farmer experimentation

Experiments are conducted to examine whether a hypothesis is valid or to determine whether something previously untried will work. In both these senses, farmers conduct experiments. This does not mean that they lay out Latin squares, have a laboratory behind their home, or apply sophisticated statistical procedures to analyse observations. Examples of farmer experimentation are given in Box 3.7.

Of course, not all farmers are experimenters. Some might even appear conservative. However, such observations should not mislead us into seeing all farmers as passive users of ready-made technology received from their forebears (or, more rarely, from extensionists). There used to be a widespread concept in development theory of passive farmers,

Box 3.7
Cases of experimenting farmers

Farmers in Chiapas, **Mexico,** found velvet bean *(Mucuna pruriens)* growing wild in the nearby jungle and noticed that it shaded out all other weeds. They tried planting it together with their maize and, in essentially jungle conditions together with judicious use of chemical fertiliser, harvested 4 t/ha of maize in the same fields year after year without the benefits of either crop rotation or fallowing (Wilken 1987).

Also farmers in northern **Honduras** tried growing velvet bean as a ground-cover crop together with maize. They now obtain maize grain yields of 2.7 – 3.3 t/ha, more than double the national average, without using chemical fertiliser. As ploughing was substituted by a no-till system and the ground cover reduces erosion, costs for land preparation and weeding could be considerably reduced. This technique spread rapidly without being promoted by any private or government agency (Milton 1989).

In a case study on farmer innovations and communication in **Niger**, McCorkle et al. (1988) found many examples of farmers' experiments normally unseen by agricultural scientists. For example, a farmer in Wazeye was experimenting with a short-cycle millet which he discovered during a trip to Filingue, where a market woman was selling it. He liked the look of the seed and the woman gave him a handful to try out back home. For two years he has been cultivating a small plot (7 – 8 m²) and is very pleased with the performance of this variety: it gives many kernels and does very well without fertiliser. He plans to continue and expand cultivation of this hardy, short-cycle millet.

the inactive victims of external oppressive and exploitative forces. More recent empirical studies of farming communities have led to the 'actor perspective' (Long 1990): a view of farmers as people who seek to optimise their advantage through careful strategies and deliberate action.

When trying to make decisions, some farmers take considerable time to explore possibilities and carefully integrate knowledge from various sources. In this process, discussions with trusted peers in comparable circumstances often play an important role. Apart from adaptations of innovations introduced from elsewhere, farmers may routinely make careful observations and small-scale trials of new ideas, germination tests of seeds, and trials of new procedures or work methods. Some different types of farmer experiments are described in Box 3.8.

Research in Africa and India into adaptive responses to drought and famine suggests that farmers' experiments increase in number and complexity after crises (de Schlippe 1956, Juma 1987, Vaughan 1987). Problems or changes perceived by the farmers, e.g. poor harvests, new pests, migration to new areas or availability of new crops, stimulate a search for useful alternatives or new options. Among the Azande farmers' experiments observed by de Schlippe (1956) after a poor harvest, some of the most ingenious were undertaken by women.

When a new crop (wheat) was recently introduced to farmers in Thailand, it was observed that farmers at some sites quickly began to investigate production technologies other than the one being officially extended. This happened spontaneously within the framework of ongoing extension programmes of both governmental and non-governmental agencies. Farmers were most innovative in devising and experimenting with various component technologies where the extensionists did not attempt to be directive in promoting a technology 'package' (Connell 1990).

Box 3.8
Types of experiments by Andean potato farmers

Andean potato farmers are among the most experienced in the world. They depend primarily on this crop for their livelihood and cash income. In the highlands of Peru (2500 – 4500 m above sea level) the potato farming system is ancient and well-defined. Here, experimentation is rarely of a radical nature. Three types of experiment can be identified.

Curiosity experiments. It is not uncommon for farmers to set up experiments simply to test ideas that come to mind. As example: a farmer in Chicche village developed the hypothesis that varieties with apical dominance would yield fewer but larger tubers than varieties without apical dominance.

To test the hypothesis he planted a row with apical dominance and one without.

Problem-solving experiments. Farmers experiment to find solutions to old and new problems. A case in point is the perceived increase in attacks of the Andean weevil (*Gorgojo de los Andes*) in improved potatoes. This led farmers to experiment with the effects of sunlight by exposing the tubers for short periods to the sun. This drove the worms out of the tubers. Tests are done on a small scale first and expanded if success appears likely.

Adaptation experiments. In the highlands, where the farmers know the environment well, adaptation involves technologies introduced from outside, particularly new varieties. These are tested in different

production zones and at different times in the growing season, and their performance is carefully monitored. Meanwhile, farmers keep the old varieties. As a result of such experiments, farmers build up stores of knowledge which allow them to talk for hours about varieties.

Adaptation experiments differ in the case of the tens of thousands of potato farmers who are 'migrating' from the highlands to the lower *'ceja de la selva'* (eyebrow of the jungle) where farming conditions are unfamiliar. Here, farmers systematically test different varieties, eliminate unsuccessful and disease-prone ones, and experiment with planting times and cultivation methods, in an effort to develop new effective knowledge for their new environment (Rhoades & Bebbington 1988).

3.5   Farmer-to-farmer communication

Farmers worldwide form part of various types of networks. Markets, funerals, festivals, school meetings and other occasions provide opportunities for local communication. Networks for exchanging seed or animals allow people from different areas to obtain new genetic materials. Women may develop their own channels of communication. Men may have societies which meet regularly and allow discussion of farm issues. For example, research in the Dominican Republic (Box 1987) revealed the existence of local networks of farmers who regularly discuss among themselves and form concepts, adapt ideas, integrate knowledge and determine acceptable action. The importance of farmer-to-farmer communication will differ according to social organisation and infrastructure (see Table 3.2).

A study in India (Feder & Slade 1985) revealed that, in areas where the Training and Visit (T&V) system was introduced, 47% of the farmers preferred fellow farmers as the primary source of information, 19% preferred the village extension worker, 16% the contact farmer and 10% agricultural radio programmes. In an area where the T&V system was not introduced, the preferences for sources of information were: 82% fellow farmers, 28% demonstrations/field days, 9% agricultural radio programmes and 2% the village extension worker.

In Thailand, the most frequently cited source of technical change in the villages was local farmers. In some instances, farmers had simply

Table 3.2 Sources from which Dutch market gardeners and Indian peasants first heard about innovations, and sources of influence on their decisions to try innovations (%)

Source	First hearing about an innovation		Trying/deciding to use the innovation	
	India	*Netherlands*	*India*	*Netherlands*
Local people (friends, neighbours, shopkeepers)	30	8	25	52
Extension service	58	14	72	47
Mass media	12	78	3	1
	100%	100%	100%	100%

Source: Van den Ban & Hawkins (1988).

worked out their own innovations; in others, they had observed them on nearby fields or during travel to more distant locations. The second most frequently cited source was the market. The government (agricultural officers and teachers) ranked third. Radio and written press were also mentioned but were not important sources of agricultural information (Grandstaff & Grandstaff 1986).

These data highlight the main factor in the diffusion of agricultural information, namely communication among farmers. As we shall see in subsequent chapters, this is an important focus for feeding local farmer experimentation, furthering exchange of results and strengthening local capacity for self-managed change.

3.6 Limits to technology development by farmers

The next obvious question is: If technology development by farmers functions so well, why are agricultural researchers, extensionists and other development workers needed? Indigenous knowledge (IK), farmer experimentation and farmer-to-farmer communication have their limitations. As we have seen in the previous chapter, production sometimes lags behind the needs of the population and, in some parts of the tropics, the technology development process is not fast enough to cope with new problems before serious environmental degradation sets in.

Limits to indigenous knowledge

IK is not uniformly spread throughout a community, and individual aptitudes for storing traditional knowledge and generating new knowledge differ. Each individual possesses only a part of the communities' IK. Specialised knowledge (e.g. treasured insights into medicinal qualities of plants) is often kept secret or known only to a select few, such as elders, midwives or healers. In any case, peasants do not document their knowledge so that it can be made available to strangers. Their knowledge may be implicit within their practices, actions and reactions, rather than a conscious resource.

Certain types of indigenous knowledge may be confined to only a select few, such as this traditional healer in northern Nigeria. *(Ann Waters-Bayer)*

Certain types of knowledge may be tied to economic or cultural roles within the community and may not be known to other members of the community. For example, studies in East Africa have shown that women usually possess remarkable knowledge about the qualities and uses of indigenous tree species, and that many of those insights are unknown to men (e.g. Juma 1987, Rocheleau 1987). Similar examples of knowledge about wild and cultivated plants, animal husbandry etc. which is limited to women are given by Dankelman & Davidson (1988) and Shiva (1988).

Different individuals or groups possess different types of knowledge, depending on their economic functions in the community. Particularly in severely disrupted social systems, farmers within an area may differ greatly in the types of knowledge they hold, and a widely shared and homogeneous knowledge system may not exist.

Farmers' knowledge is limited to what they can sense directly, usually through observation, and what they can comprehend with their own concepts. These concepts grow out of their past experiences. It may therefore be difficult for them to relate to processes which are new or affect them only very gradually or indirectly, e.g. population growth, deteriorating natural resources, external markets.

Many of the older farming traditions and the knowledge stored within them are being lost. Foreign technology, education, religions and values, the marginalisation of agriculture and other factors have led to the marginalisation of farmers' knowledge and ways of spreading it. With the loss of IK also indigenous practices, crop species, breeds, tools etc. are lost. But also the other way round, when for example certain genetic resources become extinct, knowledge about how to use them is also lost. Agrarian cultures in the Third World have not systematically stored traditional technical knowledge which may appear to be no longer relevant but may regain relevance with future changes in farming conditions.

In situations where land is limited and the population continues to grow, the traditional ways of farming may no longer be tenable. When farmers have moved (willingly or not) to new land with different ecological conditions than they knew previously, their IK may not apply there and may lead to misuse of the land.

Limits to farmer experimentation

When farmers perceive problems such as decreasing levels of soil fertility or increasing soil erosion, they may try out various potential solutions either conceived by them or known to them. However, the errors in trial-and-error experimentation may be costly, particularly in terms of time. In their responses to problems or opportunities, farmers are not aware of all the possibilities developed outside their communication network, nor can they be aware of all the repercussions of new technologies. Scientists, too, are not fully aware of all the possibilities and repercussions, but may have more systematic and wider-ranging methods of recognising them.

In his observations of spontaneous farmer experimentation with a new crop in Thailand, Connell (1990) noted the following limitations in the farmers' abilities and effectiveness in generating new technologies:

● **Undirected experimentation**. In their enthusiasm to experiment with the new technology, farmers liked to think up their own personal variants. Other technology variations occurred by chance, without the farmer being aware that she/he was doing something different from the neighbour.

● **Lack of an analytical approach**. Many of the farmers were not analytical in evaluating the techniques they tried in their fields, and were in danger of coming to false conclusions. They did not always understand the underlying reasons for a good or poor yield and attributed the success of a technique to the most obvious difference. For example, in one village, farmers compared wheat plots on the basis of whether they were broadcast or row-seeded, when the main reason for the varying stands was the extent of over-irrigation.

● **Poor experimental design**. Experimenting farmers sometimes did not design comparable units. When they tried out a new technique, the basis for comparison was what they did in a previous season or what was in a nearby field, possibly with a different soil type or management system that would invalidate the conclusion.

Connell (1990) concludes that, for these reasons, the outcome of farmers' technology development is undirected and uncertain. These are areas where farmers' abilities could be strengthened and developed. However, these limitations do not invalidate the concept of farmer experimentation. While the experiment by any one farmer might not be productive, it is very likely that some worthwhile innovation will be developed by farmers when the process takes place within a farming community or larger population with well-functioning (informal) communication channels.

Farmers' experimention may also be limited by:

● insufficient information about potential options in the search for improved technologies;

● insufficient scientific understanding of the processes involved in their experiments;

● too many variables within their experiments, rendering the interpretation of results very difficult;

● inadequate methods of measurement for reaching sound conclusions about what they want to investigate or test;

● isolation of each farmer's experimentation from that of other farmers, which means that they cannot benefit from each other's ideas, findings and interpretations.

Limits to farmer-to-farmer communication

Some of the limits to farmer-to-farmer communication are already implied by the above-mentioned limits to their indigenous knowledge and experimentation. In addition, cultural differences may prevent communication between two cultures, or the differences may be so great that the information gained within one culture may initially appear to be inapplicable within the other. Further constraints to communication between farming groups include:

● long distances, physical barriers (e.g. large rivers, mountain ranges)

Farmers have a great self-interested capacity for experimentation and local technology development. Malian women examining new seed to decide what they would like to test. *(Peter Gubbels)*

and national boundaries between farming peoples, particularly where public transport facilities do not exist or are very expensive;

● political friction between countries, between regions within a country, or between ethnic groups;

● language problems.

Also within a community, class differences or culturally determined divisions between men and women can hinder the flow of information.

On account of these limits to farmers' technology development, farmers faced with rapidly changing conditions cannot be expected to solve all their farming problems with their own knowledge, experiments and communication networks. Outsiders – either local or foreign development workers or scientists – can provide important information and skills needed to widen farmers' base for reflection and action, for example:

● stimuli and encouragement for farmers to join forces with each other to analyse their problems, determine priorities and develop improved technologies;

● basic scientific information about non-observable phenomena;

● options for testing;

● methods of designing experiments and comparing results which help the farmers come to conclusions in which they can be confident;

● ways of achieving further-reaching extension of farmers' findings to other farmers and scientists.

Farmers have a tremendous, self-interested, collective, creative capacity for local technology development. The development challenges facing the world today require that we support the development of this capacity as effectively as possible, by helping farmers make better use of their knowledge about their environment and its problems and opportunities, and by strengthening their experimental and creative powers to develop solutions and to link effectively with agricultural scientists. It is this cooperation between farmers, agricultural advisers and researchers that holds promise for successful development of site-specific techniques for sustainable agriculture.

In order to be able to work together with this aim, all actors involved need an understanding of the principles upon which LEISA is based. In Part II, these principles are outlined, and possibilities for developing LEISA systems based upon them are explored.

Principles and possibilities of LEISA

4 Low-external-input farming and agroecology

4.1 The agroecological view

Present-day ecosystems are the result of millions of years of 'trial and error' in the co-evolution of an enormous diversity of species. In the process, nonsustainable species were eliminated, possibly because they were not adapted to climatic conditions, were too susceptible to pests or diseases, were not able to secure sufficient food or energy, or were simply out-competed by more efficient species. Ecosystems are constantly changing as this natural selection process continues. Ecology, as a biological science, is the study of the relations between the organisms involved and their environment. Despite the large diversity of ecosystems which, fortunately, still exist, certain basic processes and principles have emerged. Ecology can provide some important insights for the study of agricultural systems which, by force or by choice, are likewise constantly changing and being adapted to environmental constraints. The new merger science of 'agroecology' tries to combine elements from both conventional agricultural science and ecology. A number of the principles behind this new science will be discussed here, so as to provide a better understanding of how agroecological principles can be applied to create LEISA systems.

Ecological niches for functional diversity

A central concept within ecology is that of the 'niche': the function or role of an organism in the ecosystem and the resources on which it depends, which determine its chances of survival and its positive or negative effects on other components. More than one species can occupy a niche, and each may help to create the survival conditions of the other. Alternatively, a niche may be empty, at least temporarily, which means that local resources are underutilised and opportunities exist for new components within the ecosystem.

Agroecosystems with many different niches occupied by many different kinds of species – in other words, with a high degree of diversity – are likely to be more stable than those with only one species (as in monocropping) and thus give the farmer more security. However, diversity does not necessarily lead to stability; it may even cause instability if the components are not well chosen, e.g. some trees are hosts of insects or diseases dangerous to crops; and crops, animals or trees may compete for labour, nutrients or water (Dover & Talbot 1987). If, however, functional diversity can be achieved by combining plant and animal species that have complementary characteristics and are involved in positive, synergetic interactions, then not only the stability but also the productivity of low-external-input farm systems can be improved.

Throughout West Africa, agrosilvo-pastoral systems have been developed by farmers and herders, often not of the same ethnic group but using the same land resources in a complementary way. *(Ann Waters-Bayer)*

Complementarity in agroecosystems

Within a farm system, components complement each other when they perform different functions (productive, reproductive, protective, social) and when they fit into different ecological, spatial, economic and/or organisational niches, e.g. when they exploit:

- different soil depths (superficial and deep-rooting plants);
- nutrients to differing degrees (e.g. plants that need specific elements in high or low quantities, that use residual nutrients, that take up specific nutrients more or less efficiently);
- different light intensities (light- and shade-loving plants);
- different levels of air humidity (high or low needs, resistance to wind);
- different levels of humidity in the soil (high or low needs);
- land of different quality (e.g. more or less stones, soil depth, inclination, fertility, humidity, resistance to waterlogging);
- biomass not directly useful for humans (e.g. weeds, crop residues, insects, leaves of woody plants);
- different types and periods of labour;
- different household needs;
- different markets (e.g. crops with different levels of market risks, out-of-season products, livestock).

Synergy in agroecosystems

Components of a farm system interact in synergy when they, apart from their primary function, enhance the conditions for other useful components in the farm system by, for example:

- creating favourable microclimates for other components;
- producing chemicals to stimulate desired components or suppress harmful ones (allelopathic effects of root excretions or mulches);
- decreasing pest populations (e.g. intercrops, decoy and trap crops);
- controlling weeds;
- producing herbal medicines (for humans as well as animals) or herbal pesticides or repellents;
- producing and mobilising nutrients (e.g. by nitrogen fixation or mycorrhizal symbiosis);
- producing plant biomass or waste products which serve as feed for other plants or animals;
- producing soil cover or root structure to enhance water and soil conservation;
- having deep root systems to enhance recycling of water and nutrients which have been leached out or are not within the reach of crops;
- enhancing growth conditions for other components (e.g. animal labour).

Components can also be synergetic in function, e.g. contour strips which conserve soil and water as well as produce fodder and food; hedges around fields which protect against animals or wind and also produce fuel, food, fodder or medicines. Multipurpose plants and animals that combine different functions, e.g. grasses for contour hedges and fodder production, or animals which provide manure, milk and draught power and also serve as capital reserves, are very important in this respect.

Exploiting this functional diversity to its fullest degree results in complex, integrated farm systems that make optimal use of the available resources and inputs. The challenge is to discover which combination of plants, animals and inputs leads to high productivity, high security and resource conservation, relative to the given constraints in terms of land, labour and capital.

4.2   Agroecosystems that simulate natural ecosystems

When plants and animals in a natural ecosystem which have little or no agricultural use are replaced by similar plants and animals which are more useful in agricultural terms, the result is an agroecosystem.

Where ecological and economic conditions are favourable (e.g. flat, fertile land; high market demand; availability of artificial inputs), a farm system with less perennial biomass (trees, shrubs, grasses, animals) and functional diversity than the natural ecosystem may be preferable from a production-oriented point of view. Also socioeconomic conditions, labour availability and the need for efficient use of external inputs, particularly machines, may favour the choice of a more specialised cropping or livestock production system.

Under LEIA conditions, where diverse products are needed and where perennial biomass and functional diversity are of key importance for protecting and reproducing the farm system, agroecosystems would ideally approach the climax ecosystem for the site. In the tropics, this would normally be some type of agroforestry system: in drier areas, savanna-like systems with scattered trees, shrubs and perennial grasses; and in more humid areas, systems which resemble more dense forest.

The characteristics of natural ecosystems can be used as the basis for designing sustainable farm systems. For example, the (ideal) agroforestry system shown in Figure 4.1 is designed to imitate the ways in which natural ecosystems save or accumulate nutrients against the forces of erosion, fire, leaching and volatilisation and, thus, ensure a continuous turnover of biomass. The natural mechanisms of nutrient accumulation are (Woudmansee 1984):

- continuous vegetative cover;

- litter layer on the soil;

- synchronised plant and microbial activities;

- retention of a large portion of ecosystem nutrients in living tissues, particularly in wetland systems;

- broad heterogeneity in rooting structures.

(Common) Forest ecosystem

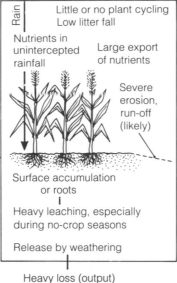

Addition by throughfall

Large canopy volume

Favourable microclimate

Little erosion, run-off

Soil surface | Litter fall

Release by root decay

Mycorrhizal association

Uptake by deep roots

Light leaching

Release by weathering

Little loss (output) from the system

(Common) Agricultural ecosystem

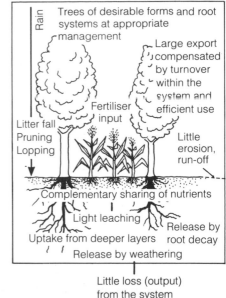

Rain

Little or no plant cycling
Low litter fall

Nutrients in unintercepted rainfall

Large export of nutrients

Severe erosion, run-off (likely)

Surface accumulation or roots

Heavy leaching, especially during no-crop seasons

Release by weathering

Heavy loss (output) from the system

(Ideal) Agroforestry ecosystem

Rain

Trees of desirable forms and root systems at appropriate management

Large export compensated by turnover within the system and efficient use

Fertiliser input

Litter fall Pruning Lopping

Little erosion, run-off

Complementary sharing of nutrients

Light leaching

Uptake from deeper layers

Release by root decay

Release by weathering

Little loss (output) from the system

Figure 4.1
Schematic representation of nutrient relations and advantages of ideal agroforestry systems in comparison with common cropping and forestry systems. (*Source:* Nair 1984)

In this agroforestry design, the natural ecosystem characteristics are combined with the needs of farming. Better soil cover is achieved by including perennial species and/or sowing cover crops. This reduces direct rain impact, traps sediments and may reduce evaporation, so that more water is available. The vegetative canopy and litter lower the soil temperature and, thus, the rates of decomposition and mineralisation. Diversity of plant species, e.g. with different rooting and canopy characteristics, can increase the available resources above and below soil level and use them more efficiently, e.g. sunlight by better canopy structuring or soil nutrient and water volume by deeper rooting and better root structuring, thus decreasing the leaching of nutrients.

In the tree savanna areas of West Africa, the traditional land-use systems retain much of the character of the natural ecosystem. *(Ann Waters-Bayer)*

Figure 4.2
The chronological arrangement of
crop components in a successional
cropping system. (*Source:* Hart 1980)

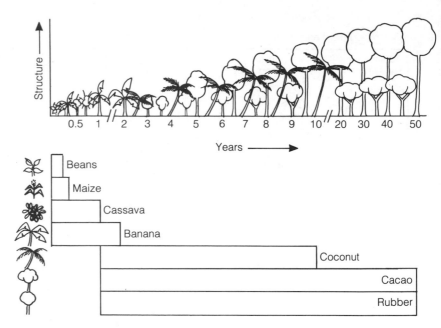

Nutrient export as percentage of total nutrient turnover within the system is lower in the agroforestry than in the cropping system shown in Figure 4.1, as the inclusion of trees increases the total nutrient turnover. Thus, the natural ecosystem characteristic of retaining a large portion of the nutrients within living tissues is included in the farm system. However, such a system can be kept functioning only if nutrient losses are sufficiently recycled and/or replaced (van der Werf 1989).

Hart (1980) has suggested an analogue approach to designing sequential food production systems: managing a farm site so as to imitate natural succession (see Figure 4.2). Beginning with annual grasses and broadleaved species such as maize and beans, the system progresses through stages of plantings to a 'forest' of economically valuable trees and understorey crops with many of the ecological characteristics of a maturing tropical rainforest.

The highly diverse and productive home gardens of Java and Sri Lanka exemplify traditional systems that simulate natural succession: each stage creates the physical conditions (light/shade, soil organic matter etc.) needed by the next. Directing succession rather than fighting it reduces the battle against weeds characteristic of annual cropping systems, lowers the energy and labour costs of establishing perennial crops, and results in an evolving farm system with increasing diversity and reduced susceptibility to disruption (Dover & Talbot 1987).

In such humid areas, where farmers have a much wider choice of species and varieties of useful crops and the competition between trees and field crops is generally lower, the ecosystem analogue hypothesis may be more applicable than in semiarid areas. Here, choice of adapted genetic resources is not so wide, and trees and arable crops compete more strongly for water. Nevertheless, farmers in semiarid areas traditionally practise savanna-like farming systems.

Indigenous agriculture presents many examples of close simulation of natural ecosystems (Wolf 1987). These could be used as a basis for developing farm systems which combine optimal use of local resources with judicious use of external inputs.

5 Basic ecological principles of LEISA

Scientific understanding of LEISA is still in its infancy. However, the insights and experience gained thus far in agroecological studies, indigenous agriculture in the tropics and ecological farming throughout the world point to some basic ecological principles which can guide the process of developing LEISA systems. Here we concentrate on ecological principles, although we acknowledge that socioeconomic, cultural and political principles play no less important a role in this context.

The ecological principles basic to LEISA can be grouped as follows:

1 Securing favourable soil conditions for plant growth, particularly by managing organic matter and enhancing soil life.
2 Optimising nutrient availability and balancing nutrient flow, particularly by means of nitrogen fixation, nutrient pumping, recycling and complementary use of external fertilisers.
3 Minimising losses due to flows of solar radiation, air and water by way of microclimate management, water management and erosion control.
4 Minimising losses due to plant and animal pests and diseases by means of prevention and safe treatment.
5 Exploiting complementarity and synergy in the use of genetic resources, which involves combining these in integrated farm systems with a high degree of functional diversity.

These principles can be applied by way of various techniques and strategies. Each of these will have different effects on productivity, security, continuity and identity within the farm system, depending on the local opportunities and limitations (above all, resource constraints) and, in most cases, on the market. In this chapter, the basic principles are discussed. In Appendix A, examples are given of site-specific ways in which they can be put into practice.

We do not go into detail here about husbandry practices for particular crops and animals. The type and intensity of husbandry will depend on the purpose for which they are used, e.g., whether cowpeas are grown primarily for the greens or the grains, or whether goats are kept primarily for milk, meat or as a savings account. In the last case, for example, it could be assumed that as little labour as necessary would be invested. When consulting standard texts about the husbandry of particular crops and animals, the reader should be aware of the functions of that component within the farm system and of its relationship with other components. Most texts refer to the husbandry of plant and animal species in isolation: sole crops and single-species herds. More information about management practices for crop mixtures or mixed-species herds, e.g. of cattle and sheep, can be obtained from practitioners of 'traditional' farming than from scientists or books.

5.1 Securing favourable soil conditions for plant growth

The physical, chemical and biological processes in the soil are strongly influenced by climate, plant and animal life, and human activities. Farmers need to be aware of how these processes are influenced and how they can be manipulated in order to grow healthy, productive plants. They must create and/or maintain the following soil conditions:

- timely availability of water, air and nutrients in balanced and buffered quantities;
- soil structure which enhances root growth, exchange of gaseous elements, water availability and storage capacity;
- soil temperature which enhances soil life and plant growth;
- absence of toxic elements.

Essential constituents of the soil

Soil is often described as consisting of solid particles, water, gaseous elements, humus and raw organic matter. An extremely important aspect which is often forgotten is that the soil is also the dwelling place for a large variety of living creatures (see Box 5.1). This soil life includes soil flora (microflora, such as bacteria, actinomycetes, fungi and algae) and fauna (microfauna, such as protozoa; mesofauna, such as nematodes and collemboles; macrofauna, such as beetles, centipedes, millipedes, ants and termites; megafauna, such as earthworms, rodents and moles). These organisms play a major role in many soil processes and soil-plant interactions, such as soil formation, creating soil structure, mineralisation to free nutrients for plant growth, building up humus, nitrogen fixation, phosphate solubilisation and uptake of nutrients by plant roots. There are strong interdependencies between roots and soil life, as roots excrete substances that stimulate soil life, which again release nutrients that are taken up by the roots (Subba Rao 1977, Lal 1987).

Soil life provides free labour for the farmer, if the right conditions have been created and organic matter with the right composition (C/N balance) has been fed at the right moment. Then, soil life can do its work effectively in synchrony with crop production.

Humus – organic matter decomposed by soil life – plays a critical role in creating fertile soils. Humus binds soil particles together into larger aggregates necessary for a stable, porous physical soil structure. Improved soil structure increases the water-holding capacity of soil. This is particularly important in (seasonally) dry areas, where raising the humus content in the soil renders the farm system more resistant to drought and permits more efficient use of available water. Humus also gives chemical structure to the soil, as nutrients are adsorbed to humus, creating a buffer of nutrients which can be made available to plant roots when needed. For example, humus can reduce phosphorus adsorption in the soil and bind micronutrients that would otherwise be leached, thus making them available for plant growth. Also elements poisonous for plant life, such as aluminium, can be absorbed and neutralised to certain levels. Humus is particularly important in tropical

---

Box 5.1
The importance of soil life

A rule of thumb is that, under favourable conditions, one tenth of the organic matter in a soil is made up of soil animals. Thus, a layer of 10 cm of a hectare of soil with 1% of organic matter contains roughly 1500 kg of soil fauna. This equals the weight of 3 – 4 cows (Dalzell et al. 1987).

soils which have a low nutrient-buffering capacity and are poor in nutrients. Leaching of nutrients is less in soils with high humus content than in those with low humus content (Lal & Stewart 1990).

Managing organic matter

Organic matter serves as a nutrient store from which the nutrients are slowly released into the soil solution and made available to plants. Organic matter in or on the soil also protects it and helps regulate soil temperature and humidity. Often, the use of organic matter is combined with other techniques with complementary functions, e.g. use of artificial fertilisers, tillage, water harvesting, shading and bunding. Organic matter management differs according to the situation and the crop. Improper management can lead to inefficient use of nutrients, nutrient losses, binding of nutrients or acidification.

Five basic ways of handling organic matter are: applying it directly to the soil, either as a surface mulch layer or incorporated into the soil; burning it (causing mineralisation); composting it; feeding it to livestock; or fermenting it in biogas installations.

Availability of sufficient organic matter is a critical point. If nutrients are replaced primarily by chemicals and farmers no longer attach high value to manuring, the soil will become poor in organic matter and buffered nutrients, and more susceptible to drought and pests. In other words, the productivity and stability of the farm system will decrease. In such cases, an initial investment in nutrients and labour will be necessary to increase biomass production to be used subsequently as fertiliser, so as to build up the farmer's working capital constituted by soil organic matter.

Some rough estimates of the levels of organic matter inputs required under different agroecological conditions can be derived from Young (1990): about 8.5 t/ha above-ground residues in humid areas, 4 t/ha in subhumid areas and 2 t/ha in semiarid areas, in order to maintain target soil carbon levels of 2.0, 1.0 and 0.5, respectively. As above-ground residues of a single crop are usually less than 3 t/ha, it is clear that, in the humid tropics, extra sources of biomass (e.g. trees, cover crops) are needed to meet this target.

In Appendix A, some technical options for organic matter management and the conditions under which they can be used are discussed.

Soil tillage

The condition of the soil can also be improved by tillage, which affects soil structure, water-holding capacity, aeration, infiltration capacity, temperature and evaporation. It reduces heat conduction and breaks capillary connections in the soil. The tilled layer dries quickly, but the subsoil moisture can be conserved better. Tillage can create favourable conditions for seed germination, and may be necessary to combat weeds and other crop pests or to help control erosion. It requires high energy inputs. These can be produced on-farm (manpower, animal traction) or can come from outside the farm (hired labour or draught animals, fuel-based mechanisation). Tillage can have negative effects on soil life and increase mineralisation of organic matter. If not done well, it can also increase erosion.

Organic matter used as a surface mulch protects the soil and helps regulate soil temperature and humidity. Maize field in Honduras mulched with velvet bean (*Mucuna pruriens*). *(Flores Milton)*

Conservation tillage and no-tillage techniques have been developed recently by scientists and farmers and are, in some places, traditional farming practices. Under LEIA conditions, zero tillage may have advantages, as the hard work of soil preparation is left to the soil life. However, as there are also limitations to this practice, the appropriate tillage (or no-tillage) technique must be carefully chosen for each specific site. No blanket recommendation can be made.

Managing soil health

Good soil health is a precondition for good plant health. Plant health is influenced directly by the uptake of certain organic compounds produced when soil organisms mineralise organic matter, e.g. phenol carboxylic acids produced when actinomycetes degrade woody plants. Actinomycetes seem to fulfil an important function in the production of antibiotic substances that can be taken up by plants (Rangarajan 1988). Resistance to infestation by *Fusarium* species and other fungi is linked to the presence of such compounds in some cereals (FAO 1977). More research is needed to discover the exact mechanisms involved.

Plant health is influenced indirectly when one soil organism suppresses the development of another one that could hinder crop growth. This can be illustrated by an example. When crops are grown, an ecological imbalance is created because the natural diversity in the ecosystem is decreased. A basic ecological reaction is to try to restore the balance by way of an attack by insects, fungi, bacteria etc. This natural self-defence by flora or fauna is labelled 'harmful', because it is detrimental to the performance of the plant being cultivated. Rather than using chemicals to counteract these attacks, they can often be prevented by adding organic matter to stimulate greater diversity of soil life. Generally, soil-borne plant diseases also decrease in occurrence when organic matter is added, because disease-causing organisms (pathogens) are hindered in their development or because their antagonists increase in number. The more varied and numerous the soil micro-organisms, the better are the chances for biological control of the pathogens.

Balanced manuring is basic to plant health. Too many or too few nutrients can make a crop more susceptible to disease and pest attacks. High doses of nitrogen fertiliser lead to high nitrogen content in the crop; the vegetative growth then becomes too abundant, resistance to pests is reduced, and certain kinds of insects multiply more rapidly. This danger is less when organic manure is applied, as organic matter releases nutrients gradually. However, the complicated relationship between organic matter and disease occurrence is not completely understood.

Organic matter management, tillage and soil health management may not be sufficient to create favourable conditions for the growth of certain crops, e.g. maize. Rainfall may be too high or too low; the water table too high; the slope too steep; or the soil too impermeable, too poor in one or more nutrients, too acid or too alkaline. This may demand investment in improvements such as drainage, water harvesting or terracing, or in external inputs such as phosphate or calcium fertilisers. LEIA farmers may not always have enough cash or time for these investments. As optimal growth conditions differ according to plant and animal species, another option to deal with these constraints is to

Farmers in the mountains of Bhutan improve soil conditions for plant growth by transporting animal manure and forest litter to their fields. *(Walter Roder)*

select crops and animals adapted to the actual growth conditions rather than adapting the growth conditions to the crop.

5.2 Optimising nutrient availability and cycling

A very important condition for good plant growth and health and, indirectly, for good animal and human health is the timely provision of sufficient and balanced quantities of nutrients that can be taken up by the plant roots. Nutrient deficiencies and imbalances are main constraints to crop production, especially in regions with poor and very acid or alkaline soils. As explained above, nutrient availability depends greatly on the general soil condition, soil life and organic matter management. However, deliberate attention must also be given to providing the nutrients required for crop growth.

There is a constant flow of nutrients through the farm (see Box 5.2). Some of the nutrients are lost or exported, e.g. by export of products, erosion, leaching and volatilisation. These nutrients have to be replaced. Where external inputs to replace them are not readily available (as in most cases of LEIA), nitrogen fixation and nutrient recycling as well as deliberate attempts to prevent losses become crucial. For example, in Zimbabwe it has been estimated that nutrients losses through soil erosion exceed application of artificial fertiliser by 300% (Stocking 1986).

If the farm system is to remain productive and healthy, it must be ensured that the amount of nutrients leaving the soil does not exceed the amount returned to the soil. In other words, over time, there must be a nutrient balance. When attempts are made under LEIA conditions to raise production for a growing local population or for more distant markets, farm systems are in danger of gradual degradation. Where more nutrients are being extracted than are being replaced by natural processes such as dust and rainfall, weathering and nitrogen fixation, techniques such as applying organic matter, recycling organic wastes and enhancing nitrogen fixation, combined with the use of artificial fertilisers (integrated plant nutrition), are necessary to maintain an adequate level of soil fertility for continued farming.

Limiting nutrient losses

Nutrient losses can be limited by:

- recycling organic wastes, e.g. manure, night soil, crop residues, crop processing residues, by returning them to the field, either directly or treated (composted, fermented etc.);

- handling organic and artificial fertilisers in such a way that nutrients are not leached by excessive rain or volatilised by high temperature or solar radiation; e.g. about 50% of the N ingested by livestock is excreted in the urine and is easily lost by evaporation and leaching; some of this can be conserved by using bedding with a high C/N ratio in the stable or kraal;

- reducing run-off and soil erosion, which removes nutrients and organic matter;

Box 5.2
Nutrient flow

Nutrients in solution are taken
from the soil by plant roots and
transported to green parts of the
plant (see Figure 5.1). There,
together with CO_2 from the air,
they are combined through a
process of photosynthesis into
complex units needed to form the
different plant parts. Energy for this
process is obtained from sunlight.

Plant tissue is consumed by
animals (herbivores, insects) and
humans, which may then be consumed by other consumers, e.g.
animals by humans; or dead
animals, humans and plants by
soil micro-organisms. These, in
turn, may be eaten by other soil
organisms. The movement of
nutrients from green plants
through plant eaters to animal
eaters is called a **food chain.**

Because consumers may use
more than one food source,
food chains interconnect with
each other and form a complex
food web. At the end of the
food chain, decomposers such
as earthworms, termites, fungi
and bacteria consume animal
excrements and tissue from
dead plants and animals,
thereby forming soil humus.
This breaks down to soluble
nutrients, which can be used
again for plant growth.

Various nutrients are involved
in this process. The most important are the basic elements
(macronutrients) carbon,
hydrogen, oxygen, nitrogen,
sulphur, phosphorus, potassium,
calcium and magnesium. But
also trace elements (micronutrients) such as iron, copper,
boron, zinc and manganese are
indispensable for plant and
animal growth.

Within the farm, nutrient flow
is more or less cyclic. However,
at different points, nutrients
enter the cycle with dust, rain,
sediment, fertiliser or concen-

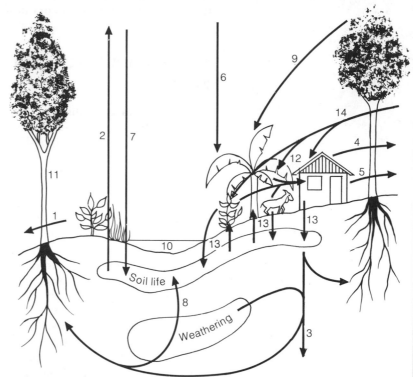

Possible losses
1 Erosion
2 Volatilisation
3 Leaching
4 Export (market/gifts)
5 Removal of wastes

Natural gains
6 Rain
7 N fixation
8 Weathering
9 Sediment/dust
10 Blue-green algae

Management options
11 Woody species
12 Feeding concentrates/
   minerals
13 Recycling (via livestock,
   compost, biogas, slurry etc.)
14 External inputs

Figure 5.1 Nutrient cycle at the farm level

trates; or leave it as marketable
products or gifts or as a result of
erosion (by wind or water),
volatilisation (diffusion of nitrogen
and sulphur components into
the air), leaching (nutrients dissolve in water and percolate to
deeper soil layers) and removal
of waste (such as deposition of
night soil off the farm). Nutrients may also be mobilised or
gained within the farm by weathering of tiny rock particles, the
action of mycorrhiza, and fixation of atmospheric nitrogen by
certain micro-organisms.

Nutrients are taken out of the
soil solution and are no longer
available to plants when they
combine chemically with other

substances in the soil or when
they are incorporated into micro-
organisms and are thus
immobilised.

Especially woody species, but
also other plants, draw nutrients
from deeper soil layers and/or
the wider surroundings and concentrate these nutrients in the
soil near the trunk.

By feeding crop residues to
animals and making compost,
biogas slurry, etc. from crop
residues, manure and similar
organic 'wastes', nutrients can be
recycled in the farm. They can also
be transferred from one place to
another, e.g. from pasture to field,
or concentrated in one place,
e.g. near the homestead.

- reducing burning of vegetation when farming is intensified, as this leads to losses of organic matter;
- reducing volatisation of nitrogen by denitrification under wet soil conditions;
- avoiding leaching by using organic and artificial fertilisers which release nutrients slowly (in synchrony with crop needs), maintaining a high humus content in the soil, and intercropping plant species with different rooting depths;
- pumping up partly leached nutrients from deeper soil layers and bringing them back to the topsoil by using litter from trees or other deep-rooting plants as mulch or green manure;
- limiting nutrient export in products by producing crops with relatively high economic value relative to nutrient content, e.g. fruits, nuts, herbs, milk;
- producing for self-sufficiency, so that as few products as possible need to be exported to the market, and using by-products for fodder and/or organic manuring.

However, nutrient export from the farm to the market cannot be completely avoided, as money is generally needed to pay taxes and school fees, to buy services and industrial products etc. It is also impossible to avoid all nutrient losses resulting from erosion and leaching.

Capturing and managing nutrients

Some nutrients can be captured on the farm by:

- fixing nitrogen by micro-organisms living in symbiosis with leguminous trees, shrubs or cover crops, or with *Azolla* ferns or some grasses; or by free-living bacteria, e.g. *Azotobacter* or blue-green algae;
- harvesting nutrients by capturing wind or water sediments from outside the farm; this can be done by vegetation or by special constructions which often function in combination with water harvesting (e.g. ponds, run-on farms), and is possible only when wind or water erosion takes place somewhere else;
- using livestock to bring nutrients (via manure) from outside the farm, e.g. from common land. A similar process takes place when mulch or fodder is carried to the farm. However, as this removes nutrients from the common land, it can be continued over a longer period only if the land is not used very intensively.

Localised nutrient deficiencies within the farm can be at least temporarily overcome by:

- concentrating nutrients on a field by, for example, localised application of manure, compost, mulch or green manure, or water and nutrient harvesting within the farm;
- growing green manures (trees, shrubs, grasses, cover crops) to make nutrients more readily available to crops. Nutrients from deeper soil layers or less soluble nutrients (e.g. phosphates, micronutrients) can

Livestock transfer nutrients from the range to the farm when they are kept overnight on land to be cropped. *(Ann Waters-Bayer)*

thus be brought into circulation. Mycorrhizal fungi and other soil life can mobilise nutrients, so that they can be used by plant roots. However, this reserve of nutrients must eventually be replenished by external nutrients to avoid 'mining'.

Recycling nutrients within a farm system can also be speeded up, allowing higher turnover in the form of products. Some organic matter, particularly mature grasses and stems of sorghum or other grain crops, has such low nitrogen content that microbial breakdown of the material – and thus nutrient recycling – is very slow. This process can be accelerated by adding nitrogen, e.g. by intercropping or oversowing with legumes; by passage through the rumen of livestock (i.e. by letting ruminants eat the low-quality material); or by burning it.

Supplementing nutrients

When replacement nutrients cannot be captured on the farm, they must be obtained from elsewhere. External sources of nutrients include:

- organic matter from elsewhere, e.g. manure from other farms, processing by-products, night soil and other compostible matter from towns;

- purchased fodder or concentrates, or human food;

- mineral fertilisers such as rock dusts, e.g. lime, rock phosphate, and bio-super (a mixture of rock dusts and micro-organisms which help mobilise the minerals) and artificial fertilisers.

Agro-industrial by-products, e.g. from sugar (molasses), oil pressing (cakes), breweries (spent malt), juice factories (peel), cotton (cotton-seed), can either be fed to livestock which produce manure and urine or can be used directly as fertiliser. Agro-industries often face problems with waste disposal. If not properly treated, these wastes become public health hazards while, at the same time, an important source of organic matter is lost for recycling back to farming.

Shortages or excesses of specific elements in the soil disturb the nutrient balance and may lead to diseases in crops and animals. The cause of such imbalances must be analysed: they may be due to Fe or Al toxicity (in the case of low pH), salinisation, alkalinisation, long-term application of mainly NPK fertilisers leading to a loss of micro-nutrients (Tandon 1990, Box 5.3), or certain elements may be naturally lacking in the soil. It may be necessary to rectify pH, salinity or alkalinity, or to apply specific missing nutrients. Using organic fertilisers with a balanced C/N ratio can improve the soil balance, the pH and the availability of nutrients, including micronutrients.

Nutrient supplementation can lead to considerable production increases not only in crops but also in livestock. Feeding mineral supplements can stimulate appetite, increase the digestibility of forage and improve animal health. It also improves crop production via the higher quantity and quality of animal excretions which can be used as fertiliser. When livestock are fed macronutrients, such as phosphate, a large part of this is excreted in the dung (Winter 1985). Also supplementation with small amounts of micronutrients, such as copper, can bring considerable benefit to both the livestock and the plants fertilised with their excrements. Similarly, mineral supplementation to crops and pastures can increase the digestibility of plant parts eaten by livestock.

In pastoral livestock-keeping systems, the value of mineral supplementation is well known. Herders bring their animals to 'salt pastures' at certain times of the year or give their animals locally available mineral supplements. For example, cattle herders in West Africa feed *kanwa*

Box 5.3
Micronutrient drain in Punjab, India

Is the green revolution killing the goose that lays the golden egg? Is it draining the very soil where high-yielding varieties grow? Scientists are discovering that intensive cropping is removing the crucial micronutrients zinc, iron, copper, manganese, magnesium, molybdenum and boron, which control various aspects of plant growth. For instance, zinc helps the plant use nitrogen and phosphorus, and iron is responsible for the synthesis and maintenance of chlorophyll.

Every crop removes macro-nutrients and micronutrients from the soil, but farmers tend to put back only N, P and K. A study at Ranchi Agricultural College revealed that applying 100 kg NPK (50:25:25) per ha led to depletion of zinc by 629 g/ha and copper by 433 g/ha. Over half of 8706 soil samples were deficient in zinc. This can depress yields by up to 4 t/ha in rice, 2 t/ha in wheat and 3.4 t/ha in maize. The next most serious deficiency is in iron. This is emerging as a limiting factor in rice production in the new rice – wheat rotation evolved in the nontraditional rice-growing areas of Punjab.

Micronutrient deficiencies affect the quality of food and, in turn, human health. Consumption of zinc-deficient grain can lead to retarded growth and sexual development, carbohydrate intolerance and poor healing of wounds, according to research by the Postgraduate Institute of Medical Sciences, Chandigarh.

Soil scientists advocate recharging the soil by adding the deficient nutrients in chemical form. However, micronutrients must be applied carefully to avoid a toxic build-up. Up to 90% of applied zinc may be converted to forms not available to the plant. Moreover, trying to remove deficiency of one micronutrient by adding it in a chemically pure form can lead to deficiency in another. In one village where farmers started applying zinc, deficiencies in iron and manganese replaced zinc deficiency. Copper deficiency has taken over in five other villages.

One solution is greater use of organic manures and multiple cropping with legumes. In a Punjab Agricultural University experiment, poultry manure, pig manure and farmyard manure proved effective in meeting zinc requirements in a maize – wheat rotation. Also cultural practices such as prolonged submerging of the field can be used to tackle iron and manganese deficiencies (Sharma 1985).

In West Africa, cattle are given *kanwa*, a traditional mineral supplement, which is often placed on old termite mounds. *(Ann Waters-Bayer)*

(called 'local potash' in English), a traditional supplement which contains mainly calcium and potassium (Otchere 1986).

Box 5.4
Flow of solar radiation

The sun transmits energy in the form of radiation to the earth. This radiation is used by plants for photosynthesis and evaporation, i.e. plants transform solar energy into chemical energy. Animals profit from this energy by eating plants, and humans use the energy and nutrients stored in plants and animals.

Solar radiation is also responsible for temperature fluctuations. During the day, the surplus radiation entering the atmosphere raises the temperature; at night, energy is lost from the atmosphere and the temperature falls. Most agricultural processes (e.g. growth of plants; activity of insects, diseases and soil life; decomposition of organic matter; functioning and well-being of animals) are sensitive to temperature and day length.

Plants can photosynthesise only a small part (2 – 5%) of the incoming solar radiation (Cooper & Tainton 1968), and only a part of this is used as food. The efficiency of solar energy use in animal production is even less: it has been estimated that only 0.02% of incoming solar radiation is reflected in live-weight gain of cattle on subtropical grasslands (Okubo et al. 1983). However, as animals can consume herbage which is not used as human food, integrating them into farm systems increases the fraction of solar energy that ends up in foodstuffs.

5.3   Managing flows of solar radiation, water and air

Different plants and animals have different needs for light, temperature, water and humidity. Some plants need full sunlight; others prefer shade. Some need high humidity; others prefer air movement. Some respond to day length, and others to temperature to induce flowering. Some plants and animals are vulnerable to extreme temperatures, heat and frost; others to a lesser degree. Insect pests, diseases and weeds may be influenced by light intensity, humidity, droughts or floods.

The growth conditions for crops and livestock are determined to a high degree by climatic conditions. These may not always be optimal for growth, may cause damage to the crops, livestock or soil, and may involve considerable risk factors. The climatic conditions on the farm are largely determined by the flows of solar radiation, water and air.

Farmers can make optimal use of these flows by choosing crops and animals that fit the specific climatic conditions. Or they may influence the spatial composition and structure of the plant canopy and soil cover to manipulate radiation, water and air flow to create microclimates that are favourable for the growth of specific plants and animals. This is called **microclimate management**.

The land form or the spatial arrangement of plants also can be chosen in such a way that the water flow is deliberately guided to increase availability of water for plant growth. This is called **water management**.

The flows of solar radiation, water and air may also cause considerable soil erosion. For the farmer, this means losses of both production and the farm's natural resource capital. By manipulating the flows **(erosion control)** the farmer may be able to minimise the risks and losses. Indigenous farmers have often developed remarkable techniques to combine the available resources of water, soil and air so as to make maximum use of radiation and water and to protect crops and livestock from damage by the different flows (see Chapter 3 for examples).

Box 5.5
Flow of water

Water may enter a farm system as rain, flood or irrigation water, groundwater or air humidity (see Figure 5.2). Plants need a certain level of humidity in the air and soil to be able to take up water and perform their life processes. Beside moisture, soil life and plant roots need air to be able to exchange gases with the atmosphere.

The availability of moisture and air in the soil depends on precipitation, water infiltration, capillary flow, rooting depth, texture of the soil particles and structure of the spaces between them. Organic matter and soil life play important roles in creating soil structure and greatly influence storage capacity and availability of soil moisture and air. Where farmers are not using organic matter to improve the soil, the farm system may be more vulnerable to drought and flood.

Changes in land use, particularly clearing of forest and other vegetation, affect the water table and local micro-climate and may even affect regional climate. Removal or change of the vegetation cover also affects run-off and infiltration rates, altering the quantity, timing and quality of overland flow, river flow and stream

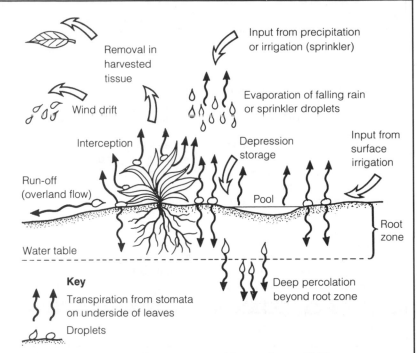

Figure 5.2 The water – soil – crop system. (*Source:* Porter 1978)

flow. Changes in stream flow and infiltration can affect groundwater levels which, in turn, can affect stream flow.

Overpumping of groundwater, as presently done in many low-rainfall areas, affects the natural vegetation and availability of drinking water, and can endanger agricultural production. In principle, a balance is needed between recharge of groundwater reserves and outflow by pumping or other means. Recharge of groundwater reserves can be

improved by reafforestation and water conservation.

Excessive overland flow of water may cause soil erosion. Valuable topsoil with relatively large quantities of nutrients and soil life is taken away and deposited elsewhere as sediment. As this is a direct attack on the natural resource capital of farmers, it is very important for them to manage water flow in such a way that it can no longer be harmful (Barrow 1987).

Box 5.6
Flow of air

Wind has both positive and negative influences on farming. It affects temperatures of – and evaporation from – soil, plants and animals, and temperature and humidity in the micro-climate. The impact of drought and cold increases with wind strength. Soil may be blown

away if not adequately protected from air flow. In situations where these effects of air flow are detrimental to farming, especially in the case of dry and cold winds and where soils are vulnerable to erosion, farmers may try to influence air flow by changing the vegetation cover or by providing shelter with vegetation strips, scattered trees or walls.

Box 5.7
Examples of microclimate management
(after Stigter 1987b)

Manipulating solar radiation:
- multistorey cropping to make optimal use of available light, e.g. in home gardens on Mount Kilimanjaro in Tanzania (Fernandes et al. 1984);
- shading, e.g. for shade-loving crops such as coffee trees or betel vines (see Box 3.4); use of cover crops and mulches to control weeds;
- exposure to solar radiation to control pests, e.g. the brown plant hopper in rice in India (Balasubramainam 1987), and to kill soil-borne pathogens;
- increase or decrease of surface absorption of radiation, e.g. mulching to lower soil temperature, painting trunks white to prevent heating;
- cover to prevent radiation loss by night;
- irrigation to influence plant temperature, e.g. sprinkler irrigation to lower temperature in groundnut crops in India (Balasubramainam 1987);
- use of solar radiation for drying crop or animal products in the field or in storage;
- maintaining trees on grazing land to provide shade for livestock.

Manipulating heat and/or moisture flow:
- mulching to regulate soil temperature and humidity;
- windbreaks to protect crops and animals;
- wind protection for ripening;
- influencing air/humidity flow by changing soil or vegetation conditions;
- using warmed air for field and/or storage drying;
- manipulating dew fall;
- wind-blow rows to allow quick drying of canopy where there is risk of fungal diseases.

Manipulating mechanical impact of wind, rain and hail:
- altering wind speed and/or direction;
- planting in lower places or pits or where deep rooting is possible;
- protecting soil from erosive flows of air or water;
- protecting crops and produce against impact of rain, wind or hail;
- using wind for winnowing.

Managing microclimates

Farmers may combine crops (multistorey cropping, intercropping, shelterbelts) with complementary canopy characteristics so that one crop creates favourable conditions (in terms of shade, wind protection, humidity etc.) for the other. This can also be done with physical structures (walls, cover etc.), mulches or irrigation. The microclimatic conditions for crop and animal production can thus be improved, and maximum use can be made of the available solar radiation. In this way, it is also possible to influence the occurrence of weeds and pests. Many examples of microclimate management can be found in indigenous farming systems (see Box 5.7).

Managing water

Differences in availability of soil water and air humidity are important reasons for differences in types of natural and agricultural vegetation and in level of biomass production. Farmers can influence the availability of water and air in the soil by improving soil structure and storage capacity (e.g. managing organic matter, tillage), increasing the infiltration capacity and decreasing evaporation (e.g. mulching, tillage), increasing water infiltration into the soil (water conservation/harvesting, irrigation) or draining excess water. In the cases of water conservation/harvesting, irrigation and drainage, special physical structures may be used to create microenvironments to exploit the synergy gained from concentrating nutrients and water.

In dry areas, irrigation reduces risk of crop failure due to drought and can greatly increase biomass production by improving growth

In the volcanic soil of Lanzarote, Canary Islands, grapes are grown by taking advantage of the cinder mulch to retain moisture and by building stone and cinder walls for wind protection. *(Chris Reij)*

conditions. Small-scale irrigation systems have been designed by traditional farmers to use external inputs of water as a complement to rainfall, water-harvesting and maximising water-use efficiency by means of organic matter management, tillage and microclimate manipulation. These systems tend to be flexible and socially acceptable (Barrow 1987, Steekelenburg 1988, Chambers 1990, Ubels 1990, Lundqvist, in press).

Soil and water management techniques are closely related, as they affect the soil-water-air system as a whole.

Erosion control

Soil erosion can be due to the impact of radiation, water or air flow, and is often caused by a combination of these. Soils are very sensitive to radiation, especially in dry climates. When soil temperature becomes too high or the soil becomes too dry, e.g. when denuded of vegetation or mulch cover, soil life is endangered, root growth (see Figure 5.3) and functioning is suboptimal, the humus of the top layer is mineralised and, as a result, the surface of loamy soils is sealed by rain impact, and sandy soils lose their coherence. This leads to increased erosion

Figure 5.3
Influence of soil temperature on root length and shoot weight of soybean. *(Source:* Lal 1975)

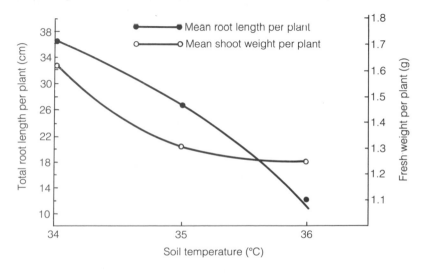

by water and wind. The negative effects of radiation and high temperatures can be reduced by preventing direct sunlight from reaching the soil surface. This can be done by covering the soil with vegetation or mulch, or by providing shade.

Erosion by water occurs where the soil is open to the impact of rainfall (splash erosion), extensive flow of run-off (rill erosion) or concentrated flow of run-off (gully erosion). Rain intensity, duration of rainfall, vegetation cover, soil texture and structure, inclination and slope length are the factors determining the impact of water flow. Erosion can be especially strong on fields with steep, long slopes and fragile soil structure where the vegetation cover is insufficient to neutralise the impact of heavy rain. Concentrated water flow has considerable erosive force. Sedimentation takes place where water flow slows down, at the bottom of the slope, where water is impounded, or in deltas.

Erosion by wind occurs where uncovered soil is open to the impact of air flow. Very sandy soils in dry, windy climates are vulnerable to this type of erosion. Sediments are deposited behind barriers which slow down air flow.

Soil erosion has been intensively studied, and many technical solutions have been offered. However, erosion control projects tackling only the technical problems are seldom successful, as the deeper causes of soil erosion often lie in socioeconomic and political problems (Blaikie 1985). Some promise is shown by integrated approaches, in which erosion control is part of a broader attempt to improve the productivity of farming and to conserve natural resources, and in which farmers are given the chance to participate in local-level planning, technology development and natural resource management. Legislation to secure land-use rights could support such an approach (Reij 1987, Hudson 1989, Shaxson et al. 1989).

Many of the technical solutions that have been suggested to overcome erosion problems in the tropics have been developed under capital-intensive, mechanised, commercial farming conditions in North America and Australia. Introduction of these 'solutions' faces the typical problems of inappropriate technology transfer. Often, the techniques do not work under LEIA conditions, as they demand too many external inputs or do not fit into the labour profile of the farm. Much greater emphasis should be placed on biological techniques and making simple improvements in traditional techniques of water and soil conservation. Such techniques are (or were) used in many more places than expected by development workers. However, where farmers have neglected these techniques in recent years, local knowledge about them is diminishing (Reijntjes 1986a, Millington 1987). In general, good vegetative cover of the soil, protection from wind, organic matter management and guiding the flow of excess water are the most important ways of conserving soil (see Appendix A for information about relevant techniques).

5.4   Minimising losses due to pests and diseases

In most farming systems today, natural mechanisms of regulating the populations of pests and other organisms have been disturbed or partly replaced by artificial mechanisms such as chemicals and drugs. Under

Box 5.8
IPM with rice farmers in the Philippines

Since the 1950s, an increasing number of farmers in the Philippines have been spraying their irrigated rice fields with pesticides. In 1965, even before the first green revolution variety IR8 was introduced, 60% of the farmers already used insecticides. This was encouraged by the fact that government services initially sprayed fields free of charge. By 1976, 90% of the farmers used chemical pesticides. Farmers felt that this was progressive, effective and essential for rice cultivation.

Farmers adopted spraying so quickly partly because credit and pesticides were available,

but mainly because stemborers multiplied with the more intensive cropping (double cropping) made possible by irrigation. As a result of the spraying, natural enemies disappeared and brown plant hoppers increased. More frequent spraying had no effect; eventually, this pest was controlled by introducing a resistant variety of rice.

Still, farmers continued to spray. There were many misunderstandings about pesticide use. Farmers sprayed whenever their neighbours did, whenever the first insect was spotted and after manuring. All fields were sprayed, even if the pest occurred only in a small area.

The recognition gradually dawned that a more sustainable

solution to pest problems lies in careful observation in the field and spraying only when necessary.

Thousands of farmers have been trained by the Philippine National IPM Programme to recognise pests and their natural enemies, to use 'action levels' in deciding when to use pesticides, and to apply insecticides only to the infested field spots. They have been taught about the effects of insecticides on natural enemies and safety precautions for applying pesticides. Farmers who have received this training appear to spray considerably less (4 – 6 times) than untrained farmers (Kenmore et al. 1987).

such 'unnatural' conditions, not using chemicals often leads to considerable production losses. Using chemicals often leads to health problems, pollution and disturbance of ecological balances (see Chapter 1). Interest is therefore growing in alternatives to chemicals in pest control and to methods of Integrated Pest Management (IPM) to decrease the use of chemical pesticides.

In IPM, in the context of the farm's environment and the population dynamics of the pest species, all suitable techniques and methods (biological, genetic, mechanical and chemical) are used in the most compatible manner possible so as to maintain pest populations at levels below those causing economic injury (Panel of Experts on Integrated Pest Control 1967). Costs of crop protection can thus be decreased, as chemical pesticides are used more efficiently, and their negative environmental effects can be reduced. IPM programmes have been developed for major commodity crops such as rice (see Box 5.8). These programmes are especially successful where conditions for commercial production are favourable and farmers are good crop managers.

However, many farmers do not yet use chemical pesticides or use them only on certain cash crops. Conventional research into crop protection and animal health care does not correspond with the needs of most LEIA farmers, as it focuses on crops and animals used in commercial farming on large estates or ranches (e.g. sugarcane, cotton, coffee, wetland rice, cattle) and largely neglects food crops grown on a small scale (e.g. sorghum, cassava, upland rice). There is a need for alternative pesticides which can replace dangerous chemicals and, preferably, can be produced locally at low cost. Often, in conventional research, the emphasis is still on combating pests and diseases rather than on prevention. It is important that integrated farm systems and techniques be developed which can minimise the need for curative measures and prevent farmers from falling victim to the pesticide treadmill.

Over centuries of experimentation, farmers in traditional systems have developed many measures to reduce the negative influence of pests such as weeds, rodents, birds and large insects. They may not recognise the influence of smaller organisms such as small insects, mites, worms, fungi, bacteria and viruses, but many measures woven into traditional farming systems prevent a massive occurrence of these pests. Examples of indigenous pest-control practices are shifting cultivation (see Box 5.9), intercropping, crop rotation, sanitary measures (removal of infected plants) and using resistant varieties. Many of these are preventive measures which do not eradicate pests but limit their population and maintain an ecological balance.

Farmers who observe pest life cycles can limit pests in crops and livestock by making use of the natural mechanisms that regulate the population dynamics of organisms. They can try to disturb the life cycle of pests so as to reduce their numbers (see Box 5.10). Many examples from traditional agriculture show that stable agricultural production is possible without using chemical pesticides and drugs. In some cases, e.g. with some resistant varieties, the farmer has to accept a lower level of production than would be possible with a combination of high-yielding varieties and pesticides. Preventive functions may compete with productive functions. However, this is not always the case, e.g. intercropping can enhance both crop protection and yield.

From traditional and biological farming practices, much can be learned about how to base measures of plant and animal protection on natural mechanisms. However, this will not provide solutions to all situations of pest attack. New scientific insights into the ecology of crop and animal pests and diseases are opening up new perspectives, where traditional pest control has failed.

Box 5.9
Shifting cultivation and pest control

Shifting cultivation not only conserves moisture, restores organic matter and nutrients to the soil, and prevents erosion and leaching, but it also controls weeds and reduces populations of insects, nematodes and various pathogens. The plots mimic tropical forest ecosystems in at least two ways that influence pest incidence. The great diversity of crops grown, sometimes as many as 40 at the same time (Conklin 1957), gives some protection, as pests are seldom able to build up to destructive proportions on the few isolated plants of each species. Also, the closed canopy consisting of some trees left standing and tall crop species such as bananas and papayas reduces the severity of pests and weeds.

Burning, rotation, intercropping and shading are practices in shifting cultivation that help reduce losses to pests and weeds. As only small plots are cleared, biological agents can easily enter from the surrounding jungle. Shifting cultivators also select for host resistance by using seed and vegetative parts from the most successful crop plants which survive in the harsh environment (Glass & Thurston 1978).

Population growth and other factors can reduce the viability of shifting cultivation systems, but many of the techniques developed by shifting cultivators may prove to be very useful for LEISA farmers of today.

Box 5.10
Using trap plants to control cotton pests

The cultivation of cotton in Nicaragua gave rise to enormous problems in pest control. The excessive use of pesticides led to poisoning, pesticide residues in human and animal tissue, decreased resistance to malaria and economic problems. By 1976, pesticides accounted for 30% of cotton production costs.

In 1981, FAO started a new cotton programme with a detailed analysis of the cotton ecosystem. On this basis, a trap crop system was developed as an alternative to frequent application of chemical pesticides to control the boll weevil (Anthonomus grandis). Wet areas, e.g. along rivers, with cotton plants that are still green, are favourable habitats where weevils can survive the dry season. At such sites, cotton plants were left standing after harvest. Regular spraying with methylparathion killed the weevils. At the onset of the rains, small areas were seeded early with cotton. These plants likewise attracted weevils, and were likewise sprayed.

In 1981, this trap crop method was tested on a cotton farm of 565 ha. As a result, 9 pesticide applications could be eliminated and yields were 15% higher. In the following years, this method was successfully extended to other areas (Daxl 1985).

Good crop protection in storage requires proper drying of the grain and well-constructed granaries. In central Nigeria, sorghum and millet are stored well in hand-moulded clay granaries. *(Ann Waters-Bayer)*

Protecting crops

Crop protection measures can be divided into the following main categories:

- sanitary measures, e.g. using healthy planting material, clean seeds, and clean tools; clearing the focus of infection;
- multiple cropping, e.g. intercropping, rotation, trap crops, decoy crops, shade trees;
- cultural measures, e.g. manuring, mulching, tillage, flooding, sowing dates, planting distances;
- mechanical measures, e.g. hand-pulling and picking, hoeing, ploughing, mechanical traps, burning, creating noise;
- biological measures, e.g. introducing or conserving natural predators, birds, insects, microbes, weeds;
- exploiting host resistance;
- chemical measures, plant-derived as well as artificial;
- storage practices.

Prerequisite for good crop protection are good basic cultural practices. It is of little use to apply pest control measures if, for example, the soil is not prepared well or soil fertility is low or unbalanced. Many crop enemies, especially fungi and bacteria, affect primarily weakened plants. It is therefore important to aim for a healthy crop by proper tillage, using seeds with good germination qualities, sowing at the right time, spacing crops appropriately and providing sufficient water and nutrients. A healthy crop can also compensate for pest damage by growing new offshoots or speeding up the growth of undamaged parts.

In LEISA, preventive methods and cultural and mechanical measures are of particular interest, as they generally involve few risks, do not create resistance or resurgence, do not endanger human health or the environment, and do not require many external inputs. They create unfavourable living conditions for pests or disturb their life cycle. Of fundamental importance in this context is functional diversity of the farming system (Altieri 1987).

When farmers start to produce more for the market, they often change their cropping system: crop diversity decreases where sole cropping is adopted, fallow periods become shorter, high-yielding varieties with low pest resistance are grown and irrigation may permit more harvests per year. This modernisation and intensification often mean that traditional methods of pest prevention disappear and losses to pests increase. A common response is to apply pesticides. The danger is great that farmers begin to see pesticides as the only solution, overestimate their usefulness and underestimate their disadvantages. Farmers need to be made aware of (or reminded about) other effective crop protection methods and instructed in sensible use of pesticides.

Protecting livestock

In contrast to plants, livestock do not have to be kept in one place. Mobility offers possibilities of avoiding diseases and disease transmission by avoiding areas of high risk. Often, herders bring their animals to certain grazing areas only for the dry season and move them out again before these areas become infested with biting flies in the wet season.

Besides such grazing strategies, many other traditional management practices also reflect a sound adaptation to the environment and help avoid or prevent animal diseases, reducing the need for curative measures. For example, in the subhumid zone of West Africa where large pockets of land are infested with the tsetse fly, Fulani cattle-keepers avoid taking their animals to graze in the severely infested areas, and minimise the time that animals spend at watering points, where tsetse flies are most likely to occur. They also delay grazing in the wet season until late in the morning, as the danger of worm infestation is high in the early morning when the grass is still wet with dew (Bayer 1986). Also the fires made at cattle camps or other livestock enclosures are a means of keeping insects away from the animals.

At the end of the dry season, transhumant herds are moved out of more humid areas to avoid disease-transmitting flies. *(Ann Waters-Bayer)*

When an acute outbreak of disease occurs, many traditional livestock-keepers take quarantine measures. Nowadays, these are usually supported by corresponding government measures. Quarantine can slow down the spread of disease but cannot stop it. Therefore, such measures have to be supported by vaccination campaigns, e.g. ring vaccination around the infected herds. Although some livestock-keeping peoples have developed their own forms of vaccination (immunisation), they generally appear to regard the modern vaccines as more effective. In the case of some diseases, lifelong protection can be achieved by one vaccination but, in other cases, vaccinations have to be repeated regularly to ensure protection. This requires an efficient veterinary service and continuous efforts to produce and – as the diseases change – to adapt vaccines. As vaccine production is based on natural processes, it is likely to be a sustainable measure, even though it requires considerable technical and organisational inputs.

In contrast, the use of chemicals to prevent infectious diseases cannot be viewed as sustainable. It may help in the short term, but long-term use of chemical drugs will inevitably induce the development of germs resistant to the particular chemical being used. As evident in the case of acaricides, the 'lifespan' of these medicines becomes increasingly shorter, the more intensively they are used. This means that new chemicals constantly have to be developed – leading to an ever-tightening spiral of chemical use and all the accompanying negative effects of chemical residues in animal products.

The drugs used in 'modern' veterinary treatment are usually imported and expensive. When national governments try to carry the costs and offer treatment free of charge, these services are often unreliable or not available in LEIA areas. Researchers are now becoming increasingly aware of the wealth of knowledge about veterinary care in indigenous livestock-keeping systems and the potential for livestock treatment at lower costs and with less external dependence than in the case of modern drugs (Mathias-Mundy & McCorkle 1989, Matzigkeit 1990, Niamir 1990).

Indigenous veterinary practices (ethnoveterinary medicine) offer a rich resource for development. In some cultures, ethnoveterinary medicine and ethnomedicine overlap: healers may treat people as well as animals. In Nepal, for example, there are at least 14 types of healers with different training and methods of treatment; all accept both human and animal patients (FAO 1984a).

Ethnoveterinarians diagnose, treat and prevent diseases in animals. Their diagnosis is influenced by the prevailing belief system and commonly relies on symptoms, postmortem inspection of diseased animals and epidemiological observations. Treatment and prevention methods include:

- herbal and other medicines;
- surgical methods such as wound care, bone-setting, blood-letting and cauterisation;
- management practices;
- vaccination.

Pharmacology (the study of medicines) is probably the most widely investigated aspect of ethnoveterinary medicine; e.g. FAO has compiled

Livestock-keepers appreciate the effectiveness of vaccines in animal health protection. In Kenya, village women have been trained to vaccinate poultry, using vaccines produced in Kenya. *(John Young, ITDG)*

lists of 140 medicinal plants in Nepal and 150 in Thailand (FAO 1984a, 1984b).

Exploiting disease tolerance in crops and livestock

An environmentally sound and very effective way to minimise pest and disease problems is to use locally-adapted plants and animals, as they are generally less susceptible to the pests and diseases than are species, breeds and varieties introduced from other areas. Sometimes, this is the only way to prevent infection with certain diseases, e.g. virus diseases.

For centuries, farmers have selected plants for their resistance to pests and diseases. The great genetic diversity of the local varieties developed by farmers also helps to reduce the risk of losses. In livestock, indigenous breeds have evolved that are resistant or tolerant to pest attacks and other environmental stresses. In part, this is the result of deliberate selection by livestock-keepers, and in part the result of natural selection under harsh conditions. In many tropical areas, the prevalence of particular diseases became evident only when exotic stock was introduced. In Nigeria, for example, the transmission of heartwater, a tick-borne disease that is fatal for exotic cattle, could not be investigated for several years because scientists could not find susceptible indigenous livestock within the area of natural occurrence of the tick (Bayer & Maina 1984).

By using crop varieties and animal breeds with time-proven tolerance of locally prevalent diseases and pests, farmers can avoid high inputs and expenses for preventive and curative measures.

Integrated measures

Most farming practices influence pest and disease control in some way. Therefore, creating healthy conditions for plants, animals and humans requires an integrated systems approach. The cumulative effect of the many different practices that have some influence on diseases and pests is probably a better insurance than a bottle of pesticide or medicine.

As pests and diseases spread beyond farm borders, community efforts and social control play an important role in suppressing pests and diseases. Laws, ceremonies, customs, and the rights and duties of chiefs (e.g. to indicate the time to sow or burn) often institutionalise health care at the community level. In the case of a measure such as removing crop residues, it is important that all farmers in the region do this, if the optimal effect is to be attained. Respecting the sowing dates is another example which asks for communal decision-making and action. Therefore, good communication and cooperation between farmers is necessary.

Improving crop protection requires cooperation between farmers and scientists. Farmers can do their own experimentation to improve crop protection, e.g. by developing multiple cropping systems, preparing plant-derived pesticides, or mechanical, cultural and sanitary measures. However, they often need additional information on, for example, the effects and dangers of plant-derived pesticides, the influence of natural enemies or instruction in recognising pest damage. Good pest control decisions are based on knowledge of the life cycle of each pest, knowledge which farmers do not always have. Other improvements, such as certain forms of biological control, development of selective pesticides and incrossing resistance against certain diseases, require specialised knowledge and equipment, and are beyond the possibilities of smallholders working on their own. Yet these techniques may also be important for crop protection. Good interaction between farmers and scientists is therefore needed. Research into traditional crop protection could help in developing crop protection measures of interest to LEIA farmers.

Usually, farmers have to deal with numerous different pests (insects, fungi, nematodes, viruses, weeds) at the same time. No one measure can deal with them all. Therefore, farmers must seek the most effective set of measures to suit local circumstances. This is not easy, as there are many internal relations in the farm system. Each change intended to suppress a particular pest may have positive or negative effects on other organisms. Moreover, many measures are taken not only to protect crops but also for other purposes, e.g. to improve soil fertility. Therefore, scientists seeking to help farmers develop appropriate combinations of pest control measures must take an interdisciplinary approach.

5.5   Exploiting complementarity and synergy in combining genetic resources

The mix of plants and animals on a farm is not just a random collection of genetic resources. Each species must fit into the biophysical and socioeconomic environment of the farm and must perform productive, reproductive, protective or social functions, or a combination of these. The species and varieties are chosen to meet subsistence needs and often also to sell, among other objectives the farm household may have (see Chapter 2). The choice of crops and livestock will depend greatly on what the household can produce and what can be obtained on the market, taking into account the quantity, quality and price of market products and services and the reliability of their supply.

To optimise the viability of farming, the family must choose and mix its crops and animals in such a way that the farm, as an integrated whole, is more than the sum of the individual organisms within it. Genetic resources are needed that perform complementary functions and can be combined so that they interact in synergy, rather than competing with each other. In most cases, deliberate choice of corresponding plants and animals results in a farm system with a high diversity of genetic resources.

Land suitability, market demands, availability of resources (land, labour, knowledge, genetic resources etc.) and inputs (fertilisers, pesticides, medicines, water etc.) may make it necessary for farmers to concentrate on certain crops or animals, i.e. to limit diversity. Creating market opportunities for products from a wider variety of crops, trees and animals would give farmers more opportunities to profit from the advantages of integrated multiple cropping systems.

As farming conditions and household needs and opportunities change constantly, the farmers must always continue the process of choosing the best mix and arrangement of genetic resources in space and time.

Box 5.11
Choosing crops for rotation

A crop rotation should meet the needs of the particular farm for which it is designed, and it should meet the requirements for sustainability.

Needs of the farm. When designing a crop rotation, some questions that should be asked regarding the needs of the farm are:

- Is there a market for crops in the rotation?
- Are the crops suitable for the soil types on the farm?
- Are the crops suitable for the moisture and climate conditions of the farm?
- Can the crops be produced with the equipment available on the farm or with minimal changes in equipment?
- Do the crops supply the on-farm feed and on-farm green manure needs as well as the farm's cash and subsistence needs?

Requirements for sustainability. The crop rotation requirements for sustainability revolve around the following principles:

- Does the rotation provide effective weed control?
- Does the rotation provide a balance of crop production and soil conservation?
- Does it contribute to soil building?
- Does the rotation include root systems that penetrate soil compaction, bringing nutrients to the surface and allowing air and water to infiltrate the soil more readily?
- Does the rotation provide for effective insect and disease control?
- Does the rotation effectively utilise available moisture? Are moisture-conserving practices included? Are high moisture users alternated with plants requiring less moisture?
- Does the rotation provide for a sufficient diversity of crops to increase stability and therefore minimise risks?
- Do the crops avoid any build-up of undesirable elements?

Probably no rotation scheme can accommodate all of these objectives, but even including several of them can help to establish a sustainable system (Kirschenmann 1988).

After outlining briefly some basic principles about interactions between different genetic resources and between these and their environment, we look at some combinations of genetic resources which can be and are used by farmers in LEISA.

Exploiting plant interactions

Plant interact in terms of time and space (both horizontal and vertical). Plant development in space, i.e. plant growth, is a process that takes place in time. During this process, the plants must take up energy, water and nutrients from the environment, but their needs for these growth factors differ according to the growth stage. During growth, climatic factors change with the season, and the plants themselves influence the microclimate (e.g. air humidity, temperature of soil and air, shade) as they become larger and use up the available nutrient and water resources. This, in turn, influences the plants' uptake of the growth factors, which affects their growth. Thus, plant growth is a continuous interactive process between changing plants and their changing environment. In multiple cropping systems, farmers strive for optimal vegetative development in which the arrangement of crops in time and space plays an important role.

Techniques related to the space dimension that can be used to achieve the desired crop development involve different plant densities, planting patterns and spatial arrangements (see Box 5.12). Techniques related to the time dimension are planting date, rotation and fertilisation.

Over time, it may be necessary to alter the combination of genetic resources to make more efficient use of resources such as nutrients,

Box 5.12
Selective use of crop species and varieties

Each species and variety has its own way of developing in inter-action with its environment, i.e. each has a characteristic need for growth factors, which is manifested in space as its morphology. Differences in morphology between crops (and, to a lesser extent, between varieties) can be used by farmers to achieve a desired impact on the interaction of the components with the environment. Not only the morphology above the soil surface but also the root pattern is important. Examples of how the morpho-logical characteristics of different crops and varieties in multiple cropping systems influence the spatial develop-ment of other components are:

- Leaf shape, density and arrangement influence light interception, and light require-ments vary during a crop's life cycle. Farmers can influence light interception by means of plant spacing and pruning.
- Crop height is also linked to light interception. Agroforestry systems have several storeys of crops. Some trees such as Acacia spp and Grevillea

Figure 5.4 Ditch between hedge and crop to avoid root interference

robusta are quite open in their growth habit and let light penetrate to the crops beneath. Others are less open, and permit the cultivation of shade-loving crops (e.g. cardamom under banana and coffee).

- Crop height can also play a role in crop protection, e.g. a tall variety of sorghum with a high cyanide content can be deliberately sown in a strip around a field with a short, high-yielding sorghum variety to attract grazing cattle and divert them from the main sorghum crop.
- Plants with different rooting patterns (horizontal and ver-tical) need not compete for nutrients and water as they can take them up from different soil layers. Deep-

rooting crops pump up nutrients that have leached into deeper soil layers. When the fallen leaves or prunings decompose, the nutrients become available to crops with shallower roots. Roots of shrubs and trees in hedges can be trained to grow deeper by regular trimming. Competi-tion in the root zone between hedge and crop can be reduced by digging a ditch between them to force the shrub roots to grow deeper (see Figure 5.4).

- As the growth of trailing and climbing crops (e.g. pumpkins, climbing beans) is not easy to control, these can easily smother other crops (negative effect) or weeds (positive effect) (Dupriez & de Leener 1988).

water and labour, to restore soil fertility (fallow) or to decrease populations of pests such as insects and weeds. Staggered planting, sequential cropping, relay cropping, rotation and succession are techniques which can be used for this purpose.

Exploiting animal – plant and animal – animal interactions

Interactions between animals and plants and between different animals can also be used by farmers to their advantage. This includes deliberate manipulation of game populations. For example, advantage can be taken of the fact that disease vectors such as tsetse flies prefer certain hosts. If a sufficiently high population of wild animals, which are preferred hosts, is maintained in the area where sheep and goats are kept, the danger of disease transmission to these animals can be reduced (Matthewman 1980).

The impact of animals upon plants can be used to manage the vegetation. With knowledge of the diet selected by different species of animals, grazing pressure can be manipulated to create or maintain a desired composition of vegetation. For example, browsing animals, such as goats, are useful in reducing encroachment of unwanted bushes into pasture. The selective grazing habits of animals can be deliberately used to control 'weeds', as in the case of grazing down grass in the early wet season to allow pasture legumes to establish well (Otsyina et al. 1987). Goats have similarly been used to weed crops: in parts of the Middle East, they are allowed to satisfy their initial appetite on natural pasture and are then put into cereal fields, where they selectively eat the herbs (Jaudas 1988, pers.comm.).

The hoof action of livestock can compact the soil and destroy vegetation if grazing pressure is very high for a long period of time. However, the impact of animals' hooves can also be used to disturb the soil surface to permit better germination of seed (Otsyina et al. 1987). This technique is used, for example, by agropastoralists in Nigeria to prepare land for growing small cereals: they concentrate cattle overnight on a small area of cleared land and then broadcast the seed over the broken soil surface the following morning. Another way of making use of this principle is to run a herd quickly over a tract of land to stimulate regeneration of natural vegetation from seed banks in the soil (Savory 1988). The impact of rapid and intense hoof action depends, however, on vegetation and soil type. Therefore, such animal-plant interactions may not benefit land-use systems at all sites (Bayer et al. 1987).

By manipulating the vegetation and changing the microclimate, farmers can improve conditions for desired animal species. Scattered trees can create shade for livestock. Trellised bean plants or high-stalked cereals sown around the farmhouse protect free-ranging poultry from predatory birds; for this reason, the cereal stalks are often left standing after the grains have been harvested. Reducing the density of woody vegetation to create a more open savanna landscape decreases the habitat for biting flies and reduces the risk that livestock will be infected by diseases the flies may carry. In subhumid West Africa, thinning of woodland has made it possible for farmers to keep zebu cattle in areas which were formerly infested with tsetse flies (Bourn 1983). The planting of scattered hedges can create a favourable environment for wild animals and birds which control pests in the adjoining cultivated fields. However, both beneficial and harmful creatures can be attracted to trees and bushes, and careful consideration must be given to the types and forms of vegetation that will tip the balance in favour of attracting creatures which benefit cropping and which can be harvested directly as food or for other useful purposes.

Maintaining diversity and flexibility

The sustainability of a farm system depends on its flexibility under changing circumstances. The availability of a wide diversity of genetic resources at the farm level contributes to this flexibility. A classic example to illustrate the dangers of genetic uniformity is the failure of the potato crop in Ireland in 1846 – 47. More recently (1971) in the USA, 15% of the Corn Belt's harvest was wiped out by a lone fungus which caused corn-leaf blight. Other reasons why it is important that

Box 5.13
**Rice varieties to suit
different needs and
conditions**

According to a Sri Lankan
farmer, only 3 – 4 rice varieties
are left in his region, whereas
he remembers the existence and
uses of 123. There were
4-month varieties and 3-month
ones for each of the two
growing seasons. There were
varieties with a high protein
content which were eaten by
Buddhist priests, who needed
food to sustain them for a long
time since they do not eat after
noon. There were varieties with
high carbohydrate content
which served as a source of
energy for those who had to
work long days in the field.
Moreover, many varieties were
available for different growth
conditions and were used
according to the prevailing
circumstances in the fields: little
or much water, poor or rich soil
etc. The farmer concludes that
some of the varieties always
grew well, whatever the problems in a particular year. The
new, hybrid varieties give high
yields in good years but, when
there is a drought, the plants
die. He adds that the droughts
are becoming worse as a result
of the disappearance of the
forests (ICDA News 1985).

farmers should be able to choose between varieties with different
properties to suit their needs are given in Box 5.13. These needs can
differ from season to season and from year to year. Not only climatic
but also economic, political and social changes have to be taken into
account. An example of a political change influencing the choice of
variety is a government's decision to stop fertiliser imports; low fertiliser
availability requires the use of other varieties.

Farmers can maintain biological diversity by using mixtures of
different species, mixtures of different varieties of the same species,
or varieties whose genetic composition is itself variable (Jiggins 1990):

● Mixtures of different species make an important contribution in
unstable and variable environments to harvest security and nutritional
balance. In some cases, they also yield a higher total usable biomass
than monocrops and increase the sustainability of the yield (Clawson
1985, Francis 1986).

● Variety mixtures offer additional diversity in the timing of germination, flowering, growth, seed-filling and harvest. In a study of
farming in a cluster of villages in central Sierra Leone, Richards (1986)
noted how farmers used this diversity by drawing on a portfolio of
different species and variety mixtures to suit the different conditions
of different sites along the slopes (see Box 3.5).

● The additional benefits which mixtures of unstable varieties might
confer have not received much attention. Their use has been widely
observed, even where stable varieties are available to farmers. Recent
surveys of variable common bean *(Phaseolus vulgaris)* varieties in
Malawi indicate that "the mixtures planted by farmers are comprised
of both the higher-yielding, but probably susceptible, and the lower-
yielding, but drought-tolerant, components. This is one explanation
why Malawian farmers grow bean mixtures. They appear to want
to maximise seed yields during good years by planting higher-yielding
types while at the same time minimising yield losses, in the event of
a drought, by including drought-tolerant types" (Mkandawire 1988).

Mixing crops

When two or more crops are grown in the same field, either at the same
time or immediately after each other, this is called 'multiple cropping'.
The term 'crops' generally refers to annual or biennial plants, but scientists are increasingly recognising that perennials (trees, shrubs, grasses,
herbs) can be combined advantageously with these crops and could also
be regarded as crops themselves. Combining arable crops with woody
species is known as agroforestry, agrosilviculture or multistorey cropping (a term emphasising the design of the vegetation canopy). Defini-
tions of these terms are given in the glossary (see Appendix B).

When humans started to cultivate plants deliberately, these grew
amidst many others. The first forms of agriculture were multiple
cropping systems (Francis 1986). During the past century, technologies
were developed (e.g. use of chemical fertilisers, pesticides, hybrid seed)
to maximise the productive functions of cropping systems by externalising the reproductive and protective functions. This led to specialisation
(sole cropping), which favoured mechanisation. The high-external-input

A multiple cropping system in Kenya which provides good soil cover and reduces farming risks compared with sole cropping. *(Chris Pennarts)*

Table 5.1 Yields and total biomass of maize, beans and squash (kg/ha) in polyculture as compared with several densities (plants/ha) of each crop in monoculture

Crop	Monoculture				Polyculture
Maize					
Density	33 300	40 000	66 600	100 000	50 000
Yield	990	1150	1230	1170	1720
Biomass	2823	3119	4478	4871	5927
Beans					
Density	56 800	64 000	100 000	133 200	40 000
Yield	425	740	610	695	110
Biomass	853	895	843	1390	253
Squash					
Density	1200	1875	7500	30 000	3330
Yield	15	250	430	225	80
Biomass	241	941	1254	802	478
Total polyculture yield					1910
Total polyculture biomass					6659

Land equivalent ratio (yield or biomass of each crop in polyculture/ maximum yield of biomass of each crop in monoculture):

Based on yield   1.73
Based on biomass 1.78

Source: Amador, M.F. (1980), Comportamiento de tres especies (maíz, frijol, calabaza) en policultivos en la Chontalpa, Tabasco, Mexico, Tesis profesional, Colegio Superior de Agricultura Tropical, Tabasco, Mexico; cited in Dover & Talbot (1987), p.36.

cropping systems which resulted are quite uniform and exhibit little diversity in terms of species and varieties.

If not forced to do otherwise, most smallholders in the tropics have continued to practise multiple cropping. During the last two decades, scientists have become increasingly aware that this is a highly suitable practice for maximising production with low levels of external inputs, while minimising risks and conserving the natural resource base. More specifically, the following advantages of multiple cropping for small-holders have been identified (Papendick et al. 1976, Beets 1982, Francis 1986, Altieri 1987, Hoof 1987):

● In most multiple cropping systems developed by smallholders, productivity in terms of harvestable products per unit area is higher than under sole cropping with the same level of management. Yield advantages can range from 20% to 60% (Steiner 1984, Francis 1986). These differences can be explained by a combination of higher growth rates, reduction of losses on account of weeds, insects and diseases and more efficient use of the available resources of water, light and nutrients (see Table 5.1).

● As several crops are grown, failure of one crop to produce enough (either as actual harvest or in terms of cash) can be compensated by other crops. This decreases farming risks.

● Multiple cropping systems, especially those with perennial grasses and trees, appear to be less prone to soil erosion (because of better soil cover and more barriers to water and air flows), make better use of available space for root and canopy growth, recycle available nutrients and water to a higher level and have greater buffer capacity for adverse periods and events (drought, pest attack, sudden high cash needs etc.) than sole cropping systems. In other words, they make better use of and give better protection to the farm's natural capital.

Integrating woody species

Woody species (trees and shrubs) can contribute to the viability of a farm system in many ways. They not only have important productive functions (yielding food, fodder, fuel, fibre, timber, medicine and pesticides); they also have reproductive, protective and social functions. Products of woody species can be used for home consumption and/or sold.

By providing products, as well as nutrient and capital reserves for adverse times and seasons, and by protecting soils and crops from harmful flows of radiation, water or wind (creation of microclimates and erosion control), woody species are indispensable to secure family subsistence in many LEIA areas. By integrating woody species, farmers can diversify outputs and spread the need for inputs (e.g. labour) over the seasons, thus reducing farming risks.

Woody species can enhance soil fertility by extracting nutrients from the surroundings and from deeper soil layers (recycling leached nutrients by nutrient pumping) and concentrating them in perennial biomass and the topsoil, by increasing organic matter in the topsoil, by interacting with mycorrhiza and soil bacteria (nitrogen fixation, phosphate solubilisation) and by capturing nutrients from air and water flow. Woody species perform these roles particularly during natural fallow and certain species can be deliberately introduced to intensify the fallow.

Woody species can create suitable microclimates for other productive components, crops or animals, within the farm system. They can help control weeds, and some species provide natural pesticides or medicines. Certain woody species can be used to decrease the need for external inputs of artificial fertilisers and biocides.

Where labour or capital resources are scarce, low-input, low-management woody species may make the most effective use of these resources. But likewise, where labour and capital are ample, certain woody species, e.g. fruit trees, may make the most effective use of these resources.

Crops and woody species can complement each other in terms of space (e.g. certain woody species can grow on places unsuitable for crops, e.g. stony, steep, temporarily flooded, acid or alkaline land; the roots of some woody species penetrate into soil layers not used by crops), in time (e.g. crop versus fallow vegetation; spreading of product and labour needs over the seasons) or in function (e.g. diversification of production; productive versus protective and reproductive functions).

Synergy between crops and woody species can be expected particularly when protective or reproductive functions of woody species with positive effects on crop growth are combined with one or more useful products. For example, improving the microclimate and controlling erosion with

Trees serve multiple purposes within farm systems. *Vitex doniana* not only yields wild plums, leaves as vegetables and fodder, and wood for making drums, it is also a tree in which Nigerian farmers like to place beehives. *(Ann Waters-Bayer)*

windbreaks can lead to higher grain production per unit area sown (which may compensate for the loss of yield on the parts of the field occupied by trees) plus production of wood, fodder, herbal medicines, 'bush-meat' etc. Contour hedges, which take up nutrients from deeper soil layers, fix nitrogen and control erosion, can lead to higher and more sustainable grain production per unit area sown plus production of wood and fodder (Nair et al. 1984, Whitington et al. 1988).

Not all of these products and functions can be combined at one site and in one species. Some species provide several products or fulfil several functions (multipurpose species) but, in general, combinations of woody species have to be sought to obtain combinations of products or functions. Each species has particular characteristics and uses and requires specific ecological conditions. Growth conditions can be such that certain functions cannot be provided. For example, in the semiarid zone, nitrogen fixation by leguminous woody species seems to be very low and roots develop horizontally instead of vertically where only a superficial zone is wetted by rain. Therefore, the function of woody

---

Box 5.14
Overcoming constraints to integration of trees

Integrating trees into farm systems faces many constraints, the most important of which are:

- Trees also have actual or perceived negative aspects: they may compete for scarce nutrients, water or light; hinder mechanisation; be hosts for pests and diseases; become noxious weeds; or contain harmful substances. Farmers must find the right place for trees in their farms, where competition with annual crops is lowest and positive effects are highest, e.g. beside the house, on fallow land, along field borders or contour strips, or on shallow, stony or waterlogged soils.
- Tree planting may involve considerable risks because of possible damage by animals, fire or pests (de Leener & Perier 1989). Also market risks are involved, as it takes many years before the intended commodity is produced and can be sold. Some of the risks can be reduced by good management, tree protection, diversification and regular harvesting.

- Trees may compete with other crops for scarce space. As it generally takes a few years until trees show positive effects on income and growth conditions, an initial reduction in overall productivity of the farm can be expected when trees are interplanted with crops. These start-up problems can be alleviated by introducing trees gradually, combining them with other components which bring advantages more quickly, or providing credit. Sometimes, for the purpose of improving soil fertility and water and soil conservation, there may be better solutions than planting trees, e.g. cover crops or grass contour lines.
- Trees are often governed by land tenure and gender-based usufruct regulations which may prevent potential tree planters from gaining the fruits of their work. Secure land-use rights are basic to tree planting, but changing rights and customs of land and tree use is a long and difficult process. Using fast-growing trees or shrubs which are traditionally not regarded as trees may be a way of getting around such problems.

- Often, insufficient attention is given by development planners to the problems of involving women in tree planting. Enabling women to derive more benefits from tree resources, e.g. by involving them in the cash economy of tree cropping, is likely to prove a rewarding strategy in tree planting (Clarke 1986, Rojas 1989, Hoeksema 1989).
- There may be legal restrictions on cutting, selling or transporting trees. In such cases, adopting legislation to favour tree cropping by smallholders and minimising administrative procedures for obtaining permits will enhance tree planting (Chambers et al. 1989).
- Markets for tree products may not exist, or prices may be unattractive. Problems may be encountered in transporting bulky products, such as fuelwood, to market. To be of any value, the products may have to be processed, which requires additional labour, capital and specialised knowledge. Creating market demand and opportunities, and improving transport and processing facilities, could promote tree growing (Hegde 1990).

species in enhancing soil fertility in the semiarid zone is probably less than in more humid zones (Kessler & Breman 1991).

Many of the constraints to integrating woody species into farm systems and some ways of overcoming these constraints are presented in Box 5.14. Formal scientific knowledge about trees and shrubs is still very limited, especially about their interactions with arable crops and animals. Questions still to be answered include: Which are the best woody species to grow together with specific seasonal crops? What are the most effective ways of exploiting trees? How can trees for forage and/or green manure be best managed and used? How should tree products be handled for processing, storage and marketing? To answer these and many more open questions, tapping indigenous knowledge and skills and further scientific research will be necessary (CTA 1988).

In many areas, woody species are disappearing, often on account of a combination of factors, such as the need for more land for crop and livestock production; continuous pressure on tree resources for fuelwood, charcoal, timber or fodder; shifts toward sole cropping and mechanisation; uncertain land tenure; privatisation and nationalisation of common lands; and restriction in the power of traditional village authorities. Nevertheless, the increase in woody species in other land-use situations indicates that, under certain circumstances, growing woody species can form part of an appropriate response to increasing pressures on the farmer's resource base. Arnold (1990) mentioned three situations in which farmers regard tree growing as an efficient use of resources:

- **Under low-intensity land use**, where farmers start shortening the fallow period. A common practice is to enrich the fallow by encouraging or planting woody species that accelerate or enhance regeneration of soil fertility, or produce outputs of subsistence or commercial value, or both; e.g. *Acacia senegal* in sub-Saharan semiarid Africa, which regenerates the soil and produces gum arabic, fuelwood, medicine and fibre; and the babassu palm *(Orbignya speciosa)* in Brazil, which provides various commercial and subsistence products,

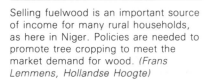

Selling fuelwood is an important source of income for many rural households, as here in Niger. Policies are needed to promote tree cropping to meet the market demand for wood. *(Frans Lemmens, Hollandse Hoogte)*

particularly oil. A next step in intensification is introducing trees as an intercrop. Numerous examples of such continuous fallow strategies can be found: maintaining *Faidherbia albida* in cultivated areas in much of Africa, or intercropping *Sesbania sesban* (a leguminous tree) with maize in Kenya. Considerable research has recently been directed towards developing a more intensively managed continuous fallow system known as alley cropping (see Appendix A1).

● **Under high-intensity land use**, when home gardens including woody species are used to supplement outputs from other parts of the farm. They also spread labour, outputs and income more evenly over the year. In the densely populated lowland areas of central Java, home gardening is the principal form of dryland farming, while irrigated rice cultivation forms the other main component of the farm system (Palte 1989). As the proportion of land devoted to rice decreases, home garden areas are cultivated more intensively, becoming mixed rather than forest gardens as annual plants are progressively intercropped to provide food and income.

● **When woodlots are established** to produce trees as a cash crop. Although trees may be grown intensively with fertiliser and irrigation, low-input, minimum-management systems of cash cropping with trees are more common. In India, for example, there has been an upsurge in tree growing in response to expanding markets for poles and other wood products (e.g. pulpwood). In rainfed areas, the main factors motivating farmers to start growing trees are low labour requirements, minimal annual operating costs in most years, greater resistance to drought and, hence, reduced risk and uncertainty.

Trees may also be cultivated to increase income-earning opportunities when the size of landholding, or site productivity, falls below the level at which the household's basic food needs can be met from on-farm production of food.

As trees yield various products, a single species can be exploited for its leaves, wood, bark, roots or fruits. However, the different uses are interdependent: a good leaf crop will not coincide with a good fruit crop. Farmers must decide what services they want the trees to perform and what products they prefer. This explains the great diversity in methods of tree management. The farmer seeking to create light shade will prune a tree differently from one who wants an abundant leaf harvest. Cultivation practices may differ depending on how the trees are to be used but, in principle, trees need to be fertilised and protected from harmful effects of air and water flows, animals, insects and diseases, just like field crops. Farmers wanting to take advantage of all the opportunities woody species offer must manage the trees or shrubs well. When woody species are left untended on farmland, they can become a nuisance by competing with crop and livestock production and creating pest hazards.

Integrating herbaceous species

Also some nonwoody species, e.g. herbs and nonwoody green manure and cover crops, can play an important supportive role in LEISA. Herbs can be used to produce natural pesticides and medicines for humans and animals and can serve as trap or decoy crops. Apart from their protective function, they can provide a significant source of income. Where medicinal herbs have traditionally been collected in the wild but are in danger of becoming extinct on account of overexploitation, farmers can start cultivating them, thus combining income generation with nature conservation.

Green manure and cover crops are important for managing soils and soil fertility and in erosion control, especially in humid areas. They can provide the necessary biomass from their leaves, branches and roots to stimulate soil life and protect the soil surface. Leguminous green manure and cover crops can fix considerable amounts of nitrogen and enhance the availability of elements necessary for crop growth. They can function indirectly when used as fodder for animals which provide manure for crops. Some green manure and cover crops (e.g. *Crotalaria ochroleuca, Tephrosia candida*) have pesticidal effects which can be used for plant protection. Sometimes, cover crops also function as 'green' or 'living mulch', reducing water losses from soils being cropped in semiarid areas (van der Heide & Hairiah 1989).

As the reproductive and protective functions of such herbaceous species may compete in the short term with the productive function of crops, they are often replaced by external inputs of chemical fertilisers, when available. However, especially in vulnerable soils, this can lead to biological inactivity, acidification, increased leaching, loss of soil structure and erosion. For farmers with very limited access to external inputs, auxiliary herbaceous species can help greatly in keeping the farm productive, efficient and sustainable.

Mixing livestock

As with multiple cropping, mixed holdings of livestock are common in LEIA systems. By keeping several species, e.g. poultry, ruminants and pigs, farmers can exploit a wider range of feed resources than if only one species is kept. In pastoral areas, camels can graze up to 50 km away from watering points, whereas cattle are limited to a grazing orbit

Keeping more than one species of livestock reduces risks and permits use of a wider range of feed resources than with single-species herds. Camels and goats at a watering point in Somalia. *(Wolfgang Bayer)*

of 10 – 15 km. Camels and goats tend to browse more, i.e. to eat the leaves of shrubs and trees; sheep and cattle generally prefer grasses and herbs, and resort to browsing only when the preferred forage is scarce. Different animal species supply different products, e.g. milk, transport and draught power can be provided by camels and cattle, whereas goats and sheep tend to be slaughtered more often for meat. Chickens often provide the 'small change' for the household, sheep and goats are sold to cover intermediate expenditures, while larger animals like cattle are sold to meet major expenditures.

Keeping more than one species of livestock is also a risk-minimising strategy. An outbreak of disease may affect only one of the species, e.g. the cow, and some species or breeds are better able to survive droughts and thus help carry a family over such difficult periods. Advantage can also be taken of the different reproductive rates of different species to rebuild livestock holdings after a drought. For example, the greater fecundity of sheep and goats permits their numbers to be multiplied more quickly than cattle or camel numbers. The small ruminants can then be exchanged or sold to obtain large ruminants.

Integrating crops and livestock

Animals can perform numerous functions in smallholder systems. They provide products, such as meat, milk, eggs, wool and hides. They serve sociocultural functions, e.g. as bridewealth, for ceremonial feasts, and as gifts or loans which strengthen social bonds. Under LEIA conditions, integration of livestock into the farm system is particularly important for:

- increasing subsistence security by diversifying the food-generating activities of the farm family;
- transferring nutrients and energy between animals and crops via manure and forage from cropped areas and via use of draught animals.

Keeping livestock to secure subsistence is particularly important where cropping risks are high, e.g. in dry areas. Livestock serve as a buffer: an animal can be slaughtered for home consumption or sold to buy food when crop yields do not meet family needs. Livestock are like a savings account, with offspring as interest. Animals are sold when cash is needed for specific purposes, including the purchase of inputs for cropping. Diversification into livestock keeping extends the risk-reduction strategies of farmers beyond multiple cropping and thus increases the economic stability of the farm system. Spreading risk by practising both crop and livestock production may lead to lower productivity within each sector than in specialised farms, but total production per unit area may even be increased, as both crop and livestock yields can be gained from the same area of land.

Livestock can enhance farm productivity by intensifying nutrient and energy cycles. Stubble fields and other crop residues, e.g. after threshing, are important sources of forage in smallholder systems. On Kenyan smallholdings, for example, an estimated 40% of annual forage energy is derived from crop residues (Stotz 1983). Weeds from cultivated fields, lower mature leaves stripped from standing crops, plants thinned from cereal stands, and vegetation on fallow fields offer additional fodder resources related to food cropping. When animals consume vegetation and produce dung, nutrients are recycled more quickly than when the vegetation decays naturally. Grazing livestock transfer nutrients from range to cropland and concentrate them on selected areas of the farm. The livestock themselves can do the work of collecting, transporting and depositing the nutrients and organic matter in the form of urine and dung.

In LEIA areas, forage is derived primarily from land which is unsuitable for cropping ('wasteland', such as areas with rocky outcrops, wayside edges and waterlogged land) and temporarily not being cropped (harvested or fallow fields). These pieces of land are often interspersed between cultivated plots and can be grazed by herded or tethered animals, or the vegetation can be cut as fodder.

Integrating fodder production into crop rotations can enhance the sustainability of a farm system, particularly to the extent that perennial grasses and legumes, including shrubs and trees, are involved. These may use nutrients and water from deeper soil layers than annual crops, help improve soil fertility, and protect the soil during periods when arable crops are not grown. Forage crops can play an important role in nutrient transfer also within the farm by providing better quality feed which, in turn, results in better quality dung, which can be used as fertiliser for crops. Part of the forage crop can also be used as green manure or mulch. Farmers are more likely to apply techniques, such as sown fallow, to restore soil fertility or prevent erosion if they can also gain immediate economic advantages in the form of fodder for their livestock.

Where animals are used for traction, some of the energy gained from grazing wasteland and temporarily uncultivated land can be exploited for crop production. Farmers can cultivate larger areas with draught animals than by hoe. Since ploughs and harnesses can normally be manufactured locally, animal traction requires lower levels of external inputs than the use of tractors. Animal power can also be used to process farm products, e.g. for threshing, and for transporting them from the fields to storage or market. However, the ecological repercussions of

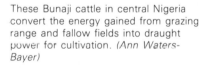

These Bunaji cattle in central Nigeria convert the energy gained from grazing range and fallow fields into draught power for cultivation. *(Ann Waters-Bayer)*

keeping draught animals are site-specific: in some cases, they may cause overgrazing and environmental degradation on pastures near the village.

Besides the more conventional livestock, such as cattle, sheep and goats, other less conventional livestock, such as rabbits, guinea pigs, ducks, bees and silkworms (see Box 5.19) can play an important role in integrated farm systems.

Integrating aquaculture

Another group of animals that deserves separate mention in this context is fish and other aquatic creatures. Natural resources available to a farm family may include water resources, such as streams, ponds and flood-prone land. The husbandry of plant and animal organisms that live in water is known as aquaculture. Integrating this form of husbandry into a farm system intensifies the use of natural resources in a sustainable manner through species diversification and nutrient recycling. Integrating fish, land-based animals, trees, vegetables and field crops within a farm is a way of maximising productivity per unit of land. By-products from one form of resource use serve as inputs for other forms, and agricultural wastes can be used to make marginal patches of land more productive.

Fish ponds can be created in wetlands (see Figure 5.5) and on homestead land where a nearby stream or spring permits. Fruit trees and vegetables planted on dikes beside the pond can be watered with pond water, which can also be used to water livestock. Manure, crop residues, weeds, tree leaves, rotten fruit and vegetables fertilise the pond. Other crop by-products, such as maize and rice brans, can also be fed to fish. Fish convert plant and animal waste into high-quality protein and enrich pond mud, which can then be used to revitalise cropland.

Smallholders in densely populated areas of Asia have already developed various integrated agriculture – aquaculture systems. Perhaps the most widely known form of integrated aquaculture is fish culture in rice fields. Detailed descriptions of various integrated systems, their biophysical elements and the interactions between them can be found in Little and Muir (1987), Edwards et al. (1988) and Dela Cruz et al. (in press). An indigenous system of agriculture – aquaculture practised in South China is described in Section 3.2.

Figure 5.5
Regenerative intensification of marginal flooded land by a farmer in India. (*Source:* Lightfoot 1990)

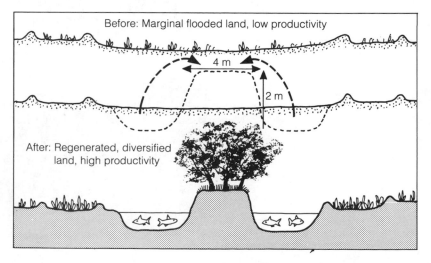

Managing limited space

<table>
<tr><td valign="top" width="33%">

Box 5.15
Making the most of scarce farm space

In many indigenous farming systems, field edges and borders, roadsides and canal banks are routinely grazed by livestock. Where space is even scarcer, these surfaces may be planted to special crops, herbs or medicinals, or to deep-rooted shrubs and trees that yield valuable products and act as soil stabilisers and windbreaks. Also waterways can be productive. Distinctive field-edge and aquatic environments also shelter wild plants and animals that can be harvested (Wilken 1987).

</td></tr>
</table>

The physical and biological structures established by farmers (bunds, terraces, tanks, contour hedges, shelterbelts etc.), field borders, pathways and trees are the infrastructure of the farm. Their shape and location are determined by slope, exposition, wind direction, resource ownership and their function to control water, wind or traffic flow, or to keep animals out of the fields.

As these structures help conserve the farm's natural resources and as their location cannot be easily changed, it is important that they be well sited. By combining functions, e.g. by planting multipurpose trees, grasses or herbs along contour lines or planting shelterbelts along field borders, a stable resource-conserving infrastructure can be created which, by virtue of positive interactions, does not decrease and may even increase the productive capacity of the system.

Land suitability of one part of the farm can be totally different from that of another part (see Box 3.5). Diversity is needed so that genetic resources adapted to the specific ecological conditions of the land (dry, wet, acid or poor soil, cold, hot or windy microclimate etc.) can be used. Also labour constraints may oblige farmers to use different genetic resources close by and far away from the homestead.

Data on the productivity of complex integrated systems with arable crops, animals (including fish) and trees are scarce, and indication is seldom given of their effects on security of farming and resource conservation. However, the fact that most indigenous smallholder farming systems contain these components in various mixtures suggest that the farmers judge their combination to be positive.

However, not all combinations increase productivity, as there are also many negative interactions between organisms. Crops, trees, animals and humans may compete with each other for land, solar energy, water, nutrients, food or labour. They may also influence each other in a negative sense (negative allelopathy, unfavourable microclimate, transfer of pests etc.). Although competition cannot be completely eliminated, it is minimised in good combinations of genetic resources. The optimal balance between positive and negative aspects of the

Box 5.16
Using vertical and horizontal space in home gardens

Full exploitation of vertical and horizontal space is found in the multistoried home gardens of Central America. Plots of less than 0.1 ha may contain two dozen or more different economic plants, each with distinctive space requirements. A garden near Atlixco, Puebla, is an example. The plot displayed a seemingly random horizontal distribution but a rather carefully arranged four-tier vertical distribution of tall trees such as mango, papaya (Carica papaya), capulín and guaje (Leucaena esculenta); numerous medium-height trees and shrubs including banana, peach, avocado, pomegranate, several types of citrus, colorin (Erythrina americana), chirimoya (Annona cherimola) and kumquat (Fortunella spp); and lower layers of high and low field crops such as maize, shrub beans, red and husk tomatoes, chili and squash, interspersed with flowers and medicinal and cooking herbs. Economic vines including beans and chayote twined into otherwise unoccupied spaces. The whole garden was fenced with additional productive and ornamental plants, such as colorin, the bamboo-like carrizo (Arundo donax) and jacaranda (Jacaranda acutifolia). Chickens and turkeys patrolled between plants and contributed in their way to this productive, three-dimensional space (Wilken 1987).

different components must be found, e.g. loss of space versus creation of better microclimate or fixation of nitrogen. It is also possible to make use of competition in a positive way, e.g. to stimulate growth by grazing, coppicing or high-density planting. Evaluation of the combinations of genetic resources evolved by indigenous smallholders would increase insight into the interactions between the different organisms, and would provide useful information for choosing the best mix.

Introducing seeds and breeds

Increased use of improved seed may be one of the cheapest and technically simplest ways for LEIA farmers to raise their productivity. However, the varieties offered to them must be adapted to LEIA conditions, and an efficient distribution system is needed to give the farmers access to these seeds (Friis-Hansen 1989).

Many governments in developing countries have started national programmes in cooperation with international agencies to develop improved varieties and seed supply systems. They have focused their activities on high-response varieties but, as mentioned in Chapter 1, the properties of these varieties often do not coincide with farmers' wishes regarding plant genetic material. Plant breeders evaluate the performance of new varieties according to a score composed of numerous criteria, the most important being yield potential. Other criteria often include response to fertiliser, resistance to pests and diseases, length of growth cycle and dietary value of the product. Sometimes, seed is also screened for suitability for mechanisation, e.g.

Box 5.17
Women's criteria for assessing cassava varieties

When FAO started rural development work in the Kwango-Kwilu area of Central Zaire in 1979, a large-scale survey revealed that nearly all food production was done by women. Clearly, any effort to develop the traditional farming sector would have to focus on women and on the predominant crop: cassava.

An inventory was made of local cassava varieties and their characteristics (yield, susceptibility to diseases and pests, etc.). Great variation was observed between varieties in disease and drought tolerance, tuber and leaf yields, taste, etc. Women themselves classify varieties according to their suitability for forest or savanna.

Discussions with the women revealed their criteria for selecting the 'best' varieties, which go well beyond the standard criterion of researchers: high tuber yield. Women's criteria include:

- tuber, and equally important, leaf yield;
- high dry matter content of tubers (for flour production);
- rapid formation of leaf canopy (reduces weeding and gives an early vegetable yield);
- taste: bitter varieties are preferred for 'luku' flour;
- dwarf varieties, from which the leaves can be easily picked and which do not suffer from wind damage in the dry season;
- late flowering, because leaves are usually harvested up to flowering, which is said to change the taste;
- shape of tubers: regular

shapes are easier to harvest, short and fat tubers do not break when the plant is lifted, and are easier to peel;
- drought and disease/pest tolerance (often bitter varieties);
- a combination of early- and late-maturing varieties to assure a continuous tuber supply and to spread risks of infestation.

It was also noted that women are very eager to experiment with new varieties and will do anything to obtain new cuttings (travel far or steal them from project fields). These discussions about cassava cultivation also proved to be valuable in helping women become aware of the possibilities of changing matters that are traditionally (and also often by development experts) considered as immutable (Fresco 1986).

Table 5.2   Traits of high-input and low-input crop types

Trait	High-input type	Low-input type
Biomass		
Amount	Small	Large
Harvest Index*	High	Rather small
Leaf Area Index**	About 3	Higher than 3
Tillering	Uniculm	Heavy tillering potential
Stem size	Dwarf	Tall
Branching (dicots)	Little	Multiple
Productivity	Yield/area/time	Biomass/area/time and external input
Environmental adaptability		
Site	Wide	Specific
Climate	Wide	Specific
Microsite	Wide	Wide
Seasonal variation	Wide	Wide
Bio-environment	Weak	Broad-horizon field resistance
Agro-environment	Specific	Specific
Competitiveness		
Intergenotypic	Weak	Strong
With other crops	Weak	Good adaptation for intercropping
With weeds	Weak	Strong
Maturity		
Time	Early	As late as agronomically possible
Type	Determinate	a) Indeterminate if environment induces seedset or tuberisation, otherwise b) Semideterminate
Root system		
Structure	Seminal	Seminal, nodal (cereals) tap roots, shallow roots
Size	Confined to topsoil	Deep (others), widely spread out
Bacterial and/or mycorrhizal associations	Inefficient	Efficient
Seed		
Size	Large	Rather small
Number/ear	High	High
Rate	High	Low (weightwise)
Genetic structure		
Inbreeders	Pure lines	Multilines, varietal
Outcrossers	Hybrids	Mixtures (synthetic) populations

*   Harvest Index (HI) = ratio of root weight to biomass
** Leaf Area Index (LAI) = an index of canopy layers

Source: Janssens et al. (1990).

mechanical harvesting. The varieties developed in such breeding programmes are generally appreciated by market-oriented, medium-to large-scale farmers, who are growing the crop in a pure stand and under relatively good growing conditions.

Smallholders look for crop varieties with a good yield which is reliable and stable over the years, and under adverse environmental conditions; varieties with built-in resistance against pests; and varieties which produce well with organic fertilisers. For this purpose, they commonly use a mixture of varieties, each of which must be compatible with the farm system (e.g. suitable for intercropping or staggered harvesting) and must fit into the labour pattern. Subsistence farmers also attach much importance to a specific taste and culinary quality, and appreciate byproducts that can be used as forage, building material, etc.

Funds for breeding programmes have often been misallocated, as the programmes were not tailored to smallholders' needs and possibilities. This is partially because varieties adapted to LEIA conditions have traits which differ from and may even be completely opposite to those of most high-response varieties. There is an urgent need to develop specifically low-input crop types with the traits given in Table 5.2.

Similarly, conventional livestock programmes have often emphasised the introduction of exotic breeds, or cross-breeding these with local stock to increase production, mainly of meat and milk. To realise their higher potential, the introduced breeds normally require high-quality feed, intensive health care and often also housing, even air-conditioned. To provide these inputs, a complex infrastructure of feedmills, breeding controls and veterinary services is necessary. Without these high external inputs, the 'improved' animals may produce less than local breeds and are not likely to survive for long. Animals with high genetic potential for production are more susceptible to stresses of all kinds (Frisch & Vercoe 1978). The introduction of exotic stock requires that the environment be manipulated to suit the needs of the animals. However, farmers operating under LEIA conditions – particularly under harsh conditions such as long dry periods, high temperatures or mountainous terrain – must largely accept the environmental constraints and are in need of animals well adapted to these conditions (Bayer 1989).

Conventional livestock breeding has concentrated on increasing food production. However, in smallholder systems, livestock are valued not only for their meat and milk, but also for their functions in providing manure, transport, fibres and skins, as a capital reserve, in sociocultural exchange systems etc. Depending on the husbandry system, farmers are looking for particular traits in the animals they select for breeding, e.g. amenability to herding, ability to walk long distances (see Section 3.2). These are traits seldom found in 'improved' livestock coming out of conventional breeding programmes.

Exploiting indigenous plants and animals

Indigenous farm systems include many local crop and livestock species, varieties and breeds adapted to the specific local conditions. Many smallholders also exploit wild plants and animals. Information on these indigenous 'unconventional' or 'underexploited' genetic resources is sketchy, but enough to indicate that they play a critical role in meeting the essential needs of people living under LEIA conditions.

Box 5.18
Women as managers of knowledge about indigenous plants

In a travelling workshop coordinated by KENGO, women farmers and scientists from East Africa shared traditional knowledge and skills of women in managing natural resources. The workshop revealed that the most important component of the 'survival economy' of the women visited in the semiarid zone, the wet highlands and the lake basin of Kenya was their knowledge of wild and cultivated indigenous plants and ways of growing, conserving and processing them. In the dry season and especially in drought years, the women gather wild fruits and vegetables (e.g. *Vigna* spp, *Amaranthus* spp, *Berchemia discolor, Adansonia digitata*) to feed their families. The berries and leaves are usually eaten fresh, but some are also systematically cultivated and preserved for periods of scarcity. These indigenous plants are important genetic resources for sustainable land-use systems, as they are resistant to drought and diseases, do not need special fertilisers, have high nutritional value and can be used not only for food but also for medicines, fodder, fuel, dyes and fibres (Hoffmann-Kuehnel 1989).

Indigenous plants provide diverse and nutritious food and may be vital during 'hungry seasons' and famines (see Box 5.18). They can also provide many useful nonfood products and sources of income. As they often grow on marginal land or form part of multiple cropping systems, exploiting indigenous plants increases land-use efficiency (FAO 1988a).

The major commercial trees of the tropics and subtropics, e.g. coffee, tea, cacao, coconut, oil palm, rubber and citrus, are well known. Lesser known are the products of indigenous economic trees such as shea butter *(Butyrospermum parkii* syn. *Vitellaria paradoxa)*, marula plum *(Sclerocarya birrea)*, African locust bean *(Parkia biglobosa)* and the 'local maggi' made from it (see Box 3.1); leaves of baobab *(Adansonia digitata)*, horseradish *(Moringa pterygosperma)* and bitter leaf *(Vernonia amygdalina)* which are used as vegetables; spices from the Akee apple *(Blighia sapida)*, red kapok *(Bombax buonopozense)* and calabash nutmeg *(Monodora myristica)*; and medicinal ingredients for treating humans, animals and plants.

Many products from indigenous trees are valuable local foods in terms of both quality and quantity. For example, one mature locust bean tree supplies about 40 kg of seeds in an average year. When fermented, these have the same nutritional value as 50 chickens and provide considerable amounts of vitamins and minerals in a period when other foods are often scarce (de Leener & Perier 1989). In the hungry season, the products of local trees and shrubs may be the basis of survival for many farm families.

Indigenous economic trees are seldom as 'wild' as they are perceived to be by outsiders. They are often integral parts of local farming systems: they may be protected and managed or even deliberately planted by farmers. *Faidherbia albida* in Africa and *Prosopis cineraria* in India are examples of indigenous trees that play a central role in land-use systems in the semiarid tropics. Such trees are often vital for the sustainability of farming, and their disappearance may lead to ecological imbalance and social and economic tension. Especially poorer farmers and women who derive an essential part of their food and/or income

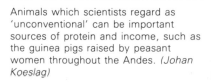

Animals which scientists regard as 'unconventional' can be important sources of protein and income, such as the guinea pigs raised by peasant women throughout the Andes. *(Johan Koeslag)*

Table 5.3   Sources of income for men, women and poor women in Uttar Pradesh, India (%)

Income	Forest and common land	Cropland	Off-farm activities	Total
Men	13	28	59	100
Women	33	35	32	100
Poor women	45	34	21	100

Source: Clarke (1986).

from trees and shrubs suffer when these become scarce (see Table 5.3).

Scientific knowledge about 'unconventional' livestock species, such as the guinea pig, camel or yak, or about indigenous breeds of conventional livestock is also very limited (Box 5.19). The species and breeds of animals indigenous to the tropics were developed under low-input conditions. Tropical breeds of cattle, for example, are adapted to forage of low quality and quantity, a trait reflected in their generally lower metabolic rate than that of animals with high genetic potential (Frisch & Vercoe 1978). Animals adapted to arid areas use water very efficiently, can drink large amounts of water within a short time and can go for three or more days without drinking. This is the case not only in camels, goats or donkeys but also in certain breeds of cattle, such as the Boran (King 1983). Under harsh low-input conditions, these adaptive traits are more important to farm families than the ability of animals to produce well under ideal conditions.

The productivity and security of farming in the tropics could be enhanced by exploiting these unconventional species to a greater extent. There is an immense need to help farmers maintain the rich genetic diversity of indigenous plants and animals adapted to low-input conditions and to make these genetic resources available to other farmers operating in climatically similar areas. To this end, it is equally important that local knowledge about how to care for and use these genetic resources be conserved and conveyed to other farmers.

Conserving and managing genetic resources and, if possible, improving them through selection is much easier and quicker in plants than in animals, as there are many times more plant species, varieties and individuals than in the case of animals. In terms of individuals from which to select, a flock of sheep is less than a child's handful of seed. Moreover, the generation intervals in large livestock species is much longer than in the case of annual crops. It is therefore particularly vital that the valuable traits in animal species and breeds selected and conserved over generations by LEIA farmers continue to be conserved, in view of the difficulty and the time (or, in the case of biotechnology, the expense) that would be involved in regaining them.

Some possibilities of strengthening local capacities to select and conserve genetic resources are presented in Appendix A.

Box 5.19
Potential of 'unconventional' livestock

More than 60 animal species contribute to human needs for food, shelter and energy, but commercial production focuses on a few so-called conventional species: cattle, sheep, goats, pigs and poultry. Many other, 'unconventional' species yield products which are vital for sustaining the human population and are particularly suitable for farming systems with limited resources. Species such as the rabbit, guinea pig, guinea fowl, duck, bee, pigeon and silkworm can adapt to a wide range of ecological conditions. The camel, llama, alpaca, yak, banteng, water buffalo, eland, oryx, deer and such small animals as capybaras, cane cutters, snails, frogs and reptiles are adapted to specific ecological niches.

Large unconventional livestock are physiologically and behaviourally adapted to harsh environments. The yak and the two-humped camel have an undercoat which enables them to tolerate low temperatures and large variations in temperature. The llama, one-humped camel, oryx and eland can live in hot arid environments because they have efficient water conservation mechanisms, long limbs and heat-reflecting coats. Some animals, e.g. water buffalo, respond to high heat loads and humidity by behavioural adaptation (wallowing, shade seeking).

Large unconventional animals can thrive on natural browse and forage. Owing to the specific morphology of their stomachs and the rumen bacteria that break down cellulose into simpler digestible compounds, they are physiologically adapted to using feed resources of very poor quality which conventional stock cannot use. Because of their different feed preferences (e.g. camels feed on thorny shrubs and salt bush, oryx on succulents and sparse grasses, eland on browse, banteng on coarse tropical grasses), they can be kept in mixed herds, thus enabling complementary utilisation of feed resources.

These animals can make use of marginal areas to produce, e.g. meat and manure (all these animals), milk (camels, yak, water buffalo) and fibres (camels, llama, alpaca, yak). In addition, camels are used for draft and transport in dry areas, llama and alpaca for transport in the Andes, the yak as riding and pack animal in mountainous central Asia, and the buffalo and banteng as a source of farm power in southeast Asia. Multipurpose animals such as these are of great importance for sustaining economic activity in harsh environments.

Small unconventional livestock are generally characterised by short generation intervals, large numbers of offspring and fast growth of young. This high reproductive capacity reduces the proportional energy needs of the reproductive unit, so that nutrients are used more efficiently in the production process. Rabbits, guinea pigs and cane cutters can digest almost any form of edible greenstuff, ranging from coarse grasses to roughages and household scraps. Being easy to house and manage, they can be integrated into farms to expand the available food resource base and can even be reared by land-less and land-poor farmers in the backyard. Also free-range ducks, pigeons and bees, as well as snails which eat decaying materials can be used to achieve more efficient nutrient recycling in the ecological chain.

Commercial production of these animals is undemanding in terms of capital investment and husbandry skills, and presents minimal economic risks. Marketing these animals can provide cash in addition to valuable protein for home consumption. The smaller quantities of meat from small animals can be consumed at once without wastage, which is an important consideration when refrigeration is not available for storing the carcass.

Research needs. Unconventional livestock species are valuable genetic resources which can contribute greatly to smallholder economies. Their successful integration into farm systems is subject to a thorough understanding of their biological potential and of how they fit into these systems. Research efforts should concentrate on improving nutrition, health and husbandry skills and selecting new, more productive species. Commercial use of game animals has little prospect at present, but 'bush-meat' will continue to be an important source of animal protein for farmers in remote areas. Therefore, the contribution of wildlife to human diets should also be evaluated (Peters 1987).

6 Development of LEISA systems

6.1 Opportunities and limitations

As outlined in Chapter 2, many ecological, socioeconomic, cultural and political processes can steer the development of farm systems away from sustainability. Some processes or constraints can be so strong, e.g. political repression, war, excessive labour migration, high prices for farm inputs and low prices for farm products, that resolving these problems becomes a precondition for initiatives at the farm or community level to enhance sustainability. To make farm systems more sustainable, integrated strategies are often needed which involve technical, commercial, legislative, motivational, educative and/or policy components. For example, to increase the viability of smallholder farming in India, Chambers (1989) proposes an integrated strategy for tree planting which comprises legal changes pertaining to tree tenure and the cutting, transporting and marketing of wood; creating and improving market opportunities; training of farmers; and developing improved methods of tree growing and management.

Some of the constraints most frequently encountered in LEIA areas are depleted, eroded, acid, alkaline, saline, waterlogged or shallow soils; steep slopes; droughts, floods, typhoons etc; serious pest or disease problems; insecure or restrictive rights to land, water or trees; commercial or transport limitations; lack of credit facilities; unreliable input delivery; restrictive gender relations etc. In order to identify the limitations and opportunities of a farm system with respect to sustainability, it is necessary to make an evaluation of the farm household's objectives and the specific technology system being applied: the genetic resources, techniques, inputs, strategies and farm layout. Some ways in which farmers can do this with the support of outsiders are outlined in Chapter 8.

Where the need and opportunity for changes of a technical nature have been recognised, suitable technologies and strategies must be sought from various sources. The options that hold the most promise for enhancing the sustainability of the farm system must be selected, tested and, possibly, adapted, or new technologies may have to be developed locally. Also in Chapter 8, experiences with such processes of technology adaptation and development are described.

The overall impact of specific technologies in the farm system can be assessed only by looking at all aspects of sustainability: productivity, security, continuity and identity. All too often, those who promote the use of artificial inputs assess only productivity, leaving assessment according to other criteria to speculation. Farmers' experience normally gives them good insight into the diverse effects of the technologies they are presently using. However, with respect to new technologies —

whether of the 'green revolution' or the LEISA type – farmers require complete, unbiased information about their impact to be able to choose wisely. The choice of technologies to enhance the sustainability of a particular farm system should not be determined by labels such as 'traditional', 'modern', 'ecological', 'biodynamic', 'LEISA' etc. These labels are often biased and are not based on critical evaluation of their effectiveness in specific situations.

Many of the processes of change mentioned in Chapter 2 put pressure on smallholders to produce more with the same or even less land, labour or cash, while having to conserve their natural resources at the same time. The necessity to make farming more productive strongly influences their choice of technology. Depending on which factor – land, labour or cash – is becoming increasingly scarce, different paths will be followed.

Dealing with land constraints

Where populations are growing rapidly, land may become a scarce factor for farming. On the same area of land, more products or higher-value products (e.g. fruits, herbs, off-season crops) have to be produced. Depending on the resources available to them, farmers will develop different strategies to raise production. With increasing land shortage, farming is intensified. Where land becomes so scarce that it can no longer provide a basis for livelihood, off-farm income must be sought; this may mean that farming becomes more extensive again. In humid areas, tree management plays an important role in such processes of intensification and extensification (see Box 6.1).

A typical example of intensification is the change from shifting cultivation to (semi-)permanent farming. The natural fallowing processes used to restore soil fertility and limit pest populations are replaced by management systems requiring more inputs for nutrient

---

Box 6.1
Farmers' response to land scarcity in Nigeria

In the humid zone of Nigeria, increasing population pressure is accompanied by decreasing farm size and declining soil fertility. As pressure on the land increases, the proportion of land under compound systems grows, as does the density of both tree and arable crop cultivation around the compound. This shift reflects farmers' perceptions that such land use, combined with increased mulching and manuring, offers the most effective way of using their resources to slow down the process of declining soil

fertility and to maintain production. Though labour inputs per hectare are no higher in the compound areas than in the fields, yields in monetary terms are 5 – 10 times as much per hectare and returns to labour 4 – 8 times as much. This higher labour productivity can be attributed to the phasing of planting and harvesting in the compound areas that reduces peak workloads, and to the better physical working conditions under the shade trees.

With increasing population density, compound areas account for up to 59% of crop output and a growing proportion of total farm income, with the proportion of income generated

from tree crops rising to a share nearly equal to that from arable crops. Livestock become an increasingly important part of the farm system, as a source of both income and manure. However, as population density continues to increase, farm size decreases further and soil fertility declines, resulting in decreased yields and returns to labour, and farmers must turn increasingly to nonfarm sources of income. As people shift to working off-farm, less labour is invested in the compound areas, leaving these dominated by trees and other perennials (Lagemann 1977).

supply and crop protection. These inputs may be produced on-farm or come from outside. For this, the farmer needs labour, knowledge and/or cash. Where these are not available, shortening of fallow periods is followed by depletion of soil fertility and results in lower production. In such cases, intensification leads to degradation of the farming system.

Burning is a technique widely used in shifting cultivation and extensive livestock-keeping to clear land; neutralise pH; mineralise phosphorus, potassium and other nutrients and make them available to crops; kill pests; and stimulate regrowth of nutritious dry-season forage (West 1965, Lacey et al. 1982, Richards 1985, Kotschi et al. 1989). When burning is well timed and controlled, the natural vegetation will respond with new, healthy growth. However, ill-timed or poorly controlled burning can seriously reduce the amount of organic matter on the soil surface and leave the soil exposed to erosion by water and wind. When land use is intensified, some of the functions of burning can be taken over by techniques such as cover cropping, mulching and applying chemical fertilisers (see Appendix A). However, these are more labour-intensive than burning.

There are no technical blueprints for intensifying land use under low-external-input conditions, but techniques that are most likely to be applicable in this context are those that involve careful conservation of soil and water, use of complementary or symbiotic genetic resources (intercropping, integrating trees and animals); taking advantage of nitrogen fixation; and complementary and efficient use of external nutrient inputs (natural or artificial). In Appendix A, several potentially suitable techniques are discussed with respect to their opportunities and limitations. These techniques will have to be combined in such a way that both short-term effects, in terms of production increases, and long-term effects, in terms of resource conservation, can be achieved.

Especially in areas with marked ecological constraints, such as semiarid areas, it is difficult but not impossible to find viable ways to intensify the farm system without degrading the environment. In all ecozones, improving soil fertility and soil protection will be central to intensifying land use. Trees, shrubs and cover crops (particularly

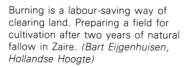

Burning is a labour-saving way of clearing land. Preparing a field for cultivation after two years of natural fallow in Zaire. *(Bart Eijgenhuisen, Hollandse Hoogte)*

legumes) but also grasses, composites and other wild plants (weeds) can contribute to accelerating the recovery of soil fertility (Egger 1987). In some cases, inputs of external nutrients, including artificial fertilisers, may be necessary to raise the level of nutrients available to plants and to compensate for nutrient export and losses. In drought-prone areas, efficient small-scale irrigation and water-conservation techniques will be of particular importance.

Dealing with labour constraints

Where production has to be increased with the same amount of (or less) labour and where farming has to compete with more attractive sources of income, methods of land-use intensification are needed that also increase labour productivity and improve labour conditions. It is particularly important for farmers in rainfed areas that seasonal labour peaks are not increased.

In many areas, particularly in less densely populated parts of Africa and South America, labour is often scarcer than land. To minimise and spread labour needs, extensive farming systems have evolved in which farmers rely on local energy sources and let nature work for them, e.g. fire to clear fields; vegetation to restore soil fertility and for fencing, windbreaks, shade or contour hedges; gravity and capillary water flow in irrigation and water-silt harvesting. To perform arduous tasks, work parties are organised (Richards 1986, Netting et al. 1990).

When the farmers want or have to produce higher levels of output, higher inputs of energy are needed. In HEIA, inputs based on fossil fuels (artificial fertilisers, pesticides, water pumps, farm machinery, transport facilities etc.) supplement or replace human and animal energy and increase the productivity of both land and labour. In LEIA areas, intensification has to depend mainly on human and animal energy, as energy derived from fossil fuels is scarce and expensive. Other optio is to improve labour productivity by making better use of locally availal le resources (e.g. by improving the efficiency of using water and nutrients) or to decrease energy needs (e.g. by preventing crop losses) must be explored (Stout 1990).

Labour productivity can be increased by mechanisation on the basis of hand-operated equipment. Mechanised milling of rice in Thailand. *(VIDOC, Royal Tropical Institute)*

Another option for increasing labour productivity is mechanisation on the basis of hand-operated and animal-powered equipment. Appropriate mechanisation for tillage and transport has received a fair amount of attention, but few appropriate machines or implements have been developed for activities specific to ecological farming (e.g. mulching, seeding into mulch, incorporating green manure, on-site water harvesting, ridging). Mechanisation of such activities under low-input conditions urgently requires more research. Examples of recently developed tools for these activities are the JAB planter for seeding in zero tillage; the tool bar for tillage and constructing ridges, e.g. for water harvesting; and the knife roller for incorporating green manures (ITDG/GRET 1991, Bertol & Wagner 1987).

The environmental effects of such mechanisation must be given particular attention. For example, if animal traction is used for tillage, this must be combined with effective ways to balance soil fertility, protect the soil and avoid local overgrazing. Special attention must also be given to changes in work loads, particularly of women, who are often responsible for labour-intensive activities, such as transport and weeding, and may not profit from labour-reducing innovations in tillage. Moreover, using animal traction for tillage precludes such techniques as relay intercropping, mulching, no-tillage and many forms of agroforestry. Using animal traction for other purposes, e.g. for transport, is more likely to be compatible with a LEISA system. Tillage with animal draught increases the incentive to expand cultivation and extensify land use. More intensive manual and biological techniques to improve the soil offer greater possibilities of making a farm system sustainable.

In this connection, the potential to improve traditional hand tools deserves mention. Many traditional implements were and still can be appropriate for low-external-input farming. An increased exchange of insights and ideas between farmers and scientists may result in ways of improving these traditional implements (e.g. shape of the hoe, length of the handle, materials used). Any attempt to improve implements, however, must be based on a proper understanding of the farmers' needs and of the constraints to local production and dissemination. Rather than introducing completely new implements, the capacity of local artisans and farmers should be strengthened to make incremental improvements to the implements they already have (Basant & Subrahmamian 1990).

Enhancing resources through use of external inputs

External inputs of organic or artificial fertilisers are indispensable to balance the nutrient flow within the farm by replacing nutrients that have been exported or lost. Nutrients from external sources may also be needed to increase biomass production or to maintain it at an acceptable level (see Box 6.2). Pesticides and medicines (natural or artificial) may be required to control severe outbreaks of pests and diseases. Artificial external inputs, such as chemical nutrients, irrigation, seeds and pesticides, can play a role in balancing the farm system, raising the productivity of land and labour, and increasing the total output of the farm, provided they are used in a way that does not damage the environment and human health.

Without external inputs, it is not possible to have open, market-oriented farming systems to provide for the needs of the nonfarming population. In countries with a high percentage of urban dwellers, external inputs are necessary to increase production to a level necessary to feed these people, as internal nutrient reserves are not high enough or cannot be cycled fast enough to attain such a high level of production. To make this ecologically and economically feasible, external inputs should be used as efficiently as possible and in combination with other resource-enhancing technologies, such as soil conservation, green manuring or preventive pest and disease control.

Artificial external inputs have to be purchased. When no other source of cash is available, it has to be obtained by selling farm products. Buying external inputs or producing crops or animals for sale competes with buying or producing consumer goods and internal farm inputs. It is therefore important for LEIA farmers to use external inputs, when available, in such a way that they enhance local resources and are recycled to the greatest extent possible (OTA 1988). Using synthetic nutrients in combination with organic inputs may be more profitable

Box 6.2
No sustainability without some external inputs

The only real cure for 'land hunger' in the overexploited parts of the West African Sahel is increased productivity of the land, both in animal husbandry and in arable farming. This will require at least imports of phosphorus (fertiliser) from outside the system, because recycling of crop residues, manure and household waste, regeneration of degraded rangeland, anti-erosion measures etc. may at best prevent further deterioration of the land resource, but are insufficient to stop nutrient depletion and to lead to improvements (van Keulen & Breman 1990).

Figure 6.1
Increased maize yields in an alley
cropping system using prunings from
Leucaena leucocephala and varying
rates of nitrogen application.
(*Source:* IITA 1985)

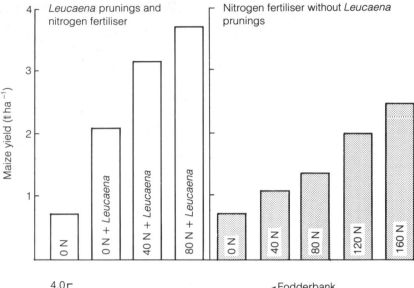

Figure 6.2
Effect of nitrogen application on grain
yield of maize grown on natural fallow
and in 'fodderbank', i.e. fallow improved
with *Stylosanthes hamata* and
superphosphate fertiliser. (*Source:* ILCA
1988)

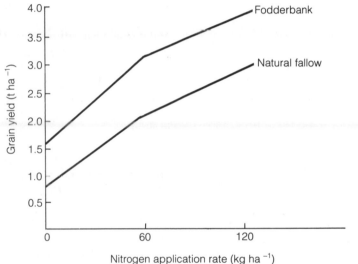

than primary dependence on synthetic nutrients to maintain or raise
fertility. Figures 6.1 and 6.2 show that, from the viewpoint of productivity, such combinations can compete well with or even surpass HEIA
systems under comparable conditions. Combining optimal use of local
resources and internal inputs with complementary use of external inputs
can be expected to result in farm systems that combine high productivity
with security and resource conservation.

More efficient use of external inputs, such as fertilisers, pesticides
and irrigation water, will increase their profitability and decrease their
negative effects on the environment. Lower losses and higher effectiveness make it possible to use less per unit area, while obtaining the same
results.

However, it is equally important to reduce the need for external
inputs – or at least to reduce increases in this need – by recycling organic
wastes back to farms, by increasing the efficiency of using internal
inputs, and by decreasing competition in the production of internal
inputs versus crop/animal production.

6.2   Strategies for transition to LEISA

Transition is the process of conversion from an unbalanced conventional or traditional farm system to an economically, ecologically and socially balanced (LEISA) one. As regaining an ecological balance may take many years, particularly when this involves growing trees and breeding animals, a transition process can be lengthy. As the conditions for farming will also be changing during transition, the farmers' capacity to adapt to these changes will be crucial for successful transition. However, it must be emphasised that transition is more than just adaptation to change; it is a conscious process to make the farm system more balanced and sustainable.

Transition involves investments in labour, land and/or money and taking risks. To gain an acceptable level of yield increase quickly, to minimise risks and to spread investments, the farm family must find an acceptable transition strategy involving specific combinations of genetic resources, techniques and inputs in a deliberately chosen sequence. The particular technologies involved and the sequence in which they are combined will depend greatly on the biophysical and socioeconomic characteristics of the farm, its historical development and present situation, and the needs and preferences of the farm household. Therefore, such strategies will be farm-specific. For example, strategies will differ depending on ecozone (arid/humid, lowland/highland), biophysical potential (high/low), resource status (nondegraded/degraded), socioeconomic status (subsistence-oriented/market-oriented) and present level of technology (artificial/natural inputs).

The process of building and maintaining LEISA systems requires awareness of and knowledge about feasible technologies, skill in applying them and a constant watch for signs of degradation and destabilisation of the farm system. As it will not be completely certain what implications the changes will have for the farm household, the strategy for transition should be undogmatic, responsive to unexpected results and open-ended. Farmers embarking upon conversion to LEISA need to be highly motivated, self-reliant and imaginative. Sustainable agriculture demands high internal inputs of good farm management.

Under LEIA conditions, where farmers may be particularly wary of the initial investments required and the risks of a temporary decrease in production, it is important that suitable 'entry points' to a process of transition be found. These are starter techniques which give good returns in the first season, involve relatively few risks and have positive effects on the ecosystem. Integration of these techniques can then lead to other beneficial changes in the farm system. For example, in northeast Thailand, the integration of fish into rice farming resulted in higher rice yields, better water conservation, lower use of agricultural chemicals, improved farm productivity, improved family nutrition, changes in village social structure, and decreased emigration. In many cases, this has also led to the development of integrated farm systems with increased production of vegetables, fruit trees and animals – all of which are components that increase the sustainability of the system (MacKay 1990).

Other examples that illustrate the importance of entry points are:

Box 6.3
Transition from HEIA to LEISA

In farms that currently depend heavily on artificial external inputs, fertilisers and pesticides, the emphasis during the transition period will be on replacing these by more natural processes and inputs. It may not be wise, however, to change too abruptly, as simply abandoning these artificial inputs before having put a regenerative farming system into place is a guaranteed prescription for disaster (Kirschenmann 1988).

Peasants in Honduras are incorporating velvet beans *(Mucuna pruriens)* as a green manure and cover crop in maize. *(Flores Milton)*

● the adoption by Kenyan farmers of nitrogen-fixing trees for livestock feed, increasing the production of milk, increasing income and manure production, and increasing yields (Chambers 1989);

● the adoption of velvet beans as green manure cover crop in maize farming in Honduras (Milton 1989); and

● water and soil conservation by stone bunds along the contours in semiarid parts of West Africa (Kerkhof 1990) and with barriers of grass and/or shrubs in various other parts of the tropics.

Some of these techniques are presented in Appendix A. No doubt there are many more simple and inexpensive innovations that could serve as an entry point or a follow-up technique in a sequential approach to achieving balanced farm systems. It is important to identify additional cases in which starter techniques have triggered off a transition process, to learn from them and to assist farmers to adapt the techniques to their own conditions (MacKay 1990).

World Neighbors' experience indicates that there are a number of widely applicable criteria that can serve as a guide in preselecting promising technologies for entering the transition process (see Box 6.4).

Legislation, marketing improvements and appropriate pricing policies can give important support to farmers seeking a strategy for transition. Many marginal farmers may not be in a position to develop sustainable farm systems unless they have access to additional income, e.g. a kind of starter credit in a revolving fund which gives them the financial space to work on their farm instead of working as farm labourers for others or going elsewhere to look for wage labour. Achieving sustainability is more easily advocated than put into practice and, in the case of many marginal farmers, it may not be possible to achieve unless supplementary sources of local, nonfarming income can be found.

The transition strategy should be determined in dialogue with the farm family, as each farm is unique and the family members are the professionals who best know the specific characteristics of that farm. Prescribed strategies seldom succeed, particularly if the situation of the

---

Box 6.4
Criteria for choosing technologies for people-centred agricultural improvement

Do the poorest farmers recognise the technology as being successful?
● Does it meet a felt need?
● Is it financially advantageous?
● Does it bring recognisable success quickly?
● Does it fit local farming patterns?

Does the technology deal with those factors that most limit production?

Will the technology benefit the poor?
● Does it utilise the resources the poor people already have?
● Is it relatively free of risk?
● Is it culturally acceptable to the poor?
● Is it labour-intensive rather than capital-intensive?
● Is it simple to understand?

Is the technology aimed at adequate markets?
● Are market prices both adequate and reliable?

● Is the market available to small farmers?
● Does the market have sufficient depth?

Is the technology safe for the area's ecology?

Can the technology be communicated efficiently?
● Does it require a minimum of on-site supervision?
● Is it simple to teach?
● Does it arouse enthusiasm among farmers?

Is the principle behind the technology widely applicable?
(Source: Bunch 1985).

farm family differs greatly from the situation under which the strategy was developed. Experiences of other farmers in the region, although their farms may differ in some aspects, can provide helpful ideas and guidance in developing a transition strategy. Outsiders can play an important role in increasing the farmers' capacity to understand and analyse their farm system; in making them aware of alternatives, together with the related potentials and potential problems; and in increasing their capacity to experiment during the process of transition.

Participatory Technology Development is an approach to developing LEISA systems in a systematic way at the farm and village level. As relatively few reports are available about transition from LEIA or HEIA systems to LEISA systems, much experience still has to be gained and documented to provide more insight into the effectiveness of different transition strategies.

Part III

Linking farmers and scientists in developing LEISA technologies

7 Actors and activities in developing LEISA technologies

As we have seen in Chapter 3, farmers are developing technology on their own, but without outside assistance they cannot advance as far or as quickly as would otherwise be possible. Scientists have also been developing technology on their own, but their impact would be greater and more beneficial if they worked more closely with farmers.

These are only two groups of actors in a complex system of generating knowledge and technology for agriculture. Besides the scientists working in basic, applied and adaptive research, many extensionists and other development workers in national agencies and in bi- and multilateral government projects are also involved in experimentation and adaptation of technology to local conditions. Nongovernmental organisations (NGOs) and farmers' organisations are likewise involved in the development of agricultural technology. Artisans and traders in the informal sector and commercial firms in the formal sector of the national and international economy also seek to develop and supply agricultural inputs and to develop ways of processing and distributing agricultural products.

As a reaction to the fact that formal scientific research has received much attention and resources, progressive 'counter-ideologies' have tended to romanticise the role of farmers as sources of innovation. This may create exaggerated expectations, as if farmers alone hold the key to solving problems which a science-based approach has failed to solve. Neither stance is helpful. It is likely that the efforts of any one group of actors in isolation will fail to provide adequate solutions to current agricultural and environmental problems. The crucial question is: how can the various actors be brought closer together into a well-functioning system of generating agricultural knowledge and technology, in which their activities complement and reinforce each other?

We will first look at the different groups of actors involved in the technology development process and consider the contributions they can make to developing LEISA systems. We will then see how they can interact synergetically in a process of Participatory Technology Development (PTD) and how this process can be promoted.

7.1 Actors in the technology development process

Farmers

As discussed earlier, most of the knowledge applied by farmers comes from their own experience in agriculture and that of their forefathers and fellow farmers. Through their 'informal' research and development (R&D) activities, farmers generate new knowledge and create new technologies. Of particular importance for LEISA development is the capacity of farmers to understand the local biophysical and cultural environment and to predict and explain the outcome of experiments under local conditions. Working together with farmers is therefore of paramount importance for creating ecologically-oriented farming systems. Indigenous knowledge (IK) is an important complement to formal scientific knowledge. As a villager in Thailand said to a visiting researcher: "We don't have the scientific knowledge to know what is possible, but the officials don't know the local conditions here. Nobody knows the local conditions better than we do" (Grandstaff & Grandstaff 1986).

Many of the indigenous farming practices mentioned in Chapter 3 – and many more, not all of which are yet known to formal science – represent at least the 'seeds' of promising LEISA technologies, e.g. composting, green manuring, mulching, multiple cropping, contour farming with bunds or hedges, water and nutrient harvesting, and ways of controlling pests. If these indigenous practices are well understood in formal scientific terms, it may be possible to improve them, e.g. by careful use of external inputs. Also many indigenous, sometimes unconventional, crop and animal species and local varieties and breeds have great potential for LEISA.

In developing LEISA systems, farmers can contribute not only their knowledge of the local ecosystem and culture but also their experience in informal experimentation and adaptation of technologies to local conditions. The innovations investigated by farmers in response to new problems and opportunities give important indications of potential improvements within their means and under the biological and physical limitations with which they have to cope. Farmers' experimental methods vary widely, but they share the following strengths:

● subjects are chosen that are relevant to the farmers;

● evaluation criteria are applied that are directly related to local values with respect to, e.g. taste and utilisation of products;

● the observations are made from within a real-life systems perspective, as they take place during actual farmwork, and are not limited to final outcomes such as yield;

● the experimentation is based on farmer knowledge, and expands and deepens this knowledge.

Elaboration of complementary research methods based on a better understanding of farmer experimentation can play a significant role in the search for LEISA technologies that can be applied in complex, diverse and risk-prone environments (Richards 1988).

If, as has so often happened in the past, IK is dismissed by formal science as irrelevant, rural people may be encouraged or forced to adopt practices that lead to inappropriate use of their resources, imbalance in the cultural or natural environment, and decline in their social welfare (Warren & Cashman 1989). As a result of such past experiences of interaction with representatives of formal science, many LEIA farmers are reluctant to express their knowledge or to discuss their practices, apologising that "we don't do it the modern way like the experts say we should ...'' They may be skeptical or suspicious about the value of foreign technologies, but hesitate to express their criticism, partly since it has been impressed upon them that the Western-educated scientists are 'right'. Farmers who have been robbed of pride and confidence in their own farming practices and capacity to innovate and manage change may feel powerless when confronted with new pressures and problems – or at least may give this impression.

It is partly through formal science that farmers have been or are in danger of being expropriated – that their particular knowledge is made superfluous, and that their labour is subjected to external interests. Many peasants have learned to draw up a line of defense to protect their own essential interests and perspectives (van der Ploeg 1990).

In such cases, the first step towards sustainable agricultural development must be to gain farmers' confidence by taking their knowledge, values and problem analyses seriously. Only when IK is recognised by scientists as valid and important can farmers interact on an equal footing with them. This creates a base for communication and cooperation. Legitimating IK 'empowers' farming communities in the sense that, through the recognition given to their knowledge and culture, they develop an increased sense of solidarity and gain in political bargaining power (McCall 1987, Vel et al. 1989, Thrupp 1987).

Artisans and traders

In LEIA areas, most improvements in farming equipment are made by the users themselves or by local artisans who respond to their needs. Often, the artisans are also farmers themselves. Also, where new implements or machinery are introduced, most of the successful adaptations to specific agroclimatic conditions are made by local farmers and artisans. Through an interactive process involving the people who use the implements, those who make them and those who sell them in the informal sector, new farming implements are modified to render them locally appropriate and old ones are modified to meet new requirements. Local technological capabilities could be enhanced and technology development and diffusion could be speeded up by strengthening decentralised artisan-based fabrication networks and creating linkages between these networks and formal R&D so as to combine local experience and scientific knowledge (Basant & Subrahmamian 1990).

Official research and development institutions

In contrast to the informal R&D of farmers and artisans, the formal R&D of international and government institutions has largely failed to produce technology directly applicable to the situations of LEIA farmers. This is hardly surprising, as most formal R&D has focused on 'high-potential' areas and forms of agriculture. From a national viewpoint, the question can indeed be raised whether the returns from farming in marginal areas and from products intended primarily for local consumption justify high investment in formal R&D. Other, less costly approaches to technology development may be the only options in such situations.

Although international and government R&D institutions are not and may never be the central agents of technology development for LEIA areas, they can have a crucial influence in both negative and positive terms. Some of the ways in which science-based agricultural technologies have contributed to environmental degradation and social marginalisation were outlined in Chapter 1. But misapplication of science in the past is no reason to deny the positive contributions that formal agricultural R&D can make in developing sustainable farming for the future. The major strengths of formal R&D lie in its:

- **Use of science.** Scientific research involves observing, identifying, investigating and developing theoretical explanations for natural phenomena. On this basis, scientists can determine the essential nutrients in the soil, produce genetically improved material by artificial means etc. Farmers can discover that something 'works', but often cannot explain the principle behind it. This means that the benefit to be derived from that principle cannot be exploited to full advantage.

- **Use of systematic procedures.** The methods developed by researchers are systematic and the processes are reproducible. This allows fast, concentrated work on a given problem and validation of results.

- **International networks and accumulation of knowledge.** Scientific knowledge from all parts of the world is systematically accumulated, stored and made available to other scientists through symposia, publications and data bases. Formal R&D institutions can draw upon this store of knowledge and can tap international networks of scientists for information and ideas. For example, a grass variety that has proven to be very effective for contour bunds in India, because it has deep roots which do not interfere with crops and does not spread as a weed, can be obtained and tested by scientists in West Africa. LEIA farmers do not have access to such a wide network.

Thus far, most formal research for tropical agriculture has focused on improved varieties and breeds that require high levels of external inputs to be productive, despite the fact that the structure of farming

Two-way transfer of knowledge between agricultural scientists and practitioners can improve formal research and extension. Discussing problems of sheep production in central Nigeria. (*Ann Waters-Bayer*)

Box 7.1
Combining internal and external inputs in Burkina Faso

A study was made of the contributions of straw, manure and compost to sorghum yields with and without the addition of small amounts of nitrogen fertiliser. The most productive organic fertiliser – compost – increased sorghum yields from 1.8 to 2.5 t/ha. Nitrogen fertiliser alone produced grain yields slightly higher than any of the organic practices. But the best result was achieved by combining compost with nitrogen fertiliser; this raised sorghum yields to 3.7 t/ha. The three organic practices increased the efficiency of nitrogen application by 20 – 30% (Pieri 1985).

and the macro-economic conditions of many developing countries more or less dictate the use of low-external-input technology. Scientific principles could be applied to exploring adaptations or alternatives to 'modern' packages which would be more economic in terms of local price ratios, availability of seasonal labour, climatic risks and yield security. This would lead to more diverse research programmes and a more realistic range of low-cost, low-risk recommendations.

Formal R&D can play a very important role in seeking effective ways of combining internal and external inputs (see Box 7.1). Potential may lie, for example, in combining artificial fertilisers with the use of local mineral resources such as natural phosphates. Indigenous agroforestry practices provide a good base for more efficient use of mineral fertilisers than in modern monocropping systems. "Governments that have modest fertiliser-promotion programmes may find that they can maximise the benefits from fertiliser by promoting agroforestry as well" (McGuahey 1986). Also the development of strains resistant to drought, pests or disease, and of plants and animals that are more efficient converters of nutrients, can contribute to LEISA.

Even where researchers have tried to develop ecologically-oriented farming technologies, such as alley cropping, efforts to introduce them which bypass farmers' experimentation have met with little acceptance. As McCorkle et al. (1988) state: "Like it or not, both research and development depend for their success on farmers' own informal system of technology validation and transfer." Formal research and extension could be made more effective by collaborating with experimenting farmers, improving two-way transfer of knowledge between farmers and scientists, and linking into farmers' communication networks. While researchers contribute their science-based knowledge and methods to the development of LEISA systems, extension staff can enlarge the opportunities for farmers to exchange experiences and systematise their process of validating and adapting technologies.

Commercial companies

A wide range of commercial companies, varying from plantations to companies that produce farm machinery, generate agricultural technology. Commercial interests may consist of exploiting local resources directly to produce a marketable commodity; selling inputs such as seed, agrochemicals, feed and farm equipment; or creating the conditions for production and marketing of a surplus produced by farmers. Since they have to live by their results, they are market-oriented and have a strong incentive to make things work. Considerations of sustainability or farmer welfare are not their prime concern. Unless farmers are sophisticated customers who have some countervailing clout to protect their interests and unless enforceable legislation safeguards sustainability, commercial companies understandably tend to follow their own interests, which can be exploitative and destructive (Röling 1990).

The companies producing inputs have a definite role to play in LEISA, because low-external-input does not mean no-input agriculture. When used in an ecologically and socially sound way, external inputs complement local resources, as shown by the example from Burkina Faso (Box 7.1). It may seem that farming with few purchased, material

Box 7.2
**Smallholder corporations
in Turkey**

The Development Foundation of
Turkey has been pioneering in
commercial development for
small-scale farmers, aimed at
the creation of farmer-run
corporations. The techniques
developed so far are for producing broilers, honey and sheep's
cheese. The added value of the
Foundation is that it creates the
marketing channel, handles
product quality, packaging and
storage, and provides highly
necessary technical expertise
and some crucial inputs, as well
as credit (Röling, pers.comm.).

inputs ('hardware') is not in the interest of commercial enterprises.
Integrated pest management, for example, reduces the number of
sprayings that farmers apply. Knowledge-intensive methods take over
from hardware-intensive methods. Yet, LEISA will also require external
inputs, such as special tools, commercially-produced natural pesticides,
appropriate genetic material, supplementary nutrients for crops and
livestock, and relevant information and advisory services. Commercial
companies can fill an important niche in this field of development. They
can also provide an outlet for farmer produce.

LEISA development work cannot afford to neglect such enterprises.
Companies capable of developing relevant technologies should be
studied and supported. The production of fish, shrimps, fruits, flowers,
cheese, woven wool, and honey – as a few examples – seems eminently
suited to such purposes (see Box 7.2). Many of these can be produced
in smallholder systems, but close communication between the companies
and the farmers in joint R&D is required to develop feasible and
commercially worthwhile production systems. In such endeavours,
cooperative enterprises may have a comparative advantage over private
companies.

If the introduction of promising LEISA techniques leads to increased
and sustainable production by small-scale farmers, there will be new
opportunities for commercial enterprises in areas where their role has
been very limited until now.

Nongovernmental organisations (NGOs)

Throughout the Third World, numerous nongovernmental organisations (NGOs) are working in LEIA areas. According to Chambers et
al. (1989), these organisations have tended to be strong in their links
with grass-root communities but weaker on the technical side of
farming.

A rapid change is occurring. Since the late 1980s, many NGOs have
been shifting their priorities, staff recruitment and training towards agriculture and have begun to move into gaps – whether thematic or spatial
– left by government services. The origins of NGO interventions in rural
communities frequently lie in concepts of empowering local communities, promoting self-reliance and alleviating poverty. Interventions may
take the form of providing relief and welfare, as in areas suffering from
ecological crisis and war; building local organisational capacity through
community education to enable local people to conceptualise and
articulate their needs within the wider political and economic structure
of the region or nation; and promoting self-reliant economic activities
through provision of credit, inputs, market facilities etc.

In their interactions with farming communities, NGOs are often
concerned with identifying, testing, adapting and disseminating locally
appropriate technology. They recruit sensitive field staff and can often
maintain them at one site for several years, giving them a chance to
gain detailed local knowledge, build up a rapport with farmers and set
up participatory development programmes. NGOs are also becoming
increasingly involved in devising new participatory methodologies for
technology development and extension specifically tailored to the needs
of small-scale farmers (Farrington & Amanor 1990). Numerous NGOs
have successfully worked with smallholders to develop appropriate

agricultural technology, e.g. World Neighbors (Bunch 1985, Gubbels 1988), KWDP (Chavangi et al. 1985), Oxfam (Kerkhof 1990), IIRR (Gonsalves 1989), FASE (Sautier & Amaral 1989), PRONAT (ENDA 1987).

The efforts made by NGOs, largely in response to the shortcomings of governmental research and extension services in difficult farming areas, have addressed three aspects of agricultural research policy (Farrington & Amanor 1990):

● The creation of a philosophy of agricultural development which is critical of conventional development and seeks to promote a more sustainable agriculture rooted in 'ecological' perspectives. Actions are concerned with preserving local farming systems, local cultivars and genetic diversity, minimising expensive external inputs, and building upon indigenous knowledge systems and low-external-input alternatives to chemical fertilisers, pesticides and modern varieties.

● The development of new research and extension methodologies to adapt technology to the needs of local people. NGOs have often been impressed by local people's knowledge of their environment and of their innovative abilities, and have attempted to develop local institutions and training facilities to strengthen these capabilities, in which they act as facilitators and catalysts to local initiatives.

● The building of research capabilities and fostering of linkages with national agricultural research services (NARSs), to provide technical options which the NGO with participating farmers or farmer groups can try out and adapt to local conditions.

Farmer organisations

Strong pressure is usually exerted on technology development agencies to serve rich, commercial farmers. These farmers command a high

By forming local organisations, LEIA farmers can exert pressure on development agencies to focus on their problems and develop relevant technologies. Low-external-input dam building in Thailand. *(VIDOC, Royal Tropical Institute)*

proportion of the country's productive resources and usually produce a large part of the marketed surplus for urban consumption and export. Where they are also well organised, they can exert considerable pressure on governments to make sure that research institutes pay due attention to their problems.

Traditionally, LEIA farmers have lacked the power and capacity to exert similar pressure. Their access to research information has been restricted and their ability to articulate their needs has been poor. If they are to exert more 'demand-pull' on NARSs to focus research on the problems in LEIA areas, the farmers in these areas must be enabled to better organise themselves.

The influence of technology users through farmer organisations is such a crucial factor that one can expect more benefit from projects that increase the influence of LEIA farmers over formal technology development systems, than from investments in laboratories, staff training, cars, megaphones etc. to improve the intervention power of research/extension services (Röling 1990).

7.2   Participatory Technology Development (PTD)

With increasing recognition of the value of and need for working with local communities to identify, test, evaluate and disseminate new agricultural technologies, various 'participatory' approaches have been taken, mainly by NGOs but also by some government organisations (e.g. Rhoades 1984, Chambers & Jiggins 1986, Farrington & Martin 1988, Haverkort et al. 1988, Chambers et al. 1989). Some approaches leave most of the decision-making control in the hands of those who live and work outside the local community. Others are much more based on the priorities expressed by local farmers and consciously seek to support and develop local capacity to manage change.

In LEIA areas, where high inputs of research time and funds from formal R&D agencies and commercial companies cannot be expected, the approach must aim at self-sustaining development: helping farmers become more effective technology developers themselves. The term 'Participatory (or People-centred) Technology Development' (PTD) refers to approaches that aim at strengthening local capacities to experiment and innovate. Farmers are encouraged to generate and evaluate indigenous technologies and to chose, test and adapt external technologies on the basis of their own knowledge and value systems.

PTD in agriculture is not a substitute for station-based research or scientist-managed on-farm trials. It is a complementary process which involves linking the power and capacities of agricultural science to the priorities and capacities of farming communities, in order to develop productive and sustainable farming systems.

The general approach

PTD is a process of purposeful and creative interaction between local communities and outside facilitators which involves:

● gaining a joint understanding of the main characteristics and changes of that particular agroecological system;

● defining priority problems;

● experimenting locally with a variety of options derived both from indigenous knowledge, i.e. from local farmers and from farmers elsewhere, and from formal science; and

● enhancing farmers' experimental capacities and farmer-to-farmer communication.

PTD is a path to LEISA. It builds upon farmers' knowledge and agricultural practices and encourages the optimal use of locally available resources, complemented by external knowledge and external inputs, where applicable and available.

This approach to technology development is closely linked with a process of general community development on a self-reliant basis. The activities involved in PTD – critical analysis of community-managed changes in the agroecological system, identification and use of indigenous technical knowledge, reconstruction of successful local innovation, self-organisation and self-implementation of systematic experiments with selected options – all foster the awareness, self-respect and self-confidence as well as the diagnostic and experimenting skills of the farmers involved.

PTD approaches also try to foster development of a network of village organisations, to intensify communication on local experimentation and to improve linkages with supporting organisations and institutions. The development of village groups and independent relations of exchange encourages the maintenance and further spread of good local practices and enhances the prospects for ongoing experimentation, improved resource management and self-sustained agricultural change. The implementation of PTD thus 'walks on two legs':

● development of technologies and the agroecological system;

● development of the 'social carriers' of the technologies and system.

In other words, it not only enables the generation of technologies adapted to local environments, but also develops the local capacities, sociocultural structures and organisational linkages necessary to sustain the process.

PTD and conventional R&D activities differ in three essential ways:

1   PTD does not attempt to generate results that can be generalised across wide areas, although it may well do so. The specific techniques and farming-system adaptations generated by the process are primarily of very localised validity. However, the underlying ideas may find a much wider applicability.

2   The PTD process follows a different approach to collecting, codifying, interpreting and utilising knowledge and information. PTD practitioners work with the rich store of local knowledge and information to describe and explain problems and relationships and to test possibilities. PTD is concerned mainly, though not exclusively, with extrapolating from local knowledge and experience (including locally available scientific knowledge and experience) to describe, explain and test technical options with local validity. This concern, in turn, guides the choice of methods used in implementing

Development of techniques to propagate *Buddleia* trees in Peru started with the knowledge and experience of the peasants in making traditional cuttings. *(David Ocana Vidal)*

the approach. Rapid, low-cost, locally manageable methods are appropriate, generating results that are reliable, although at some cost in accuracy and precision when compared to station-based experimentation and to large-scale surveys designed to generate statistically valid data. Nevertheless, repetition of experiments among hundreds of farmers or even thousands of farmers can eventually produce results of great statistical significance (Bunch, pers.comm.).

3   PTD practitioners can be based in any rural and agricultural development service or project as well as among members of a local community. They may be experienced agricultural fieldstaff from governmental or nongovernmental agencies, retired government extension or community development workers, or village leaders with agricultural training, but may also be research scientists.

Documentation of PTD experiences has appeared only recently, e.g. reports from Central America and West Africa by World Neighbors (Bunch 1985, Gubbels 1988), from Indonesia by Propelmas (Vel et al. 1989) and from Peru by Grupo Yanapai (Fernandez 1988). The broad geographical spread of these activities suggests that there is a widely-felt need and a wide potential for this kind of approach. Experiences with PTD are probably far more than those documented. Not all PTD practitioners have the skills to analyse and write up their experiences, and they are frequently too occupied with their fieldwork to find time to record what they are doing. Besides, their reference group is the rural population and not the international scientific community.

The general sequence of PTD activities

The documented experiences suggest that, despite the intrinsic diversity in the methods used, there is a certain sequence in the way PTD is usually done. In a workshop on PTD organised by ILEIA in 1988, some 200 methods were analysed and clustered into six categories of activities (see ILEIA 1989):

● **Getting started** – how PTD practitioners from outside a community choose an area, introduce themselves, build up a good relationship with the local people, analyse the existing agricultural situation and form a basis for cooperation with networks of farmers to start the process of technology development; this includes widening the understanding of all involved about the ecological, socioeconomic, cultural and political dimensions of the current situation;

● **Looking for things to try** – gathering information for detailed analysis and prioritisation of locally felt problems and identifying promising solutions, in order to set up an agenda for experimentation;

● **Designing experiments** – developing patterns and methods of trying things out, which give reliable results and can be managed and evaluated by the farmers themselves;

● **Trying things out** – carrying out, measuring and assessing the experiments, and simultaneously building up farmers' experimental skills and strengthening their capacity to conduct and monitor experiments;

● **Sharing the results** – amplifying spontaneous diffusion of the results of farmer experimentation by building up a programme to improve farmer-to-farmer communication;

● **Keeping up the process** – activities that lead to replication of the PTD process and create favourable conditions for ongoing technology development for LEISA.

In the following chapter, each of these sets of activities is described in detail, and examples are given of field-tested methods within each set of activities. These are meant to inspire PTD practitioners to design their own methods of working with farmers, adjusted to the local sociocultural and agricultural situation. Just as the methods of PTD are not uniform, also the results are not uniform technologies that can easily be introduced elsewhere. Given the great diversity in environmental and social conditions of LEIA farmers, an equally diverse array of LEISA technologies adapted to the local conditions will emerge. PTD will benefit from the fruits of agricultural research – both conventional and participatory – in other parts of the world, but attempts have to be made to develop site-specific technologies that have built-in flexibility and adaptability (see Box 7.3).

In farming, technology development is a never-ending process. No innovation will ever be permanent. ''A productive agriculture requires a constantly changing mix of techniques and inputs. Seeds degenerate, insects and pests spread and develop resistance, market prices fluctuate, new inputs appear and old ones become expensive, agricultural and economic laws change and temporarily successful technologies become less profitable as their spread forces market prices downward'' (Bunch 1985). It is therefore of prime importance to build up the capacity to continue innovating rather than to produce a static technology.

7.3   Case examples of Participatory Technology Development

The above discussion of PTD is very theoretical, quite in contrast to PTD itself, which is being developed by practitioners in the field. The approach becomes clear only in practical, real-life examples such as the following. Further examples of PTD processes can be found in the ILEIA reader entitled *Joining Farmers' Experiments* (Haverkort et al. 1991).

PTD Case I: Weed control in the Philippines

The Farming Systems Development Project – Eastern Visayas developed a participatory method to identify farmers' priority problems, analyse farming systems, elaborate farmers' hypotheses and implement farmer-led experiments. The on-site research team, comprising two economists, one livestock specialist, one agronomist and one extensionist, was supported by senior staff from the Department of Agriculture and the Visayas State College of Agriculture.

Getting started. In group meetings, farmers were asked about their current topics of conversation. From these topics, farmers chose

---

Box 7.3
PTD strategies and tools are site-specific

The complex strategy that is necessary for effective rural development will vary according to local conditions but should combine technology development with awareness and community organisation. Farmers should be supported politically and in executing activities by some sort of local institution. The strategy must be flexible enough to encompass activities outside agriculture, in case the most severe constraint in fighting poverty does not lie in the agricultural practices. The processes involved should be iterative with increasing complexity of development activities undertaken by farmers and increasing organisational strength of farmers' groups.

The ideal strategy for working to improve farmers' lives can only be found by bringing together the knowledge and experience of farmers, field workers and scientists. In this effort we must use tools that are designed not as products of our own preconceptions, but rather according to the realities in each area (Vel et al. 1989).

declining soil productivity as one they would like to elaborate further. During visits by the researchers to the fields, farmers showed what they were talking about. In subsequent group discussions, the farmers reached a consensus that the problem they wished to address first was the infestation of infertile marginal uplands by cogon grass *(Imperata cylindrica)*.

During the process of identifying farmers' problems and discussing with key informants, the researchers gained sufficient knowledge to work out guide topics for further study, such as farming decisions and processes (plot selection, cultivation procedures, cropping sequences) and issues pertaining directly to the problem (e.g. why cogon is present, why farmers cultivate these areas, what constraints they face).

In an informal survey of 24 randomly selected households from a total of 150 in three upland villages, the guide topics were discussed over several sessions. The survey responses provided information on the biophysical causes and socioeconomic constraints that farmers perceived surrounding the cogon problem. On a blackboard, each of these points was written in a box with arrows leading to the central problem box, and key informants explained the relationships between them. The boxes were redrawn into concentric rings around the problem, each box forming one segment of a circular systems diagram (see Figure 7.1). The size of each segment was determined by the proportion of farmers who responded to that point. A meeting of all

Figure 7.1
Systems diagram of economic constraints and biophysical causes concerning *Imperata* problem as perceived by farmers in 24 households in Gandara, Philippines (showing proportion of farmers who selected each factor in each half of the two concentric rings). (*Source:* Lightfoot et al. 1988)

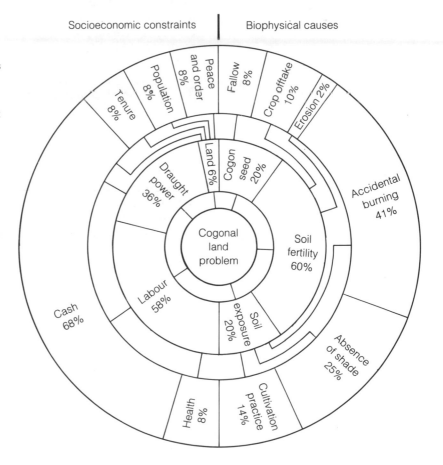

Typical landscape in Eastern Visayas, Philippines. *(Clive Lightfoot)*

respondents was then held to obtain agreement that the diagram indeed reflected what was happening on their farms.

Looking for things to try. First, the experiments, ideas and knowledge of the farmers were elicited. Most farmers knew that cogon did not germinate in covered soil or grow in shaded areas. Several had observed that it was shaded out or suffocated by vigorously vining plants. In group meetings attended by farmers and researchers, the farmers' observations were supported by formal research findings. Farmers also expressed other ideas for controlling cogon, e.g. ploughing and planting cassava or sugarcane. To supplement this list, researchers brought forth the idea of using herbicides.

At technology screening meetings, key informants and researchers presented various options to farmer groups for debate of the pros and cons. The systems diagram was used to focus the debate: pros became potential benefits with respect to biophysical causes and cons became potential conflicts with respect to socioeconomic constraints. For example, farmers judged that ploughing would require too much labour and draught power (already in short supply) and herbicides would be too costly, but cash and labour constraints did not appear to conflict with shading out cogon by planting trees or vining legumes.

Several farmers wanted to try shading out cogon, but some wanted to see the trees and vining legumes growing before deciding whether to test them. A field trip was arranged for them to see *Leucaena, Pueraria* and *Centrosema* species growing at a research station and on farms.

Designing and implementing trials. In group meetings, farmers decided to test vining legumes (*Pueraria* and *Centrosema* spp) for rehabilitating cogon-infested land. They expected that the legumes would shade out the cogon and improve soil fertility, and that soil covered with low legumes would be easier to cultivate than soil with tall grasses and shrubs.

Farmers chose plot locations and sizes from a researcher-defined range of 500–1000 m². Researchers set the number of replicates or farms. Treatments were limited by availability of legume seed. Farmers and researchers demarcated the plot, and researchers provided the legume seed. Farmers prepared the land, developed their own methods to establish the legumes, and measured parameters of interest to them. Researchers recorded cogon stand densities and labour requirements, and took soil samples. Periodically, farmers and researchers visited the plot together to note progress and take biological measurements.

Farmers continually analysed the experiment. Researchers analysed the data on labour for establishing legumes and the percentage cover of legume and cogon. These data and analyses, along with farmer responses, were discussed in regular farmer group meetings.

This method encouraged the use of systems logic in identifying problems, analysing systems and defining experiments. The resulting trials differed from typical crop trials, which stress maximising grain yield or cash income per hectare. The priorities of the farmers were, rather, the long-term rehabilitation of cogon-infested land and saving labour. Conventional trials usually focus on one or two crops. The wider view of the farming system gained through the participatory approach revealed that upland farmers cultivate many agroecological zones using a crop-fallow rotation and are therefore interested in the management of not only cropped but also fallow land (Lightfoot & Ocado 1988).

PTD Case II: Farmer-centred development in Mali

Since 1983, World Neighbors (WN) has been working in West Africa, trying to strengthen farmers' capacity to identify, test and adapt new agricultural technologies by undertaking small-scale experimentation. The approach evolved from WN's 30 years of field experience, particularly in Central America (Bunch 1985). The first West African programme was established in Togo, followed by programmes in Mali, Chad and Burkina Faso.

Problem analysis by farmers. To prepare for problem analysis, WN tries to gain a practical working knowledge of the local farming system. The purpose is not to understand the system in all its complexity in order to diagnose constraints for the farmer, but rather to learn only what is essential for guiding farmers in undertaking their own analyses.

After gaining this general information from local research and extension services, a survey is made by a team of one or two experienced WN animators and a resource person who speaks the local language. In selected villages in the programme area, the team arranges informal gatherings of local leaders, family heads and women representing a cross-section of the population. To generate reflection and analytical debate, questions with three basic themes are asked:

● Comparing the farming practices in the time of your parents and grandparents with the agriculture you practise now, what major changes have occurred? What will happen in the 5 – 10 years to come?

● What are the major problems or difficulties you face that limit your agricultural production? Why have these problems occurred? What are their root causes?

● What different ideas or new techniques have you tried in recent years to cope with these problems? How successful have these new ideas been in solving these problems?

In village meetings, each farmer examined some of the new seed and considered the merits of trying it out. (Peter Gubbels)

For example, the farmers in the Mali programme area regarded their main problems as decreasing soil fertility and low, irregular rainfall. They had already applied various indigenous innovations to cope with these problems, e.g. identifying local varieties of sorghum and millet with short growing cycles, and improving the use of low-lying land where better water retention makes it less risky to grow crops not resistant to drought. However, this informal R&D was not systematic. Individuals innovated on their own and might be observed by neighbours. The meeting initiated by WN was the first time that villagers had met to reflect on past agricultural changes and to analyse farming problems as a group. There had been no concerted, organised community effort to determine priorities and seek solutions.

Identifying options. WN establishes links with agricultural research stations and extension services, especially those close by, and finds out from other NGOs and informal contacts whether farmers elsewhere have already developed innovations to solve problems similar to those faced by the farmers in the programme area. Also, during the initial survey and analysis, innovative farmers in the area are identified. The innovations are screened in advance according to certain key criteria: above all, simplicity, accessibility to the resource-poor majority, low risk and likelihood of generating significant results quickly.

WN arranges for village delegates to make field visits to research stations or other parts of the country to inform themselves about the innovations. In this way, they can talk directly with people who have concrete experience and – even more important – they begin to establish independent links with various potential sources of innovations.

Testing innovations. After two years of abnormally short rains, the Malian farmers were interested in testing short-cycle varieties of food crops. WN identified a range of short-cycle millet, sorghum and cowpea varieties, some recommended by researchers, others of local origin suggested by individual farmers. In meetings in four villages in Sanando District, each man and woman was given a handful of each new seed variety. They were asked how it compared in appearance to local varieties. This stimulated lively discussion, particularly when the WN animator said the seeds were reputed to yield in only 3 months. The farmers asked many questions about type of soil required, spacing etc.

The WN animator admitted knowing very little about the varieties and asked the farmers what they wanted to do with the seed. Some farmers proposed planting a whole field with the new varieties. Others thought it would be risky to grow such a large area of unknown seed. Eventually, the groups decided to test the seeds on a limited scale.

The animator posed a set of questions designed to stimulate the farmers to work out the basic steps on their own:

- Should all farm families try out the new seeds, or just a few?

- If a few, by what criteria should these individuals be chosen?

- Should all the trials be put together in one big field, or should they be on the land of each selected farmer?

- Is one single test enough for each variety of seed? Why or why not? If not, how many test plots of each seed variety should be made in the village in order to be more confident of the results?

- How can the production of these new seeds best be compared to the local varieties?
- Should the test plot be on a special field or in the field in which the farmer grows his local variety of the same type of crop?

In each of the four villages, the community assembly chose individuals (pilot farmers) to undertake the trials on their own land. The plots varied from 10 × 10 steps to 30 × 30, depending on the experiment and crop. The communities decided that each trial should be replicated at least 5 – 10 times in each village, as they wanted to test the varieties under the major variant conditions prevailing in the village. 'Control plots' with the local variety were sown next to the trial plots with the new seed, in order to compare results under the same conditions (sowing date, land preparation, weeding, spacing etc.).

Immediately before the growing season, WN invited the pilot farmers to a short training course on basic principles of conducting field trials, e.g. staking out plots and making field observations. It was made clear that the pilot farmers would manage their plots and set the production variables according to their best judgement and traditional practice. Whatever the plots produced would be theirs after harvest.

During the growing season, WN staff regularly visited each pilot farmer. Additional important data that the farmers could not yet record or measure themselves were collected by WN staff for subsequent programme-level interpretation and analysis.

Evaluating results. Some months after harvest, village-level evaluation meetings were held. Each pilot farmer reported to the assembly. Interpretation of results covered a wide range of criteria, including yield, taste, drought and pest resistance, conservability and marketability. In response to the questions "Why did Farmer A have better yields than Farmer B?" and "How do you account for the differences between the replications?", the farmers generated a wealth of data and information.

In a 3-day session, delegates from all four villages came together to compare results and discuss conclusions. They examined each innovation tested in turn and decided to either:

- reject it entirely;
- test the innovation again with more replications and under different conditions; or
- recommend the innovation for widespread extension.

For the innovations selected for extension, the farmers recommended cultural techniques (date of sowing, type of soil, plant density etc.) appropriate for local conditions. These recommendations were not derived from scientifically rigorous experimental data but rather from the farmers' experience and observations. Their validity became clear during the next two years, when the adapted innovations spread rapidly to 10 new villages through a village-managed extension effort.

Institutionalising the process. In order to consolidate this process and make it self-sustaining, WN gives additional training to 'farmer-experimenters' recruited by their communities. These volunteers learn about simple, small-scale experiments – not only involving variety trials

The farmers reported the results of their experiments to the community and discussed reasons for differences between the trial replications. *(Peter Gubbels)*

but also testing intercrops, plant spacing, rotation, tillage, maintaining soil fertility, conserving soil and water, and managing pests. Since almost all the farmers in the WN programme areas of West Africa are illiterate, the training includes a functional literacy component to enable them to measure accurately and to record additional relevant data such as rainfall, planting dates and plant stands.

If farmers have successfully learned to organise their own on-farm research, WN has found that a village-managed extension programme, based on volunteer farmer trainers, can easily be organised. Such an approach is very cost-effective compared with traditional extension methods because the innovations have already been tested and adapted to local conditions by the farmers themselves, and it is not necessary to provide transport and hire paid staff.

WN seeks to institutionalise this process at both village and inter-village level so that farmers will continue meeting to analyse problems, identify and test innovations, evaluate results and extend the proven technologies on their own, after WN has phased out its support (Gubbels 1988).

7.4   Promoting participatory processes

Involving formal Research and Development (R&D)

A major challenge in promoting participatory processes of technology development is to create a greater capacity within national research and extension institutions to facilitate effective interaction between scientists and LEIA farmers. This will require shifts in research policies and priorities away from HEIA packages to LEISA options, changes in organisation and management of these institutions, and the development of strong links between them and LEIA farmers, either directly or via NGOs and local organisations which draw on grass-roots membership.

Direct links with farmers. If LEISA is not to remain an illusive ideal, it is important to find out what stimulates formal Research and Development (R&D) staff to collaborate with LEIA farmers in technology development, and to provide the appropriate incentives. The staff must be made aware of the possibilities and advantages of working closely together with farmers in creating sustainable agricultural systems.

Scientific and ethnoscientific information represent two kinds of knowledge systems and expertise, and combining them can enhance the quality of research carried out by both farmers and researchers (McCorkle et al. 1988). Some ways in which the formal research system can gain from collaboration with experimenting farmers include:

● Researchers can gain greater insight into the problems and potential solutions from the farmers' viewpoint and may discover possibilities not previously contemplated, which can then be further investigated in scientist-controlled trials.

● Researchers can become acquainted with the concepts and methods of farmers' trials and are thus better able to communicate with farmers on the planning and evaluating of trials of relevance to the farmers, both on-station and on-farm trials.

- Researchers can gain greater awareness of the differences between the priorities and objectives of the farmers and those of the scientists and can adjust the content, design and evaluation criteria of scientific trials accordingly.

- Researchers can gain better understanding of local agroecological and socioeconomic conditions and how introduced technologies can be better adapted to them.

- By observing farmers' modifications to introduced technologies and by discussing these modifications with the farmers, scientists can identify technology components that must be studied more rigorously to validate results or to explore possibly better alternatives.

In a participatory process of developing technical options and adapting them to different farm systems, questions requiring more basic research become evident. In tackling these basic issues under more controlled conditions than possible in farmers' fields, scientists can help explain farmers' results and seek potential solutions to problems encountered by farmers. This information and these alternatives can then be fed back to farmers for incorporation into their own research (Waters-Bayer 1990).

Improving the flows of information among farmers, extensionists and scientists would speed up the process of technology development for LEISA systems. It could also help to focus formal research on questions of practical importance for agricultural development and to make more effective use of the scarce time and expertise in NARSs. Perhaps the most important experience that researchers and extensionists can have is to experience that it is both satisfying and exhilarating to work together with farmers in developing truly useful technologies.

Links via NGOs and farmer organisations. As a means of reaching disadvantaged farmers in marginal areas and making more effective use of their funds and human resources, official development agencies have recently been seeking closer contacts with NGOs and farmer organisations. The Overseas Development Institute (ODI) is currently conducting a study of NGO involvement in agricultural technology development in Latin America, Africa and Asia and the scope for closer linkages with public sector R&D (Farrington & Amanor 1990).

In this study, it was noted that some NGOs avoided collaboration with national agricultural research services (NARSs) because of misgivings about the latter's top-down approach and the rigid disciplinary and commodity orientation of the research. However, links between NGOs and NARSs are increasing in several countries, partly in response to cut-backs in local agricultural services. NGOs are moving in to fill vacuums caused by the decline of extension services, and are assuming greater responsibilities in identifying and distributing required inputs, such as suitable seed, and providing technical support services. Some NARSs, aware of their financial limitations, have begun to actively seek collaboration with NGOs. For instance, eight NGOs in The Gambia carried out on-farm adaptive trials with a variety of technologies (mainly seed varieties, but also animal traction and fertilisers) provided by government services. Farmers and farmer groups participated in on-farm testing and evaluation and they expressed willingness to continue the trials. The programme is set to expand (Gilbert 1990).

By observing how farmers change introduced techniques, scientists can learn of interesting alternatives to explore. This innovative farmer is improving an 'improved pasture' technology by inserting live fenceposts. *(Ann Waters-Bayer)*

There are also numerous instances in which government services, though not collaborating directly with NGOs, have been influenced by the approaches and methodology – especially of a participatory and rapid appraisal type – which NGOs have pioneered. The rapid monitoring and evaluation methodology developed in agroforestry by CARE and Mazingira provides an example (Buck 1991).

In view of the complementarity between the capacities of NGOs to stimulate self-help community development and the capacity of NARSs to provide specialist scientific information, improvement of the interactions between these actors in the technology development process could bring considerable benefits to farmers trying to create sustainable livelihood systems in LEIA areas.

Reversing attitudes of researchers and extensionists. It is often assumed that the need for farmer participation in technology development is so self-evident that it is sufficient merely to train researchers and extensionists how to work in this way. However, if professionals are to adopt participatory approaches, they must also be able and willing to do so. There are many reasons why researchers cannot, or do not want to, engage in participatory processes, especially not with LEIA farmers:

- Researchers often see themselves as scientists, not technology developers. They see their work as a scientific activity which does not require much exposure to farmers' practices. They are rewarded for their publications and aim at recognition among their scientific peers.

- Researchers may not see themselves as performing a complementary function in a technology system of which the performance should be measured in terms of technology utilisation by farmers. Hence, many researchers have never worried about their contribution to farm development or about the mechanisms that determine whether they contribute (Röling 1990).

- The training of most researchers is solely technical or scientific. They tend to regard agricultural development in 'objective' terms, and do not realise that the attitudes, ideas and perceptions of farmers, however 'nonscientific' they might seem to a formal scientist, can be very real in their consequences.

- Participation takes much time and patience, and requires scarce resources, such as means of transport. Most national agricultural research agencies in developing countries devote only small proportions of their budgets to operating costs. Devoting any of these resources to participation takes great conviction and determination.

Therefore, if the PTD process is to become embedded in national institutions, attitudes of many researchers and extensionists will have to be reversed. They will have to learn:

- to reverse their sense of themselves as persons of higher status and expertise and as possessors of special disciplinary knowledge to which others have nothing useful to contribute (Chambers 1983);

- not to rush into a situation with a predetermined view of the nature of the problem or of the solution;

To be able to adopt participatory approaches to technology development, researchers must learn to listen to farmers. *(Ann Waters-Bayer)*

● to put much time and effort into developing a good understanding of the cultural identity of the people and their sociocultural structure, and building up a relationship based on mutual confidence;

● to keep constantly in mind that the main aim of PTD is to strengthen the local communities' capacity to experiment and innovate.

The attitudes and interests of agricultural development professionals are greatly influenced by their education and training. Most of the curricula and textbooks used at agricultural colleges and universities are biased toward HEIA technologies and market economics. They generally have a reductionistic focus, assume economic behaviour that externalises environmental and social costs, and suggest technologies geared towards specialisation and the use of external inputs rather than towards the use of locally available resources and natural processes. In many Third World countries, textbooks are being used which appear to be mere translations from Western texts, with no or little adaptation to the tropical context, not to speak of any concern for sustainability, indigenous crops and animals and indigenous farmers' practices.

Thus, professionals are trained within an outdated and inappropriate paradigm, and impart their knowledge and attitudes, e.g. via farmer training centres, within the same paradigm. Often, the training methods are far from participatory. As a result, inappropriate agronomic concepts and authoritarian attitudes are being acquired.

Agricultural professionals, including farmers, will need a different type of training if they are to develop and apply LEISA technology. In this training, emphasis must be placed on holistic concepts, intuitive behaviour, cooperative attitude, respect for nature, and respect for local farming systems and indigenous knowledge (see Box 7.4).

Box 7.4
Learning systems agriculture

To meet the needs of sustainable agriculture, radical rethinking about agricultural education is urgently required. Most fundamental is the need to re-establish universities as communities of learners. Students must be given greater learning autonomy, so that their responsibility, leadership, innovation and creative skills are enhanced rather than stifled.

This necessitates the development of flexible learner-centred as opposed to teacher-centred curricula. More focus must be placed on applying knowledge to real problem situations and in working with people to reach agreement about the existence and nature of the problem. Assessment procedures must be altered to give greater responsibility to the students, and to encourage them to understand the real world better, rather than teaching them how to pass examinations. Curricula must focus on 'praxis' — practice informed by critical theories and achieved through the conscious commitment to methodological enquiry.

In the late 1970s, Hawkesbury College in New South Wales, Australia, reorganised to make praxis the focus of agricultural education. The curriculum is aimed at training 'systems agriculturalists' within a people-centred framework, and is based on concepts of experiential learning, action research and systemic approaches to problem solving.

Systems agriculture is concerned with the identification of problem situations by the participants in which the researcher facilitates the learning process. It focuses on agriculture as a **human activity system** and employs methodologies of problem solving and situation improvement suited to the nature and shared perceptions of the problem. The equation **problem solving = research = learning** leaves little room for extension and the hierarchical organisational structures of which it is a part and which preserve the teaching paradigm (Ison 1990).

Also outside the schools, colleges and universities, in-service training of researchers, advisors and project staff will be needed to make them aware of the potential and limitations of LEISA and to prepare them for their changing roles.

Additional modules for the diverse curricula, new types of textbooks, training guides and corresponding staff training are needed. Probably the best way to start producing these materials is within a process of training trainers, using participatory methods in developing training materials based on accumulation and analysis of the trainers' and students' own learning experiences.

Networking for LEISA development

As LEISA and PTD have not previously been part of the mainstream approach to agricultural development, they have been little documented in conventional forms. In order to heighten awareness of the validity of this approach and to reveal the achievements made thus far, it is extremely important that existing and new field experiences in LEISA and PTD be well documented and widely exchanged. This can be done through national and regional newsletters and information centres, and via national and regional workshops and seminars.

The learning process in developing LEISA systems can be accelerated by establishing networks to create and strengthen links between individuals and groups with similar interests and aims. Many individuals and groups have already built up a wealth of relevant experience, but are often not aware of each other's existence. If they can be identified and linked, they can pool their resources and information and can help each other develop LEISA technologies and participatory methods of technology development.

A 'network' can be of a formal nature, with an executive committee, a schedule of regular meetings, and joint activities, such as a newsletter, workshops, training courses and conferences. But it can also be a more informal system of exchange, which emerges spontaneously as the needs and opportunities present themselves, such as a credit and savings group. Networks can operate on different levels, from very localised groupings of farmers to international networks with telecommunication linkages. Very broadly classified, networks for LEISA can be found on three levels:

1 **Farmer networks**. Farmers generally belong to several spontaneous and informal networks. A good example is the network of cassava growers in the Dominican Republic as described by Box (1987). The farmers adopted a system for exchanging information, planning and evaluating experiments, and exchanging planting materials and market opportunities. Their networks function independently from the networks of extension services, researchers and traders. Market days, public wells, festivals and traditional ceremonies provide the opportunities for farmers to meet. Farmer networks can also be induced, as is done by PRATEC (Proyecto Andino de Tecnologías Campesinas) in Peru. PRATEC documents relevant techniques developed by farmers, and makes this information available to other farmers. Similarly, World Neighbors organises village meetings at

The learning process in developing LEISA systems can be speeded up by strengthening networking between groups of farmers. Informal networking in Indonesia. *(Laurens van Veldhuisen)*

which farmers can discuss and improve the way they do experiments (see Box 8.18).

2   **Regional or national networks**. Many NGOs and farmer organisations form coordinating bodies, through which they pool resources to share technical services and transport facilities; assess, evaluate and disseminate technical information; and create linkages with government to foster cooperation and enable them to articulate their views. Examples are FASE (Federation of Organisations for Social and Educational Assistence) in Brazil, SAN (Seed Action Network) in southeast Asia and ACDEP (Association of Church Development Projects) in northern Ghana.

3   **International networks**. To coordinate their activities and to be able to wield greater influence on national and international policy, international networks involving organisations, individuals and more localised networks have been formed. Examples are IFOAM (International Federation of Organic Agricultural Movements), PAN (Pesticide Action Network), ELCI (Environment Liaison Centre International), ENDA (Environnement et Développement du Tiers-Monde), and EULEISA (a network of European networks supporting the development of LEISA in the Third World). By exchanging information and documents, and by adopting similar and compatible systems for documenting and storing their information, such networks can increase the flow of relevant information, avoid duplication of efforts and facilitate linkages between networks on different levels. The addresses of these and similar networks can be found in Appendix C.

Major functions of networks for sustainable agriculture comprise:

● offering a forum for the exchange of experience, information and, in some cases, other inputs (e.g. seed);

● providing a basis for joint planning and cooperation;

● giving moral support to members in their efforts to achieve something which is not yet part of mainstream activities;

● exercising influence on political decision-makers through lobbying.

Networks gain their greatest strength through their deliberate use of synergy: through participation in a network, members can achieve that which they would not be able to achieve alone (or would only achieve with much greater difficulty). Where the various persons and organisations trying to promote sustainable agriculture speak with a common voice, they can more forcefully advocate necessary changes in policy and formal institutions of education and technology development.

8 Participatory Technology Development in practice: process and methods

by Janice Jiggins and Henk de Zeeuw

This following description of PTD in practice is not meant as a recipe book. Rather, it is a collection and classification of field experiences and methods used by development workers attempting to help farmers develop LEISA systems. It is intended to give encouragement and inspiration to other development workers and to stimulate their creativity.

The six basic types of activities in the PTD process, and examples of methods related to them, are outlined in Table 8.1. The methods are not the only ones currently in use, but they are among the most common and best documented. Some require a certain level of literacy and numeracy; others are appropriate for those without reading, writing or numerical skills. Interested readers are urged to write to the given sources (see Appendix C) for more information about specific methods.

In this chapter, we discuss the nature and purpose of each type of activity in the PTD process and summarise examples of the relevant methods listed in Table 8.1. Within each type of activity, any single method will probably not be sufficient to operationalise a PTD approach. PTD practitioners will need to combine several methods in a mix which is appropriate to the cultural context, their own skills and resources, and their previous experience. Often, this will also imply developing completely new methods and/or adapting existing ones.

8.1 Getting started: networking and making inventories

The way that PTD practitioners from outside a community choose an area to work in, introduce themselves and negotiate with the community about how they are going to work together, depends in its details on the context. The institutional home of the PTD workers, whether they will be breaking new ground or working in the context of an existing project, and the agricultural development objectives of their organisation, are all factors that will influence their approach.

Whatever the setting, once the initial choice of area is made, the PTD practitioner will need to begin a round of introductions, explaining the proposed way of working together and introducing the people from outside who will be involved. The purpose of this is not simply familiarisation. It is also to identify and build a set of relationships between people who have a mutual interest in supporting and participating in PTD activities (Scheuermeier 1988). Such people might include: village leaders, local school teachers, members of local produce-trading associations and seed (stock) exchange circuits, men and women

Table 8.1   Six types of activities in Participatory Technology Development

Activity	Description	Examples of operational methods	Examples of output indicators
Getting started	Building relationships for cooperation Preliminary situation analysis Awareness mobilisation	Organisational resources inventory Community walk Screening secondary data Problem census Community survey	Inventories Protocols for community participation Core PTD network Enhanced agroecological awareness
Looking for things to try	Identifying priorities Identifying local community and scientific knowledge/information Screening options and choosing selection criteria	Farmer expert workshop Techniques to tap indigenous knowledge Study tour Options screening workshop	Agreed research agenda Improved local capacity to diagnose problems and identify options for improvement Enhanced self-respect
Designing experiments	Reviewing existing experimental practice Planning and designing experiments Designing evaluation protocols	Improving natural experimentation Farmer-to-farmer training Design workshop Testing alternative designs	Manageable, evaluable, reliable experimental designs Protocols for monitoring and evaluation Improved local capacity to design experiments
Trying things out	Implementing experiments Measurement/observation Evaluation	Stepwise implementation Regular group meetings Field days/exchange visits Strengthening supportive linkages	Ongoing experimental programme Enhanced local capacity to implement, monitor and evaluate experiments Enlarged and stronger exchange and support linkages
Sharing the results	Communicating basic ideas and principles, results and PTD process Training in skills, proven technologies and use of experimental methods	Visits to secondary sites Farmer-to-farmer training Farmers' manuals and audiovisuals Field workshops	Spontaneous diffusion of ideas and technologies Enhanced local capacity for farmer-to-farmer training + communication Increasing number of villages involved in PTD
Keeping up the process	Creating favourable conditions for ongoing experimentation and agricultural development	Organisational development Documenting the experimentation Participatory monitoring of impacts on agroecological sustainability	Consolidated community networks/organisations for rural self-management Resource materials Consolidated linkages with institutions

who head indigenous problem-solving or work groups, local experiment station staff. It also includes the gathering of data required for a preliminary analysis of the sociocultural and agroecological situation and as a basis for a 'pre-partnership dialogue' (Czech 1986) with the community.

Two criteria for the final selection of villages that seem valuable in most settings are that:

- the community is aware of the direct and wider implications of a declining natural resource base, and understands the importance of an ecologically oriented approach in order to sustain their livelihood;

- in the village there is sufficient scope for sustaining the PTD process (potentials for development of adequate local leadership, no strong opposing factions that hamper action and organisational growth, etc).

In contrast to the classical contents of agricultural development programmes (e.g. artificial fertilisers, high-yielding varieties, chemical plant protection, mechanisation), appropriate and sustainable agriculture demands that farming communities are aware of and identify with the conditions of their environment. Such an awareness and understanding is a prerequisite for proper diagnosis of problems and identification of viable solutions. Therefore, awareness mobilisation and developing a higher (collective) understanding and knowledge of their agroecological system, the relations between soils-plants-nutrients-water-climate and the activities of the households and third parties, is necessary during the starting phase.

It is important that, as soon as practicable, PTD practitioners establish a clear basis for the proposed collaboration and programme of agricultural self-management. The actual form this takes will depend to some extent on the cultural context but experience suggests that some form of written or verbal mutual contract is advisable, which simply sets out the proposed process, the role of participants, potential outcomes, and a mechanism for resolving disputes. Where community assemblies already exist, these might form appropriate bodies with which to negotiate such contracts. The outcome of these activities should be:

- a clear perspective and protocols for the cooperation between the local communities and the PTD team;

- a basic understanding of the sociocultural and agroecological situation in the selected villages;

- a core network of persons, groups and organisations that can play a role in strengthening and sustaining the local experimenting capacity.

Examples of methods that can be applied during the start-up period are as follows. Many of these are basically techniques of Rapid Rural Appraisal (RRA) which lay particular stress on the participation of the local people in analysing their present situation. Further RRA techniques are outlined in Hildebrand (1981), Conway et al. (1987), McCracken et al. (1988) and the RRA Notes issued since 1988 by the International Institute for Environment and Development (address given in Appendix C).

Organisational resources inventory

A good starting point for networking is making an inventory of all groups and organisations that might wish to be involved, and exploring their complementary capacities and resources. The inventory might include: local NGOs already working with farmer groups of various

The PTD process can be carried out by existing organisations, such as women's groups. Bean experiment of a women's group in northern Kenya. (Ann Waters-Bayer)

Box 8.1
Organisation inventory in the Indian Himalayas

In the Dhaulapur Project in the Himalayan Mountain region of India, the first step in promoting social organisation and agro-ecological development was to make a list of all types of existing social groups and organisations in the area of operation, regardless whether officially registered or not. It was recorded how many of the various organisations exist and in which villages.

Discussions were organised in the villages with members of the various types of groups and organisations to understand fully all aspects of the traditional systems of organisation: objectives, social standing/perceptions of the organisations, their social and ethnic profile, how communication takes place, formal and informal leaders, activities, plans etc.

The inventory resulted in a profile of each organisation or group indicating the fields in which it may become an active partner, the participation it may mobilise and to what degree it might contribute to sustaining the process of technology development and exchange in the longer term (Czech 1986).

memberships (e.g. cassava peelers, pig keepers, spice growers, sheep herders), religious centres with long experience of trying out new crops and varieties, schools running special projects such as school gardens or planting indigenous trees, research station staff running trials on local crops and crop varieties, field staff of nearby agricultural projects etc. All of these are potential sources of information and experience, and potential collaborators in the PTD process.

Special attention should be given to the potential of existing traditional and 'modern' forms of organisation in the village (village councils, water-users' associations, women's groups, producer and processing associations). How can the PTD process be built on existing organisations? Should new forms be developed? Which of these organisations and their leaders can function as main coordinator ('carrier') of the PTD process? Meetings with representatives of all these groups and organisations will form a starting point for developing the PTD network in the villages and the outside support system.

Community walks

Another good way to get acquainted with the community and to gather basic information is the community walk. Members of the community are asked to take the PTD facilitators on a series of guided tours of the area. The role of the community guides is to point out features of interest, problems and attempted solutions (residential clusters; water sources; cropping patterns; postharvest crop handling; soil fertility and erosion control measures; water harvesting, storage and use; the role of animals, fish and birds in production and food cycles etc.). The role of the facilitators is to stop, look and listen (Bunch 1985).

When getting acquainted with the community, it is crucial to avoid bias. There are four common types of bias which can occur at this stage (Chambers 1983; Poats et al. 1988):

- road bias – confining preliminary exploration to those fields, households and roadsides that are easy to reach;

- elite bias – restricting contact to the better-off farmers who are already known to the fieldworker;

- gender bias – meeting only male farmers and not women farmers;

- production bias – concentrating only on production and neglecting post-harvest preservation, processing and food preparation.

The danger of such bias is that PTD practitioners fail to understand the range of variation and diversity of patterns that typically exist in any area. These biases can be avoided by, for example, including a woman colleague in the PTD team, inviting women to act as community guides, developing a rough map of the community together with the community guides and visiting each quarter of the map, and one in every ten households, during the community walks.

Screening and discussing secondary data

Before the village diagnosis of the situation, the PTD team often critically screens the secondary data available in literature and reports,

Box 8.2
Community walks in the Philippines

In the Bicol Rainfed Agricultural Programme in the Philippines, schoolchildren were asked to make a spot map of the village area. A first walk-through was made mainly to get to know the people in different residential clusters and the layout and variation in the agroecological system. Second and subsequent tours following transect lines depended on the complexity of the farming systems and agro-ecological zones of the village area. Along the way the guides invited household members to join in and discuss the observations on natural vegetation, crop associations, cultural practices, changes in land use and land type etc. The 'tour guides' also assisted in organising and implementing problem census meetings with various key informants selected during the walk-throughs (Veneracion 1987).

maps, aerial photos, records of village health posts, censuses etc. The exercise helps identify in a preliminary way some key patterns and changes in the agroecological system (e.g. comparing old and new maps can lead to hypotheses about changing land use or call attention to land degradation) and the sociocultural structure of its population, and gives valuable 'fuel' for initiating dialogue with the communities.

Visualisation of the screening results in the form of maps and diagrams facilitates discussion of the data with members of the communities. This discussion leads to validation or correction of the information gathered, generates additional information, and triggers off local awareness of changes and key problems in the village's agro-ecological situation. This may lead to the decision to jointly analyse certain topics more deeply in order to make a better diagnosis of the situation and identify opportunities to alleviate the problems.

Problem census

This is a structured method for helping community members identify, describe and clarify problem areas and priorities and the ways of dealing with them that have already been tried by the community (Werter 1987; Crouch 1984, reprinted in Haverkort et al. 1991). It works best with small groups of 6 – 10 villagers, partly because it is important that different members of the community have a chance to express any differences in priorities and points of view. The greater the diversity of farming patterns and household types within the community, the greater the importance of working with each type separately. For this reason, it is often useful for different groups, e.g. of women and men, to conduct separate as well as joint censuses.

The problem census is often performed in at least two rounds: one to identify the main problems and a second one focused on the detailed analysis of each of these problems. During the first round, the group members list and rank the farming problems they face. The facilitator assists the group by asking probing questions and by writing key words on newsprint, or using a flannelboard, or placing symbolic objects for an identified problem (e.g. a cracked pot for water problems) where they can be seen by all participants.

The next step is more detailed discussion of each priority problem (what, where, when, how, why, implications, relations with other problems?), with the facilitator asking probing questions. This step also involves discussion of the ways in which group members have already tried to deal with the problem and the results they have achieved (why good, why bad, what hampered, what helped?).

To encourage participants to express their feelings about and experiences with a certain problem, projective techniques ('projective' because they help project or express ideas) can be used: posters, flannelboard, slides, drama (skits), audiotapes and other visual or oral prompts are useful. Such prompts, combined with well-chosen questions and group discussion, help clarify a problem by focusing it and encouraging all participants to express their opinions about the elements and background of a problem. The prompts and related questions also help structure the outcome of discussions so that it can be better remembered.

Box 8.3
Using audiovisual prompts in problem analysis

The use of drama as a prompt may be as follows: first the group identifies and discusses an issue or problem in agriculture and postharvest activities of particular concern to them. Then they split up into subgroups that make a short role play (10 – 15 minutes) on that issue, which each group presents in turn to the others. Each presentation is followed by discussion, with members of the community and the PTD facilitators acting as 'reporters' who keep a record of the main points raised (Rocheleau 1987).

In a research and development programme of the Centre for Agricultural Research (ICA) in southern Colombia, a PTD team used slides related to a single topic, in combination with four sets of questions, to stimulate identification and discussion of problems, causes and options.

Each slide depicted one basic aspect of the topic. The slides had been taken in the community so the people and setting shown could be recognised by the discussion group. The sets of questions were:
- What do you see? Do you have this/do this?
- Is it important? Why/why not?
- Do you have problems with it? When/where/why?
- What have you tried to do about the problem(s)? What happened? (Paul Engel, pers.comm.).

Tapes and pictures drawn by a local artist were used in the Grain Storage Project in Tanzania. All preliminary discussions in ward (neighbourhood) meetings on problems encountered in village and household food supply and storage were recorded on battery-run tape recorders. The tapes were analysed by the project staff together with the village-appointed storage committee.

In the next round of ward meetings, the parts of the tapes that focused on key topics were played back and discussion was prompted by means of prepared questions based on the extracts from the tapes. This method helped focus definition of problem areas, analysis and discussion of options, while also allowing villagers who had missed the first round of discussions to catch up and participate fully in the second round.

Pictures drawn by a local artist facilitated the dialogue. Each picture represented the point discussed in the extracts from the tapes. The use of pictures gave added stimulation to the participation of those who were not used to expressing problems and options in an abstract way (CDTF 1977).

Another well-known example of problem census and analysis with the help of visual prompts is the GRAAP method, in which the flannelboard is the central tool for facilitating the dialogue (GRAAP 1987).

Community-led surveys

A community-led survey can be defined as the gathering and analysis of information by the villagers themselves about topics identified by the community in foregoing discussions on the (changes in the) situation in their agroecological and livelihood system. The community survey teams do not need to develop a formal questionnaire. However, it is very helpful to have a group discussion, prior to the information gathering itself, about the types of information to be gathered, the possible ways and places to get certain information, the main questions to be asked and the main things to be observed. A division of tasks among participants can also be made.

8.2   Looking for things to try: developing the research agenda

This set of activities involves gathering information for detailed analysis of priority problems and identification of promising solutions.

Farmers are continuously experimenting and adapting their farming practices. Farmer-initiated technological change does not occur through

Box 8.4
Community-led surveys in Sri Lanka

The Change Agents Programme in Sri Lanka takes a participatory approach to community surveys. Facilitators identify, train and work with village youths (male and female) as change agents. In one case, in Matikotamulla, initial contacts revealed that the problems and prospects of betel farming were priority issues. (Betel is a vine grown in Sri Lanka for its leaves, which are chewed as a mild stimulant.) Villagers each knew something about part of the problem but felt that more information was needed to develop a full picture. So a women's group volunteered to collect information on the production aspects. They formed small teams which visited producer households and collected information on the area cultivated, plant material, production methods, labour and other inputs used, and income.

Meanwhile a boys' youth group investigated the marketing aspects, talking to producers about their experiences with traders, visiting markets to observe how traders behaved and interviewing a few, and trying to follow the product through the marketing chain up to the big town-based export traders and even went to the capital city to gather more information on wholesale pricing and the export end of the industry. Each group reported back to the village the results of their investigations for further discussion (Tilakaratna 1981).

Farmer expert workshops include visits to the fields where the farmer experts can explain their techniques. *(John Connell)*

simple trial and error: there is a farmer-based method which is, in many ways, similar to the scientific method. Farmers, too, follow the steps of identifying and analysing a problem, formulating a testable hypothesis, testing the hypothesis empirically, and validating or invalidating it (Rhoades & Bebbington 1988).

The PTD practitioner coming from outside tries to link up with and strengthen the farmers' research process by making technology development a collective process and by developing the research process with farmers explicitly and systematically, while strengthening farmers' analytic capacities, awareness and self-confidence.

One important lesson of recent PTD experience to take into account when developing the research agenda is that farmers living in diverse and variable environments seek out a range of options rather than a package of techniques. They are interested in developing and extending the portfolio of choices available to them, to be used as the climate and other physical conditions allow, or economic opportunity and family circumstances indicate. It is unusual that communities, and the different groups within them, suggest a single topic or crop for experimentation. Another lesson is that farmers seldom experiment with new techniques to replace existing ones but rather search for techniques that can usefully be combined with known techniques and included in the existing system.

The process of looking for things to try includes:

● determining the range of topics of concern and curiosity;

● screening indigenous technical knowledge and past experimentation for likely options for further testing or for unresolved problems;

● gathering promising ideas from outside the villages (from farmers elsewhere and from scientific knowledge);

● reaching agreement on the research agenda (selecting priority problems, developing selection criteria and screening options, precise formulation of the hypothesis to be tested).

The expected outcome of this process is:

- an agreed research agenda;
- improved skills of farmers to diagnose a problem, to detect promising options and to develop testable hypotheses;
- increased socioecological awareness and self-confidence;
- an improved organisational basis for systematic local experimentation.

Some methods that can be applied during the process of developing the local research agenda are as follows.

Farmer expert workshops

Once a range of problems/topics has been identified, workshops can be held with those who are directly concerned with the problem and/or have a functional knowledge of the topic. The idea is to probe deeper into the problem and to mobilise all relevant indigenous knowledge. It is important that the team recognise that they are not playing the role of 'experts': they are soliciting the expertise of community members. The organisation of the workshops should reflect this role.

Box 8.5
Workshop of Peruvian livestock keepers

Staff members of Grupo Yanapai in Peru attempted in vain to discuss livestock husbandry improvement with committees, the members of which were appointed by the community assemblies. These committees were made up of men who repeatedly showed more interest in cropping than in livestock-related problems. At one point, members of one of the committees made it clear that to start work on livestock husbandry one would have to talk with the women. This remark, as well as observations that women were the ones who spent most time with the animals, led to the conclusion that interest in this area would be found more among women than among men.

As attempts at bringing women together through the community assemblies proved difficult, it was decided to invite the women of the community to come together separately to discuss their farming problems. At the first meeting held in the community of Aramachay, 23 women prioritised their problems as:

- internal and external parasite control in sheep;
- producing forage for use in the dry season;
- improved management of communal rangelands;
- animal selection criteria;
- seed selection and storage techniques;
- criteria for determining seed density at planting.

Having defined and prioritised their problems, the groups began identifying possible solutions, following these steps:

- gathering among group members biological and technological knowledge on the problem;
- discussion with external specialists who helped complement and analyse the information gathered by the women;
- evaluation of possible alternatives proposed by both group members and outside specialists;
- planning of tests (on their fields and using their animals) for the alternatives selected for their probable viability;
- implementing and evaluating the tests planned.

The groups carried out tests on parasite control with local plants and sowing legume forages in fallows. Plans were made to begin a communal range improvement programme and to test modified guinea pig-keeping systems at the household level.

Not only were the women interested in working toward solving these problems; they also decided to organise themselves into a Women's Agricultural Production Committee. Three committees are now working in three communities. Each of the groups of 20 – 35 members has its own directorate and is recognised by the community as a legal entity within its organisational structure (Fernandez 1988).

Useful tips include: choosing site and time for ease of access and convenience of community participants; announcing clearly the purpose and intended participants; presenting all participants when the workshop begins; giving sufficient time for the farmer experts to describe and explain; focusing on probing rather than challenging or rejecting local experts' explanations and experience.

Finding farmers who are acknowledged experts on a certain topic can take some time but is not particularly difficult. For example, a group that has done a problem census can usually identify an acquaintance who is considered especially knowledgeable about the identified problem, or who has a reputation for experimenting with new ideas and materials related to it. This process of identification, known as 'peer group referral', can be stimulated by asking such questions as: Who do you consider to be an expert with respect to the problem? Can you tell me of someone in this neighbourhood who has tried out a new way/material to deal with this problem?

Interaction between the farmer experts will reveal areas of consensus and difference, generate confidence in the validity of the information and knowledge displayed by cross-checking responses, and move the group towards agreement on options for follow-up experimentation. The groups may meet only once, meet several times and then dissolve or gradually develop into 'design groups' and 'option-testing groups' (Norman et al. 1989) and more permanent platforms for exchanging experiences, experimentation and innovation in a certain functional area.

Other names for the farmer expert workshops (with slightly varying intentions and procedures) are:

Key informant group interviews. The main emphasis is bringing together those persons who are 'key' sources of information with respect to a specific area of knowledge and experience. A common mistake is to assume that a key informant is a person with public status – such as an official, or a community or family leader, who is literate or comparatively well-off. Farmers cultivating rich, deep soils usually know little about the experimental behaviour and experience of farmers with less fertile land. The head of the household usually has only imperfect knowledge of the subenterprises run by other family members. If women are responsible for daily care of stall-fed livestock, it is not particularly helpful to approach men, as owners of the stock, for information on hygiene, feeding, stock behaviour etc.

Innovator workshops. The main aim here is to learn from those community members who have experience with application/adaptation of a specified solution to a problem and/or are known in the village for continuously experimenting with methods and materials. The concept 'innovator' should not be identified with those farmers who are known by the extension worker as 'quick adopters' of recommendations based on research station findings. What is sought are the experiences of those farmers who are actively experimenting – maybe on a very small scale and with only minor changes of technology – under the conditions faced by the majority of farmers, developing indigenous ideas and adapting/incorporating ideas coming from outside (e.g. UNESCAP 1979, FAO 1985).

Key sources of information about small stock are often the women and children who care for the animals, rather than a male 'household head'. *(Ann Waters-Bayer)*

Focus groups. The emphasis here is on the strong interest of the group members in finding better ways of dealing with a specific problem. Focus groups will tend to develop from *ad hoc* groups into more permanent groups, especially if they are small and membership is fairly homogeneous (e.g. Norman et al. 1989, Fernandez 1989).

Tapping indigenous knowledge

Various techniques have been developed to gather and discuss indigenous knowledge on specific topics. Some techniques that individually or in combination can be used as means to implement farmer expert workshops, or can be applied in another context are:

Making diagrams. This is simply the visual presentation of data and causal relationships. Its usefulness, as an aid in helping community experts describe and explain, can be sharpened if initial attempts become the focus for discussion and refinement at subsequent group meetings. Diagrams include representations of space, such as sketch maps of land use around a village or homestead (Gupta et al. 1989, Rocheleau 1988) and transects (Conway et al. 1987, Budelman 1983); representations of time such as cropping and other seasonal calendars (Conway et al. 1987); and representations of causal relationships (see Figures 8.1, 8.2 and 8.3).

Diagrams can be physically drawn, at their simplest, with a stick on wetted earth, but broad brush and newsprint, notepads, portable blackboards and flannelboards give clearer images. They are drawn on the spot: e.g. a transect can be drawn as part of a community walk through the landscape, or a herding calender during a focus group discussion of sheep management. The resulting diagrams can then be transcribed to a more permanent format for future discussions with other groups.

Case histories. Facilitators help farmers describe their experiences with trying out and adapting a particular technical innovation or a practice in specific production or postharvest activities (Box 1989). Farmers describe what was the original problem, what alternative ideas were considered and where these originated, what steps the farmer took to try out each of these ideas and why they chose that way. What failed and what worked well and why? If they had to do it again, what would they change in their experiments?

Such case histories are particularly effective in revealing trends over time, important disruptive events, sources of new ideas or materials, and records of tried and proven (or failed) experiments. They can be used as starting points for wider discussion of the same technology or practice. They can also be used to review and discuss the experimental methods and designs applied by the experimenters.

Critical incident. The aim here is to focus discussion on a key event in the recent past, in order to explain changes that have occurred, such as in environmental management, crop husbandry or disease control. Having identified a subject of particular concern (e.g. pest control in rice), the participants identify a critical happening (e.g. a devastating infestation of brown rice hopper 3 years ago) and, in response to probing

Figure 8.1
Map of peasant farm in the Pananao Sierra, Dominican Republic. Maps can thus be used to learn quickly from rural people by tapping their collective local knowledge. The shared analysis facilitates communication and helps outsiders gain insights into social patterns of land use. (After Rocheleau 1987)

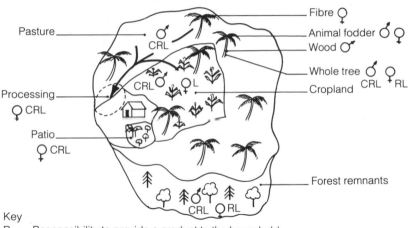

Key
R = Responsibility to provide a product to the household
L = Labour input for establishment, maintenance and harvest
C = Control over resources or process
♀ = Women
♂ = Men

Figure 8.2
Transect of a village in northern Pakistan. The PTD practitioners (outsiders and local people) walk along a transect through a village, catchment, region etc. to explore differences in land use, vegetation, soils, cultural practices, infrastructure, water availability etc. The transect diagram produced is a stylised representation of a single or several walks. (*Source:* Conway et al. 1987)

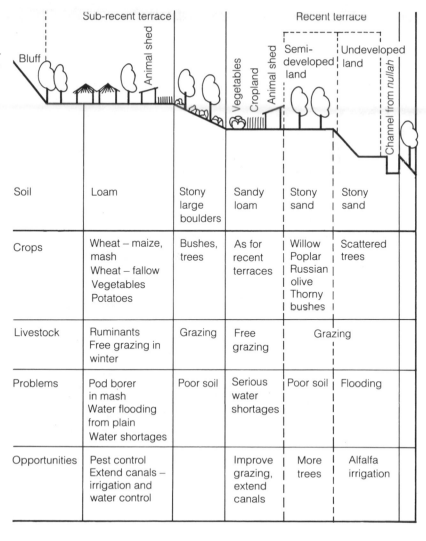

	Sub-recent terrace		Recent terrace			
Soil	Loam	Stony large boulders	Sandy loam	Stony sand	Stony sand	
Crops	Wheat – maize, mash Wheat – fallow Vegetables Potatoes	Bushes, trees	As for recent terraces	Willow Poplar Russian olive Thorny bushes	Scattered trees	
Livestock	Ruminants Free grazing in winter	Grazing	Free grazing	Grazing		
Problems	Pod borer in mash Water flooding from plain Water shortages	Poor soil	Serious water shortages	Poor soil	Flooding	
Opportunities	Pest control Extend canals – irrigation and water control		Improve grazing, extend canals	More trees	Alfalfa irrigation	

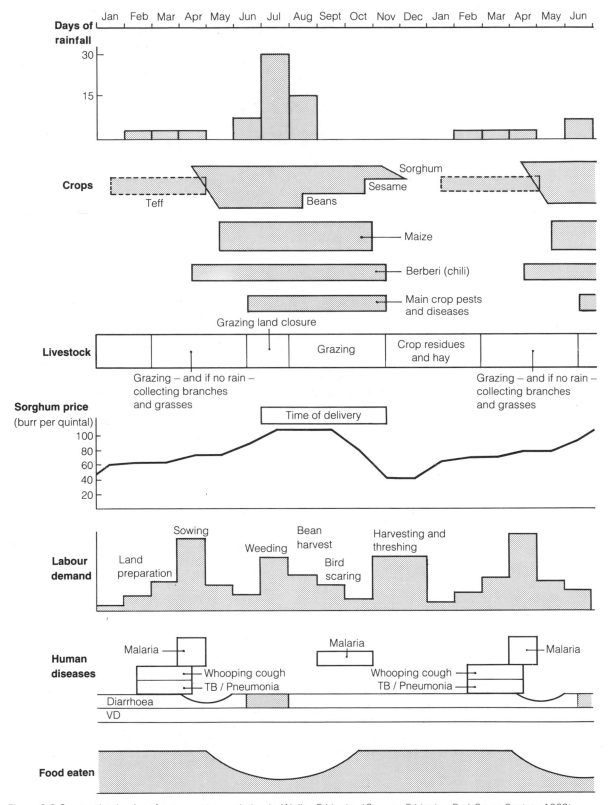

Figure 8.3 Seasonal calendar of a peasant association in Wollo, Ethiopia. (*Source:* Ethiopian Red Cross Society 1988)

questions, trace the events that led up to it, the nature of the incident, responses to it and its consequences. Group analysis of the incident leads to identification of ideas for further testing (or maybe a decision to visit a neighbouring rice project or agricultural station to find out more about it). The exercise can be repeated with different groups until nothing new comes out of the discussions; through iterative use of the technique, the sum of community knowledge is accumulated and needs for further investigation are indicated (Flanagan 1954).

Preference ranking. This can be done in at least two ways: pairwise ranking and matrix ranking. They both aim to quickly explain farmers' (or other interest groups'/individuals') own criteria for preferring one species (or crop, animal, soil type etc.) over another. The technique has been applied with success to analyse, e.g. under what conditions outbreaks of certain pests and diseases occurred and what could be done to prevent severe attacks (e.g. Barker 1979); to make inventories of locally available varieties of certain weeds, their utilities and growing requirements (Richards 1979); and to determine the relative importance of indigenous browse species in the nutrition of pastoral herds (Bayer 1990). The results are useful in:

- making inventories of available resources and identifying possible solutions or priority problems;

- helping outsiders understand farmers' decision-making;

- pinpointing desired characteristics, varying according to user types, to be taken into account when selecting materials for field trials;

- establishing criteria for evaluating the results of field trials and other experiments.

One form is **pairwise ranking**: farmers are asked to compare certain species or breeds two by two and indicate why they prefer (in terms of usefulness) one over the other (see Box 8.6).

Ranking techniques are also useful for discovering the differing needs and priorities of different categories of people within an area or community (men, women, young, old, richer, poorer etc.). For example, Scoones (1989) describes how pairwise ranking was used to compare preferences for tree species between different groups (settled residents and displaced immigrants near Khartoum, Ethiopia). In addition to the differences in priorities expressed by the actual ranking of the species, differences in the decision-making criteria used by the different groups could be demonstrated in their lists of 'good' and 'bad' properties of each tree.

Another form of preference ranking is **direct matrix ranking** (Conway et al. 1988). The group chooses a topic or class (such as weeds in maize, cassava varieties, sheep breeds). Sometimes it is possible (e.g. with weeds in maize) to collect samples of the items mentioned from the field so that the group can observe and refer to them during the discussions. They then 'brainstorm' to identify the ones they regard as most important and list them across the top of a board (e.g. on newsprint). For each in turn, the PTD team asks questions – what is good about it? what is bad? – until nobody can think of any more criteria. All the criteria mentioned are listed down the side. Negative or undesirable criteria (e.g. grows new plants from pieces of stem) are reworded to

Box 8.6
Farmers' ranking of useful trees in Ethiopia

In a farmer expert workshop in Wollo Province, Ethiopia, participants wanted to study the virtues and drawbacks of some tree species used for reforestation. The six most widely used reforestation species were selected and their names were written on squares of cardboard. Some farmers, known for their knowledge of trees, were presented two of the squares, *Eucalyptus camaldulensis* and *E. globulus*, and were asked to collectively chose which was 'better in terms of usefulness' and to explain why they had chosen that species over the other. They were also asked whether the less preferred species was superior to the preferred in any respect. Finally, they were asked whether there was anything

else they could tell about the pair. The comparison of different pairs of squares was continued until all the possible combinations had been considered. The final ranking was made by examining all the pair combinations, laying out the squares of paper in a line so that each species was above all those to which it was preferred. The ranking, characteristics and uses were:

1 African olive. Diverse implements: digging sticks, yoke and other parts of ploughs, hoes, axe, handles, sticks); house construction (not attacked by termites; firewood (no smoke); incense from leaves.

2 *E. camaldulensis.* Easy to split; strong for construction; durable; straight; easy to make charcoal.

3 *E. globulus.* Good for holding nails; high elasticity,

bends easily; difficult to make charcoal; farming implements; firewood.

4 Juniper. Window and door timber; chair making.

5 White acacia. House building.

6 Croton. Door construction; smoky as firewood.

The farmers were then asked whether there was any characteristic or potential tree missing from this list. After some discussion, the farmers said they would like a hard furniture tree like *Podocarpus* which would be better then Juniper.

These outcomes were compared and combined with outside knowledge about these tree species and the exercise resulted in recommendations for on-station research with respect to the characteristics and variation of the tree species sought (Conway et al. 1988).

form positive or desirable criteria (i.e. does not grow new plant from pieces of stem). The group then considers which of the items listed across the top of the matrix is best, judged by each of the criteria in turn. The judging process can be prompted by questions such as: Which is best? Which is next best? Which is worst? Which is next worst? Of the remaining, which is better? If items still remain, choices can be forced by questions such as: If you could only have/grow one of these, which would you choose? The answers are recorded directly into the matrix.

The direct matrix ranking technique can also be used to identify how local people classify land types and according to what criteria they make decisions about land use (see Table 8.2).

Inventory of farmers' indicators. Collecting the local terminology and indicators used in a certain technical field and trying to detect the meaning underlying these terms has proven to be a very valuable technique. A well-known example is the use of indicator plants to determine the qualities of the soil (fertility, water retention, crust-forming, power needed to plough etc.) and its utility for certain purposes and users (see Box 8.7). By applying the indicators farmers use for their decision-making, classifications of the natural vegetation, land suitability maps etc. can be developed with farmers on the spot (rather than having to rely on time-consuming and costly surveys).

Box 8.7
Farmers' indicators of soil types in Botswana

A researcher from the Environment and Rural Development Programme in Botswana asked elderly farmers with extensive farming experience to name the types of soils in each of the fields they had access to, and when they had acquired that field. It turned out that these farmers deliberately tried to get access to the three main soil types (*seloko:* clay; *motlhaba:* sand; *mokata:* loam), if possible within one field, otherwise by acquiring fields with the missing soil type.

Next, farmers were asked to indicate and explain their preferences for each of these soil types, resulting in a list of both advantages and disadvantages (with respect to soil fertility, water-holding capacity, ploughing characteristics, suitability for certain crops, resistance to erosion etc.) of each of the main soil types. Checking revealed that the advantages/disadvantages mentioned by the farmers were backed by data from scientific literature.

Finally, farmers were asked to explain how they recognised the different soil types and their qualities. It turned out that most farmers used trees and grasses as indicators for the suitability of soil for cultivation, e.g. *Acacia erusbescens (moloto)* and *Eragrostis rigidor (rathatwe)* as indicators for good arable soil of the *motlhaba* type, whereas *Mellifera (mongana)* and *Cymbopogon plurinodis (mosagwe)* indicate good soils of the *mokata/seloko* (loamy clay) soil type (Arntzen 1984).

Table 8.2 Direct matrix ranking of land types in Papua New Guinea (1 = very good, 5 = very poor)

Land type (local name)*	Poiem	Sepiem	Sunem	Erisonde	Tipso	Poi
Fertility	1	1	2	1	1	4
Slope	2	3	5	3	5	1
Vegetation easy to clear	3	4	5	5	5	5
Cleared vegetation makes good compost	2	2	2	2	2	4
Soil is easy to work	4	3	2	3	3	3
Well-drained	5	5	5	4	2	5
Productive for sweet potato	1	1	2	1	1	5
Productive for mixed vegetables	1	3	2	1	1	4
Good for pig foraging	3	4	4	4	1	1
Good for gathering nuts	5	5	5	1	1	5

*Key to local names:
Poiem	Gardens on alluvial or drained swamp land
Sepiem	Gardens from grassland (mainly sweet potato)
Sunem	Gardens from steeply sloping grassland (sweet potato and mixed crops)
Erisonde	Gardens from forest or secondary regrowth ('greens' and mixed crops with sweet potato)
Tipso	Lower montane forest
Poi	Wetlands (alluvial or swamp, undrained)

Source: Mearns (1988).

Study tours

A simple but effective way to gather ideas for things to try in order to solve the identified problems is to arrange study tours. Groups of persons are selected by the community to visit local research stations, neighbouring agricultural projects, or farmers and farmer groups in other villages known for their experimentation. The groups observe and discuss trials and experiences with adopting and adapting innovations in production technology and processing and storage practices, in order to pick up ideas (about both the technology and the experimental methods) they could try out in their own environment. However, since there is not likely to be agreement within the community on a single topic of overriding concern or on particular problems within it, the PTD process must take into account divergent needs and views within the community.

Box 8.8
Study tour by Filipino farmers

A group of farmers in Claveria concerned about soil erosion control on steep slopes sent six members to visit upland farmers on Cebu (a neighbouring island), where World Neighbors had been assisting farmers to lay out contour bunds, plant hedgerows and induce natural terracing. The Cebuano farmers trained the visiting farmers in laying out contours using an A-frame, bunding-ditching to establish strips and control erosion, and hedgerow planting of fodder grasses and legume trees. The following year, the Cebuano farmers paid a return visit to the Claveria farmers to observe the bunds and crops, discuss the adaptations made by the Claveria farmers and share ideas about how the system might be further developed (Fujisaka 1989).

Box 8.9
Systems diagrams in the Philippines

The Farming Systems Development Project in Eastern Visayas (see PTD Case I in Chapter 7) systematised the outcome of situation diagnosis with the help of systems diagrams:
- The informal interviews and group meetings provided information on the biophysical causes and socioeconomic constraints surrounding a problem. Each cause was drawn in a separate box with arrows leading to the centrally placed problem box.
- A group of key persons discussed the relationships between the boxes and the problem, and the boxes were redrawn with box sizes determined by the 'explaining value' of that cause assigned to it by the participants.
- The resulting diagram was discussed in group meetings to obtain agreement that it faithfully represented the problem and its causes, and to develop understanding of how each problem interacts in the whole farming system (see Figure 7.1).
  One priority problem analysed in this way was the cultivation of marginal uplands infested with *Imperata* grass.
  After the search for potential solutions, the diagram was used to discuss the potential of identified solutions (e.g. using herbicides, shading out, ploughing) by relating them to the biophysical causes of the problem and the socioeconomic constraints shown in the diagram. This helped the farmers decide what they wanted to test, in this case, shading out *Imperata* with vining legumes (Lightfoot et al. 1988).

Option screening workshops

During the process of diagnosing the situation and seeking solutions, many ideas are identified and discussed. At a certain moment, the community or group has to define its 'research agenda'. During one or more screening meetings, the pros and cons of the available options are discussed systematically and hypotheses for testing are developed. The screening meetings build on the results of earlier activities and techniques. They should clarify to all involved which options should be tested and why each test will be done (how this option relates to their problem and what is expected to happen). A technique that helps the group take into account all aspects of the complex problem situation when selecting options is the use of systems diagrams (Box 8.9).

8.3   Designing experiments: building on local experimental capacity

Aims of this activity are to develop experimental designs that suit the farmers' purposes and to strengthen their capacity (skills, organisation, self-confidence) to design experiments independently. Here, there is a fine balance between supporting and developing local experimental capacity, and imposing trial designs and experimental concepts that have greater validity in research-station practice. There are no firm rules to follow, but the basic idea is to improve, reinforce and add to farmers' experimental practice. The major steps are:

- reviewing farmers' experimental practices (what do they try out? how do they do it? why do they do it in this way?);

- planning and designing the selected experiments with the farmers (who will actually do the experiments? do we need controls? how many times should an experiment be repeated and under what conditions? according to what criteria do we select locations? how large do we want the experimental plots to be? what treatments do we want to apply? how do we lay out the trials? what inputs do we need and how can we obtain them?);

- developing protocols for evaluating the experiments (what will be our criteria for evaluating what happens in the experiment? what/when do we need to observe/measure/record/discuss and how do we organise this?).

An important experience of early PTD practitioners to keep in mind when designing experiments and evaluation criteria is that an experiment with certain technical options may be perceived by different categories of the population in a different way, depending on their role and tasks in production and processing, the end use they have in mind and the resources they control. The expected utility of the option for different users and situations should be clarified in advance as much as possible.

The expected outcome of this activity is:

- experimental designs that are reliable, evaluable and manageable by farmers;

- improved skills to design experiments;

- monitoring and evaluation protocols;

- existing networks, or newly established PTD interest groups, prepared to implement and manage the process of experimentation, monitoring and evaluation.

The following methods have proven effective in helping farmers systematise and improve their experimentation.

On-the-spot improvement of natural experimentation

At its simplest, the PTD team can increase the likelihood that 'natural experiments' produce evaluable and reliable results by giving fairly minimal support, as the example from a PTD team in Butare, Rwanda, shows (see Box 8.10).

Farmer-to-farmer training

PTD teams can put farmers in touch with farmers in another area who have already conducted experiments with the technology/idea they want to test. 'Hands-on' training of a community-selected group in establishing and managing the experiments, farmer-to-farmer, is convincing and practical, allowing time for discussion of the problems

---

**Box 8.10
Supporting farmers'
experiments in Rwanda**

When we learn that the cooperative is planning to harvest a clump of sesbania (a green manure crop) and incorporate the biomass in the raised bed, we ask if we can attend. Together, we take a yield sample and do a few quick calculations: 30 t/ha of biomass should contain about 300 kg nitrogen. The entire bed is about 4 times as large as the clump; thus, if the farmers spread the biomass equally, also on the land that supported the sesbania, they ought to obtain roughly 60 kg/ha of nitrogen, which we say sounds about right for the cabbage/maize intercrop they are planning. Were they to fertilise only the area under sesbania, as one farmer suggested, in order to avoid mining the soil, we tell them there would be risk of excessive growth, lodging and diseases. But why not watch the growth of the crops in the area where the sesbania stood compared to the rest of the bed? And how about leaving a small area untreated so that we can all see the effect of the incorporation? They agree and together we mark off the three sections (Loevinsohn 1989).

that the trainer farmers encountered in implementing the experiments, and of the differences between the two sites.

Design workshops

Those farmers who have a strong interest in trying out a certain option and have a stake in solving a certain problem may be brought together to discuss the setting up and organisation of the experiments they will be involved in. The group makes use of the information gathered during 'farmer-to-farmer' training, study tours, innovator workshops etc. with respect to organising and managing experiments. The facilitator may make use of the following techniques to facilitate decision-making:

Case histories. When discussing farmer expert workshops, it was mentioned that this technique can also be used to describe and discuss local expertise in 'adaptive research'. This local experimental knowledge and practical experience can form a sound basis for the development of experimental designs which can be managed by the farmers.

Slides and videos. Simple slide series and videos of farmers' experiments in other areas can be used to stimulate discussion of experimental design. World Neighbors has developed a slide set which shows farmers the basic principles: how to lay out a simple experiment, testing one factor at a time, on a small scale, with enough replications to provide convincing results.

Prompting questions. With the aid of prompting questions, the PTD facilitator can guide group discussions by farmers about the basic steps in experimenting and help them make their own decisions about how to undertake trials. The facilitator does not impose a set of pre-determined experimental procedures preferred by the outsider. Rather, she/he asks the farmers a set of carefully prepared questions designed to find out what they themselves think about how to undertake an agricultural experiment, and to stimulate them to elaborate the basic steps of the process on their own (Gubbels 1988). Examples of such questions are given in PTD Case II in Chapter 7.

Testing alternative designs

At the beginning of the PTD process, it is not always easy to draw up a design that will produce valid research results but also fits into existing work patterns and plot layouts. One or more seasons of experimentation in the design itself might be necessary to detect where problems arise and how the design can be improved. This monitoring and improvement of experimental methods is also an important aspect of the farmers' learning process and the development of local experimental capacity.

8.4   Trying things out: implementing and evaluating experiments

This includes not only actually doing the experiments and related activities, such as measurement and evaluation, but also development of

A farmer who has already experimented with a new technique – as here in biological pest control – can train other farmers in this technique. *(Henk Kieft)*

Box 8.11
Adjusting experimental design in Peru

When Women's Production Committees in Aramachay carried out experiments, the following problems arose during the first year:

- If the usual type of spatial trial blocks were insisted upon, the participating farmer would have to prepare the land with a pick instead of an ox-team. This requires more time and effort and more people.
- Traditionally, planting is done as a group working in close coordination. The ox-team goes ahead, opening the furrow, followed by the person who places the seed in the ground and then by the one who applies the fertiliser. If any of the three takes longer at his/her task than is allotted, the work of the rest is disrupted. For this reason the distribution of small amounts of seed of any variety or fertiliser treatment ran into problems, because it required that the sower/applier must reload her *quipi* (basket) which takes additional time.
- Participating farmers were found unwilling to leave portions of the plots unplanted, as this represents a waste of utilisable land. Besides, unplanted areas encourage weed growth. If there is a need for spacing, it is only acceptable in the case of grains, the last crop in the rotation cycle.

In view of these problems and incorporating the suggestions of the participating farmers, adjustments were made in the experimental design which permitted control and measurement and, at the same time, normal conduct of the planting and harvesting work.

In order to avoid unused spaces between blocks and

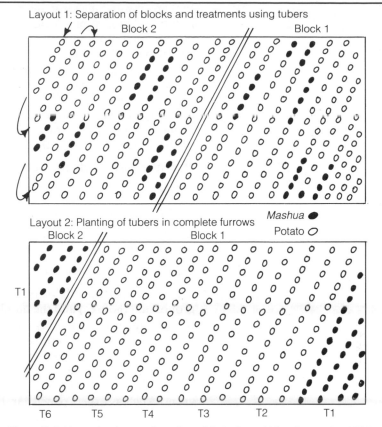

Figure 8.4 Alternative layouts for tuber trials in Peru. (After Fernandes 1989)

treatments, the farmers tried planting *tarhui* (an Andean legume) as a divider, but at harvest it was observed that the ripening of potatoes and *tarhui* did not coincide and that the ox-team killed the *tarhui* when loosening the potatoes. In the second year, the *tarhui* was substituted with *mashua* (an Andean tuber), often grown together with potatoes. This process consisted of planting an entire *quipi* of potatoes of the variety or treatment desired in the prepared furrows and then a few *mashua* before the next treatment. This avoids time loss at planting and permits simultaneous harvest (Layout I in Figure 8.4). In subsequent trials it was found that 5 *mashua* tubers are the minimum to guarantee that the treatments will not be mixed when the ox-

team loosens the tubers at harvest. The only special care that must be taken is to note carefully the direction in which the planting team has worked (up one furrow and down the next), so that the treatments can be identified.

A second alternative for avoiding unused space is to plant the tubers of each treatment in a series of complete rows (Layout II in Figure 8.4). If this is done, the number of rows must coincide with the quantity of tubers in a *quipi*. For potatoes, this implies a minimum area of 20 m^2 and requires a larger quantity of seed for each treatment and block than for the first alternative. This method can also be used for experiments with maize, peas, *quinua* and *tarhui* (Fernandez, in press).

the capacity to implement and monitor experiments (skill development, group building, strengthening exchange and supportive linkages with other communities and organisations active in the area). Such activities will also develop a structure for processes of dissemination of PTD experiences and the replication of PTD processes in other villages.

Experience with PTD suggests that isolated, individual experimenters trying out possible solutions to a problem do not build up this capacity as strongly as when groups agree on a research agenda. The experiments themselves may still be conducted on single plots by individual members of the group. The advantages of the group process include:

● greater ease of observation and discussion among collaborators;

● greater confidence and willingness to try things out;

● greater diversity of things tried;

● greater ability to control 'interference' such as bird attacks, poor water management;

● greater potential for mobilising cooperative effort.

Of course, in the early years, the social scale (numbers of participating groups and the area) and the numbers of different types of experiment should remain serviceable by the PTD team: no hard and fast guidelines can be given for how large a scale this might prove to be because it depends on factors such as the resources of the PTD team and the nature of the terrain. However, experience suggests that the more a PTD team works with groups (*ad hoc* groups such as focus groups, permanent groups such as expert panels, and existing community organisations), the more likely that the social capacity to sustain agricultural self-development survives the departure of the PTD team.

The expected outcome of this activity is:

● a growing number of experiments, with technologies of increasing complexity, are implemented and evaluated systematically;

● the PTD network within and between villages is developed, as well as the institutional linkages;

● the practical skills of involved groups of farmers to implement and evaluate experiments systematically are strengthened;

● growing active support of outside organisations and institutions.

In addition to the methods noted in the previous section, there are numerous other activities that can further support, consolidate and improve local capacity for sustained agricultural self-development:

Stepwise implementation

Trying out may already start in a relatively early stage of the partnership. One 'starter-activity' may be singled out early. Such an activity should be relatively simple, require few inputs, be attractive to the majority of the villagers, produce good possibilities for follow-up and bring quick results (Vel et al. 1989). The planning and successful implementation of this activity will help enhance the motivation of the participants, and develop their relationship with the PTD team.

**Box 8.12
Stepwise testing by
Indonesian farmers**

In the Propelmas Rural Development Project in Sumba, Eastern Indonesia, growing green gram was selected as an 'entry-point activity'. Green gram is a crop that can be readily consumed or marketed, farmers were very interested in growing it and the input requirements are low.

Evaluation of the first year's experience revealed many technical problems as well as important socioeconomic and cultural dimensions of the local community, e.g. the poor farmers cultivated green gram mainly on steep hills and did not prepare the soil well before planting. Discussion revealed that they grew the crop on this seemingly unsuitable land because they knew that, with low labour input, green gram would have low but acceptable yields. Further inquiry revealed that poor farmers need to reduce labour because, in the same period, they must join working groups in the rice fields, with strong sanctions for non-participation – a relict from feudal times. During the first year, it also became clear who was really interested and who not, which participants were suited as group coordinators, and how they traditionally organise and run their groups.

The insights gained were used when planning new trials, each building on and enforcing the previous one, e.g. planting *Leucaena* hedges on steep hills to combat erosion and provide fodder, and mulching of green gram. Once the groups had implemented these steps successfully, and awareness, group decision-making and leadership had been enhanced, a start was made with stall-feeding of beef calves and organising group marketing and supplying inputs such as animals and seeds (Vel et al. 1989).

Furthermore, it gives the PTD team a chance to learn much about the sociocultural system and how local organisations function. It helps the communities develop a clearer understanding of the PTD concept and stimulates discussion among themselves and with the outsiders, meanwhile implementing a concrete and relevant activity and strengthening experimental capacities. In the next round, new tests with other and possibly more complex options can be planned and implemented.

Regular group meetings

The farmers involved in testing one or more potential solutions to a certain problem gather regularly, preferably in the fields where an experiment is conducted. Such meetings have proven to be very important for discussing problems encountered, further clarifying or improving research protocols, and observing, measuring and discussing

---

**Box 8.13
*Farmers' focus groups in
Botswana***

In Shoshong, Makwate and Makoro villages in the Mahalapye area of Botswana, the Agricultural Technology Improvement Programme (ATIP) assists groups of 10 – 20 farmers strongly interested in a certain theme. The groups select a chairperson and set their own meeting date and place. They meet once a month. Each group member implements one or two experiments falling within the focus of the group. Technologies being tried include an animal-drawn row planter, double ploughing, the use of fodder crops, contour ploughing and short-season crop varieties.

Together with ATIP staff the groups prepare a topical agenda for each meeting. A meeting starts with each farmer reporting on problems and observations. The most animated discussions take place when group members interact on the basis of differing personal experiences in managing their trials. Data collection is limited to very basic information that can be collected at one point in time through middle and end of season measurement of a few simple variables (Worman et al. 1988).

Box 8.14
Farmers' field days in Botswana

In the villages mentioned in Box 8.13, people from outside the group (including participants of experimenting groups in other villages) are invited when the crops reach maturity, to see how the new techniques and practices have performed. At these field days, several fields where farmers have conducted their own experiments are visited, and the farmer concerned describes what he did, explains problems encountered and discusses the results with the visitors. The groups are becoming more and more independent of the ATIP staff and are developing an innovative network between the groups and with collaborative staff of government extension services and local NGOs (Worman et al. 1988).

certain topics during the growing season. The periodical collection and discussion of data and observations also improves farmers' understanding of how and why experiments are monitored.

Field days/exchange visits

Methods such as preference ranking (see above) can be used to generate criteria for evaluating experimental results. Where different groups have contrasting preference rankings (e.g. male producers prefer characteristics which have a high value in cash markets, while female processors prefer ease of transformation, good storability and short cooking time), care must be taken to ensure that both groups have a chance to evaluate results.

One way to do this is to help experimenters organise field days, both during and at the end of trials, so that others can observe and discuss. Experience suggests that such field days can lead to the outright rejection and elimination of particular practices, plant materials etc. Field days seldom result in a consensus that a single option is 'best'. They commonly lead to identification and acceptance of a number of options, each with clearly defined parameters for use and incorporation into existing systems.

Another method is to hold exchange visits, with experimenters at one site visiting the trials of experimenters at another.

Strengthening supportive linkages

As experience and skills develop, other service centres that will remain permanently in the area, such as research stations, schools, tree nurseries and church missions, can be drawn more closely into the process. PTD teams can help community representatives make contact with such organisations, to solicit their aid in providing back-up support to farmer experiments. Networking earlier in the process should facilitate this development. For example, forestry staff working in tree nurseries can assist in acquiring the tree types farmers want to try out, participate in trials for multiplying types with the characteristics farmers prefer, and participate in evaluation of farmer experiments.

8.5   Sharing the results: communication, dissemination and training

The rate and pattern of diffusion of technologies proven successful through farmer experimentation has not been formally studied, but PTD practitioners report that spontaneous diffusion occurs as experimenting communities share results with friends and spread seed (stock), and as new products gain recognition along trading routes (Gubbels 1988, Budelman 1983). These effects can be amplified through building a programme to share results with others. An important component of such a programme is the mobilisation of the networks developed during earlier phases, as channels for communication and dissemination.

The emphasis in the diffusion will be partly on the locally realised outcomes (cultural practices, seed etc.) of farmer experimentation.

**Box 8.15
Extension by folk drama in Kenya**

In Kakemega District, the Kenya Woodfuel Development Programme (KWDP) employed local actors to stage an amateur drama incorporating local sayings and songs. The experiences gained in intensive cooperation between KWDP staff and a limited number of farmer groups with respect to tree planting and management were fed into the drama, which is played on market days. The drama provides a reflection on the woodfuel problem and encourages participants to develop their own ideas as to how the situation can be improved. An outline of the plot of the drama in comic-strip format is distributed after the performance so that the audience can discuss the subject among themselves and with others who did not attend (Chavangi et al. 1985).

However, the main emphasis will be on the basic ideas and principles underlying these experiments and the diffusion of the methodological aspects of the PTD process to other communities: diffusion of both promising options to experiment with, as well as ideas and experiences about how to experiment, i.e. innovative concepts, skills and forms of organisation.

The outcome of this activity should be:

- enhanced farmer-to-farmer diffusion of ideas and technologies;
- an increasing number of villages involve themselves in processes of organised technology development, making use of the experiences of other communities;
- a farmer-managed system of inter-village training and communication.

Some methods that have proved effective in enhancing farmer-to-farmer communication, dissemination and training are as follows.

Visits to secondary sites

PTD practitioners often report that neighbouring communities approach the PTD team for similar help in improving their agricultural self-management. The team can assist members of experimenting groups to visit their neighbours on request, and to explain and discuss their experience of the PTD process. A two-way dialogue, sharing experience and results, can be facilitated by using audiovisuals developed during earlier phases of the process, by socio-drama and by using indigenous forms of communication, such as puppet shows. Return visits to observe experimenters' trials are proving to be a useful follow-up activity.

Farmer-to-farmer training

An important way to spread results and information about PTD methods is farmer-to-farmer training. This may have different forms:

- **Informal individual peer teaching**. Participants in the experimenting groups teach farmers in their direct neighbourhood who become interested during implementation of the experiments. Sharing of materials (e.g. seed) with the trainees is often part of the transfer process.
- **Informal group training**. Farmer experimenters act as trainers for a visiting group of farmers from other villages (study tour, field days) or participate in innovator workshops in other communities, acting as the key presenters (e.g. Simaraks et al. 1986).
- **Formal group training**. Farmer experimenters act as trainers in farmer training courses. This often involves additional training of the cooperators and joint planning of the training courses (content, methodology). Thus, the agricultural self-development capacity is further enhanced and a local network of experimenters/trainers is formed.

Box 8.16
Training farmer experimenters

In West Africa, World Neighbors gives training in small-scale experimentation to farmers recruited by their communities. The farmers learn how to do simple variety trials and how to test techniques of intercropping, plant spacing, pest management, tillage, soil conservation etc. Functional literacy training is given so that they can accurately measure and record relevant data, such as rainfall and planting dates (Gubbels 1988).

In the MASIPAG (farmer-scientist partnership for rural development) programme in the Philippines, 44 NGOs and scientists at the University of the Philippines at Los Baños are, among other activities, helping field-based technicians in PTD projects provide training to farmer representatives in local rice hybridisation and selection techniques (Medina 1988).

Farmers' manuals and audiovisuals

As mentioned under 'visits to secondary sites', audiovisuals can play an important role in spreading the experimentation process. To enhance the self-management capacity of the farmers, PTD practitioners encourage them to document their experiences and make these accessible for other villages and areas. PTD practitioners have both reinforced the use of folk means of communication and trained farmer representatives in producing and using modern mass media.

Field workshops

PTD practitioners can also train other potential members (local key persons or outside facilitators) of PTD teams in managing and implementing PTD processes. Organising field workshops can be an effective way to begin this process of team replication. Inclusion of farmer experimenters in the workshop as trainers of researchers and extension workers has proved to be a very effective way of reorienting (deschooling) them for the appropriate role performance in PTD. Such workshops can also help sensitise senior officials and generate required institutional support.

8.6 Keeping up the process: embedding local technology development

Developing agricultural self-management capacity involves more than initiating a particular process or introducing a set of skills and methods. Its aim is to leave communities with ongoing capacity to implement an effective and reliable PTD process. PTD teams thus have to be concerned with organisational development and the creation of other favourable conditions for ongoing experimentation and development of sustainable agroecological systems. This will include such activities as:

● assisting PTD groups (and other organisational elements that evolve in the course of the implementation) to consolidate;

● strengthening consolidation of inter-village cooperation, e.g. by stimulating linking up with existing, or developing new, farmers' organisations at area level;

● consolidating the institutional support for local PTD processes by promoting farmers' participation in formulating and assessing formal research programmes, providing training possibilities for staff of those institutions, promoting policy-level support and integration of support to PTD in area-development strategies and institutional mandates;

● developing locally manageable systems for monitoring the experimentation and diffusion process and its impact on the agroecological system and the livelihood of the communities involved.

The expected outcome of this activity is:

● consolidated community networks/organisations for agricultural self-management and a more supportive institutional environment;

**Box 8.17
Colombian farmers'
manuals**

Assisted by agricultural extension
staff of DIAR (an integrated
rural development programme),
the villagers of El Tigre, Choco
Region, Colombia, developed
improvements in the construction of the traditional *azotea*,
which normally consisted of an
old canoe put on poles to protect it against the regular floods,
and which served as a garden
mainly for medicinal herbs and
spices. They also experimented
with cultivation of a limited
number of vegetable crops.

The group members met
regularly to discuss progress and
problems. At the end of the
second growing season, they
reviewed the process they had
gone through from problem
diagnosis to consumption and
storage of products. The discussions were tape-recorded and a
local artist, together with the
farmers, made drawings to illustrate each step.

This resulted in a simple manual
in which El Tigre farmers explain
to other villages in the area why
and how they conducted their
azotea experiment. The manual
was distributed to the neighbouring villages. El Tigre farmers
used the manual to explain the
experimental process to visiting
farmers and to train them in
construction and management
of the *azotea* and related husbandry and processing practices.

In other villages in the area,
small groups of farmers organised
themselves around other focal
interests. One group of women in
Tagachi developed an improved
system for raising chickens
along the same lines as the
farmers in El Tigre. To share
their experiences with other
farmers, this group produced a
wall newspaper about the process
and results of their experiment,
including *coplas* (traditional form
of poem/song) and drawings. The
wall newspaper was copied
several times and hung at
central places in neighbouring
villages (Mazo 1986).

In an Indio village, Embera, the
participants in a farmer seminar
illustrated the results in drawings. The project staff made the
drawings into a series of slides
which was used by representatives of OREWA (a regional
Indio organisation) to start a
similar process in other areas
(Cardono & Orozco 1987).

- documented and operationalised PTD approach and resource
materials;
- ensured relevant services and input supply.

Activities oriented towards improving the sustainability of the process
of technology development must be incorporated into the PTD process
from the very beginning. Several of these activities have been mentioned
in previous sections. However, during implementation, the outside
facilitators will have to give more attention to the consolidation of what
has been set in motion. Some additional methods are the following.

Organisational development

Experience suggests that more permanent organisational structures
emerge from the organisational elements that evolve in the course of
implementation (e.g. innovator workshops, design and testing groups,
networks of farmer experimenters/trainers, and the participation of
local institutions, such as tree nurseries, schools and local research
stations). We have seen the example of the women of Aramachay in
Peru, who developed from *ad hoc* participants in a research activity
into members of their own Agricultural Production Committees (see
Box 8.5).

Members of already existing organisations, such as cooperatives or
village assemblies, might also develop new functions as PTD practitioners. It does not seem important whether one community organisation assumes a leading role or whether a number of organisations in the
community pursue agricultural self-management, as long as the diversity
of technological needs in the community is adequately represented.

The PTD team can facilitate consolidation of emerging organisational structures in the villages by:

- helping the groups develop adequate leadership and appropriate joint decision-making and evaluation procedures;

- encouraging the experimenting groups to function increasingly independently of the outside facilitators and to maintain all relations with third parties themselves;

- helping groups from different villages to gather regularly to discuss results, exchange views on next year's programme and plan together outward/upward-directed actions;

- linking the newly emerged organisational structures in the villages with existing farmer organisations at higher levels and/or with a supportive NGO;

- institutionalising the linkage between farmer experimenters and formal research, e.g. to arrange for farmer experimenters or other community representatives to become permanent members of research programme reviews and evaluation procedures at local research stations; the PTD team may also promote that research institutes use 'expert panels', i.e. groups that function as more or less permanent expert representatives with respect to a particular problem area (Norman et al. 1989), with the roles of focusing the research agenda on a specific topic or problem, testing experimental designs, implementing, observing and evaluating, and reporting results back to the community and network with similar panels at neighbouring sites in their own and neighbouring communities;

Box 8.18
Linking farmers to other farmers and researchers

Two years after the last outside facilitator of World Neighbors left Bassar, Togo, 12 village communities were continuing to meet annually to analyse and evaluate the previous season's experiments and to schedule the research for the coming season. At the meetings, they also chose a limited number of delegates to make the rounds of various agricultural development programmes and research stations, actively seeking new ideas and technologies, to report back to the community (Gubbels 1988).

In Eastern Indonesia a number of NGOs involved in rural extension have set up a joint consulting service, consisting of researchers with a 'PTD attitude' who maintain close links with governmental researchers and scientists and provide assistance to local farmer groups and organisations (Vel et al. 1989).

In Tabacundo, Ecuador, the radio school service organised a weekly programme called 'Mensaje Campesino' (the peasants' message), which is produced by farmers for farmers, with the help of 'radio auxiliaries': volunteers are given brief training in how to operate a cassette recorder. The farmers recorded material of their own choice. This simple initiative has encouraged the communities to increase exchanges amongst groups and communities. Twenty villages have formed local associations since the beginning of the programme (O'Sullivan-Ryan & Kaplun 1979).

In Pasto, Peru, the Agricultural Research Institute hired 5 minutes weekly from a local commercial station for a programme called 'Minga' (the local name for the traditional form of group labour on a rotational basis). Staff of the on-farm research and development programme collected messages and questions from farmer groups. Within a few months, Minga was the most popular programme of the region and intensified farmer-to-farmer exchange and influenced strongly the orientation and activities of the programme personnel (P.Engel, pers. comm.).

- encouraging institutions collaborating in the support network to send representatives to attend annual research reviews conducted by local communities, so that roles and responsibilities can be coordinated and resources allocated for the coming season;

- strengthening linkages between the experimenting groups and mass communication media, such as radio and newspapers oriented towards popular participation and adult education and using local languages.

Documenting the experimentation

Good documentation of what is happening in the field, the results (successes and failures) of farmer experimentation, the monitoring of adoption and diffusion processes and rates, and the impact on the livelihood of the farmers and the sustainability of the agroecological system is essential for strengthening the PTD process. The following methods may be applied.

Filing. Files that document the experimentation process at household and village level are an important asset for the local technology development process: they are the mirror of what is going on in the field and in people's minds. The PTD team may keep files per topic including information on the reasoning behind the experiment, its design and organisation, the working hypotheses, the results of measurements and farmers' evaluations, recommendations based on experiences (adaptation/further testing, rejection, wider diffusion, required supporting activities, alternative/additional ideas to be tested). The periodically up-dated files document the practical experiences of the farmers in developing indigenous technology or trying out new ideas coming from outside. It documents what people thought, what they tried and how it turned out. The files form a rich source of ideas and practical information for farmers from other areas, as well as for researchers and other new outside supporters of the community (Scheuermeier 1988).

Developing resource materials. Replication of the PTD process can be facilitated and accelerated by developing and providing resource

The PTD process can be stimulated by appropriate resource materials, such as this poster of rice – fish – duck integration used in workshops in Thailand. *(Media Center for Development)*

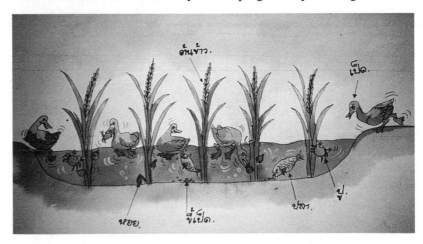

materials. There are three main sets of potential users: communities seeking to develop their own agricultural self-management capacity; PTD practitioners; planners and policy makers. Some useful materials may be developed in the course of implementing a PTD process, including training materials. As shown in previous sections, farmers can contribute greatly to the development of resource materials. It has also been found useful to involve other members of the support network, e.g. school teachers and field extension workers, in the tasks of recording what was done and preparing materials adapted to the different needs of the three main sets of users. Alternatively, the PTD team might want to draw on the specialist skills of a development resource organisation (e.g. GRAAP), university, college or NGO for help in preparing resource materials.

Monitoring impacts

The focus on mobilising local resources suggests that the PTD process will not depend to any great extent on the price and availability of external resources, such as chemical pesticides nor primarily on the skills of government scientists and extension staff working to fulfil the objectives of centrally-determined programmes. Nonetheless, the possibility remains that the community might select options that fulfil short-term goals to the detriment of longer-term agroecosystem stability and sustainability. Agroecosystems may be threatened by innovation, either directly through damage to the natural resource base and the environment or indirectly through inequitable spread of benefits and access to resources. These possibilities cannot be completely avoided in a participatory process, but PTD teams can introduce monitoring methods (FAO 1988b) which draw attention to potential areas of concern and conflicting goals, and support actions which conserve, replenish and regenerate renewable resources.

Farmer groups can and should participate in monitoring technological innovations and assessing their impact on agroecosystem properties. Some of the methods used to analyse the agroecological system and screen potential innovations can also be applied to assess the impact of certain technological innovations in a qualitative way, e.g. participatory diagnosis and systems diagrams.

Appendices

Appendix A   Some promising LEISA techniques and practices

Farming techniques encompass all human activities on a farm that aim at enhancing agricultural production. These activities can involve skilful management of farm resources, assets, inputs and/or outputs. They combine human knowledge, insights and skills with technical means and are mainly oriented towards managing the physical and biological components and processes of the farm. Examples are specific ways of ploughing, manuring, weeding or caring for animals. Farming practices involve numerous interrelated farming techniques. Examples are farm-specific ways of combining plants and animals and of managing soil and water. A technology system involves the whole complex of techniques used in a farm system.

A LEISA technology system is a combination of deliberately chosen techniques oriented towards sustainability. LEISA techniques have productive, protective, reproductive and/or social functions, and complement each other. Under the complex and diverse conditions of LEIA farmers, techniques are farm-specific and depend on the local availability of skills, assets and inputs. In view of the often high variability of climatic conditions, an important characteristic of LEISA systems is flexibility in the choice of techniques.

Promising LEISA techniques and practices are those which LEIA farmers in specific areas have found to be effective in making their farm systems more sustainable. Promising LEISA techniques are not blueprints. Therefore, PTD deliberately directed towards developing LEISA is necessary to select appropriate techniques and to adapt them to the specific conditions of individual farm systems.

Here, various promising LEISA techniques and practices are presented. Many are traditionally used somewhere in the tropics. Some have recently been improved by farmers or scientists; others are relatively new. Their effectiveness is demonstrated by cases, and information is given as to where practical documentation about them can be found. In the publications listed under the corresponding topic in Appendix C, information can also be found about additional techniques which have the potential to make farming more sustainable.

Appendix A1   Soil and nutrient management

The following techniques of managing soil and nutrients can enhance sustainability of farming by increasing the organic matter content of the soil and promoting soil life. They also contribute to nutrient recycling by increasing and balancing nutrient reserves and making nutrients available for plant growth.

Manure handling and improvement

Although farmers generally do not dispute the importance of manure to improve soils, many farmers use it primarily for other purposes or handle it inefficiently. This means that a valuable, locally available resource is not used to its fullest advantage for agriculture. Improved collection, composting, storage and transport of dung and urine can reduce nutrient losses. The nutrient content of manure is directly related to the species, sex and age of the animals and the quality and quantity of the feed and bedding material. The quality and quantity of manure can be improved by:

- choosing appropriate animal species (consumers of roughages, consumers of quality feed) and adjusting animal numbers to the available feed resources;

- improving feed by balancing the protein/energy content (e.g. by feeding concentrates, increasing the quantity of legumes in the ration or treating crop residues with urea) and providing good bedding material and housing for livestock.

Manure may be managed less than optimally for many reasons. If there are large distances between the places where dung, feed and bedding material are collected, the places where manure is stored and the fields where it is applied, manure handling will require much time and energy. Under LEIA conditions, this is normally human labour,

---

Box A1
Traditional use of nutrient inputs in Bhutan

The traditional farming systems of mountainous Bhutan integrate cropping, livestock-keeping and use of forest products. Because soil fertility is low, the farmers must rely on external inputs of plant nutrients. Through continuous use of nutrients from forest, grassland and biological nitrogen fixation, they have been able to sustain their agricultural production without the use of artificial fertilisers.

Livestock which graze forests, grassland and fallow fields by day and deposit manure in overnight enclosures are the main agents for collecting and transporting plant nutrients. More nutrients are added to the animal excreta through the lavish use of litter (leaves, ferns or needles) from the forest. This bedding allows most of the nutrients in the urine to be retained, and is highly valued by the farmers. Individuals have recorded rights to certain forest tracts to collect litter, typically of blue pine and oak species.

As the enclosures are often far from the fields, carrying manure – usually the task of women – is a major labour input in cropping. Given the transportation problems and since phosphorus is the primary limiting element in the shifting cultivation system at higher altitudes, manure is sometimes burned and only the ash is applied. The farmers thus lose most of the nitrogen and organic carbon, but the quantity to be transported is reduced substantially. Animals are also tethered on cropland overnight to reduce labour for transporting manure.

Fuelwood for cooking is collected from the forests, and the ash is used mainly to fertilise kitchen gardens. Partly decomposed leaf material from broadleaf forests is widely used as a source of organic matter and plant nutrients for fruit trees.

Nitrogen inputs are mostly through biological fixation by a number of native legumes, such as the wild vetch species which are often found growing together with wheat, barley, maize and millets. Soybean, *Phaseolus* and *Vigna* species are widely used as intercrops with maize and millet. The nitrogen requirements of improved farming systems in Bhutan can largely be covered by biological fixation through legumes grown for seed, fodder or green manure. However, these systems will require moderate inputs of phosphorus fertilisers (Roder 1990).
Contact: Agricultural Research Centre Yusipang, PO Box 212, Thimpu, Bhutan.

perhaps with the aid of animals for transport. The financial means or labour to invest in better housing or better feeding of livestock may be scarce. Also cultural taboos can make handling of manure unacceptable. Nevertheless, there are numerous examples of manure management practices by LEIA farmers which could form the basis for developing improved techniques (see Box A1).

Nutrient cycling within a farm system can also be improved through the use of manure from small animals, such as poultry. By making simple night-housing for chickens that run freely by day, smallholders can collect manure, which can then be applied on small plots. In central America, smallholders are also using the manure from leaf-cutter ants and the manure and blood from bats as fertiliser (Bunch, pers. comm.).

Composting

Composting is the breakdown of organic material by micro-organisms and soil fauna to give a humus end product called compost. It is an important technique for recycling organic waste (weeds, crop residues, waste from postharvest processing, dung, nightsoil, urine etc.) and for improving the quality and quantity of organic fertiliser. Compost is a slow-release organic fertiliser which stimulates soil life and improves soil structure. It also has beneficial effects on the resistance of plants to pests and diseases.

Normally, composting is done in heaps. In dry areas or dry periods, it can also be done in a shaded pit. Good quality raw materials and proper handling are decisive for the quality of the compost obtained. By mixing in mineral additives, e.g. rock dust, rock phosphate, urea fertiliser or chalk, the nutrient content of the compost can be improved (see Box A8).

Constraints to composting may include the availability and quality of raw materials, transport, labour and water, and cultural taboos. Composting is mainly done in connection with gardening, if labour is

---

Box A2
Composting in a semiarid environment

Sessions of 2 – 3 days are organised by ENDA in response to requests by Senegalese farmers' groups for training in the use of compost. After consultation with the farmers, technical personnel from research institutes are invited to make presentations, which are supported by a slide show produced by ENDA in the Wolof language. Sometimes, farmers with practical experience in composting techniques give hands-on demonstrations to their fellow farmers. After

training, farmers return to their villages and pass on the knowledge gained to others. In 1985, ENDA published an illustrated brochure on composting in both French and Wolof. Farmers who know how to read one of these languages can follow its directions for making a compost heap. The most common substances used for making compost are dung, peanut oil, fish scraps and miscellaneous plant matter.

Sahel farmers have generally not been inclined to adopt composting for several reasons, of which the most important are:

● Compostible matter is not

easily available. In much of Sahel, dung and vegetative matter is scarce. Plants grow quickly in the short wet season (2 – 3 months) but dry out just as quickly in the dry season and disappear after grazing and trampling by animals. The dung is often burned as fuel.
● Digging a hole in which to bury the compost is hard work.
● Stirring and turning over compost is repugnant to peasants, for whom compost and dung are the same thing (Thiam 1987).
Contact: ENDA-PRONAT, BP 3370, Dakar, Senegal.

cheap and water availability is not a problem. Some of the difficulties encountered in a semiarid climate are illustrated by ENDA-PRONAT, an organisation which, among other things, has been promoting composting to improve crop yields and soil productivity in Senegal for several years (Box A2).

Alternatives to heap or pit composting are direct application of the raw organic material as mulch or worked into the soil, e.g. by ploughing or by termites (see Figure A1); *in situ* composting, e.g. worked into ridges (see Figure A2) or planted trenches.

Other important techniques of organic waste management are biogas production and composting of night-soil. Proper handling of night-soil is also important for improving the hygiene and health conditions of farm households and villages.

Figure A1
Water harvesting in soil depressions in which manure is deposited to attract termites and cereal is sown, practised in Yatenga region, Burkina Faso. (*Source:* Wright 1984, in Pacey & Cullis 1986)

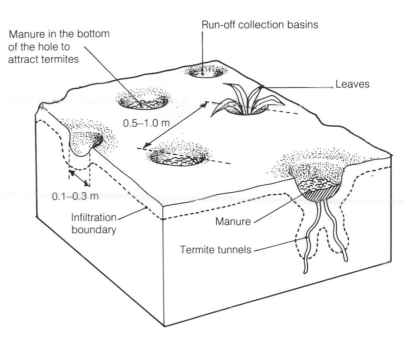

Figure A2
Working vegetation into ridges to provide organic manure. (After Fresco 1984)

Green manuring

Trees, shrubs, cover crops, grain legumes, grasses, weeds, ferns and algae provide green manure, an inexpensive source of organic fertiliser to build up or maintain soil organic matter and fertility. Green manure crops can contribute 30 – 60 kg N per ha annually (Greenland 1986) to the subsequent crop. The cumulative effects of continued use of green manures are important, not only in terms of nitrogen supply but also with regard to soil organic matter and other elements such as phosphate and micro-elements which are mobilised, concentrated in the topsoil and made available for plant growth.

Deep-rooted green manure crops in a rotation can help recover nutrients leached to the subsoil. Under high rainfall conditions, especially at the start of the wet season, permanent deep-rooted systems, as in some trees, are needed for recycling. Most food crops have shallow roots, which develop too slowly to intercept the mineralisation flush when the soil is first wetted. Some leguminous cover crops, such as *Centrosema, Pueraria* and *Crotalaria*, also appear to be able to develop deep root systems on acid soils in the humid tropics.

Particularly where land becomes scarce and fallow periods must be shortened, persistent weeds such as *Imperata* species proliferate, creating grasslands that are difficult to recover for cropping (Srivastava 1986). Shifting cultivators who have been obliged to shorten or abandon the fallow period often try to suppress weeds by using cover crops, e.g. yam beans *(Pachyrrhizus erosus)* in Southeast Asia or *Mucuna utilis* in West Africa (Akobundu 1983). Rapid establishment of a cover crop is crucial to its potential to suppress weeds. Akobundu and Poku (1984) observed that, within 19 weeks, *M. utilis* could completely cover plots infested with *Imperata cylindrica*. If a cover crop is being introduced into a crop rotation, good potential to suppress weeds is a key to farmer acceptance, as beneficial effects from weed control can be observed immediately, while effects of improved nutrient supply may occur only in the longer run (van der Heide & Hairiah 1989).

Forms of green manuring. Green manure crops can be planted in different combinations and configurations in time and space:

- improved fallow, i.e. replacing natural fallow vegetation with green manure crops to speed up regeneration of soil fertility and permit permanent cultivation (see Box A3); these green manures may be left to grow for one or several years, or only during the dry season;

- alley cropping, a form of simultaneous fallow in which quickly-growing trees, shrubs (usually legumes) or grasses are planted in rows and are regularly cut back; the prunings are used as mulch or worked into the soil in the alleys between the rows;

- integration of trees into cropland, as is found in several traditional farming systems, e.g. in West Africa *(Faidherbia albida)* and in Costa Rica, where tree legumes (usually *Erythrina poeppigiana*) growing among the crops are regularly cut for mulch material to maintain soil fertility in plots of coffee and other crops (Russo & Budowski 1986);

- relay fallowing by sowing bush legumes among the food crops after these have established and, in the dry season, using the cut green

Box A3
Experiences with improved fallow

In experiments in southern Nigeria, one-season fallowing with *Crotalaria juncea* (sunnhemp) led to increases of 12% in the organic carbon content in the soil, 4.3 kg/ha in available phosphate and 45.3 kg/ha in exchangeable potassium, corresponding to the increases in a 4-year natural fallow (Agboola 1980).

Experiments with *Pueraria phaseoloides* (tropical kudzu) on an ultisol in the Peruvian Amazon produced nutrient gains per ha per year of 59 kg N, 14 kg P, 66 kg K, 53 kg Ca, 28 kg Mg, 113 g Cu and 283 g Zn after 14 months' fallow and burning the plants on site. Preliminary results suggest that soil regeneration during a one-year kudzu fallow equals or surpasses that of a 15–20 year forest fallow (NCSU 1980).

Grasses can also improve soil fertility, e.g. *Cynodon nlemfuensis* sown on impoverished soils that had been cultivated for many years was able to mobilise 109 kg N, 22 kg P, 156 kg K and 33 kg Ca per hectare and year, and to increase the organic matter content of the soil considerably (Juo & Lal 1977).

Results with *Azadirachta indica* (neem) reveal the importance of trees in long-term fallow. On red, acid, sandy soils in northwest Nigeria, neem increased the pH of the A-horizon (0–15 cm) from 5.4 (natural fallow) to 6.8; the organic carbon content rose from 0.12% in the control to 0.57% under the tree, the concentration of exchangeable cations from 0.39 meq in the natural fallow to 2.40 meq/100 g soil, and base saturation from 20 to 98% (Radwanski & Wickens 1981). Inhibition of nitrification by neem seeds could also play an important role.

Even more interesting are the direct increases in yield after sown fallow. In Malawi, a 3-year fallow with *Cajanus cajan* led to an increase in maize yield of about 1 t/ha (from 2.07 to 3.21 t/ha) as compared with bush fallow (Kwatpata 1984). Agboola (1980) found that one-season fallows of *Cajanus cajan*, *Calopogonium mucunoides* and *Vigna radiata* increased the yields of subsequent maize crops by 10-30%.

In Nyabisindu, Rwanda, comparison of the different components to maintain soil fertility revealed that a 10-month legume fallow with *Tephrosia vogelii*, *Crotalaria* spp and *Cajanus cajan* more than quadrupled maize yield as compared with the control, doubled it as compared with stable manure (15 t/ha/year) and did not significantly differ from an application of 120 kg N, 100 kg P and 100 kg K per ha. The highest yield was obtained by combining improved fallow with application of farmyard manure (Prinz 1986).
Contact: Dieter Prinz, Institute for Irrigation, Kaiserstrasse 12, D-7500 Karlsruhe, Germany.

biomass as mulch or working it into the soil; examples are *Tephrosia vogelii* in Cameroon (Prinz 1986), *Sesbania rostrata* in Southeast Asia (IRRI 1988) and *Mucuna pruriens* in Honduras (Box A5);

- live mulching, in which the rows of food crops are sown into a low but dense cover crop of grasses or legumes, e.g. *Centrosema pubescens, Pueraria phaseoloides, Arachis prostrata*; strips of the cover crop are removed by hand or killed by herbicides when the food crops are to be sown, thus reducing soil tillage operations to zero (see mulching);

- shaded green manures (in fruit orchards, coffee plots, multistorey kitchen gardens etc.);

- azolla and blue-green algae.

Green manuring is particularly important for humid areas. In semiarid areas, where green manure and field crops compete for water, positive experience with green manure is still limited to traditional trees such as *F. albida* and *Acacia senegal*. However, the possibility that new options emerge, e.g. cover crops combined with water harvesting, cannot be excluded (see tied ridging, Box A11).

Table A1   Fallow plants for improving soil fertility

Calliandra calothyrsus	*Phaseolus aureus*
Calopogonium mucunoides	*Phaseolus lunatus*
Canavalia ensiformis	*Phaseolus mungo*
Canavalia gladiata	*Psophocarpus tetragonolobus*
Centrosema pubescens	*Pueraria phaseoloides*
Crotalaria juncea	*Sesbania bispinosa*
Crotalaria lanceolata	*Sesbania cannabina*
Crotalaria ochroleuca	*Sesbania grandiflora*
Crotalaria retusa	*Sesbania rostrata*
Desmodium ascendens	*Sesbania sesban*
Desmodium hetrophyllum	*Stylosanthes guianensis*
Desmodium intortum	*Stylosanthes hamata*
Dolichos lablab	*Tephrosia bracteolata*
Gliricidia sepium	*Tephrosia candida*
Grevillea robusta	*Tephrosia vogelii*
Indigofera sp.	*Vigna aconitifolia*
Leucaena leucocephala	*Vigna radiata*
Mimosa invisa	*Vigna umbellata*
Mucuna pruriens	*Vigna unguiculata*
Mucuna utilis	

Source: Copijn (1985).

Choice of green manure species. Various nitrogen-fixing leguminous and nonleguminous species – particularly trees, creepers and bushes – can be used as green manures (see Table A1). Some guidelines for selecting green manure species for improving fallow land are given in Table A2. Using grain legumes for green manuring brings quick economic benefit but, as they tend to accumulate nutrients in the grain, which is then harvested, their positive effect on subsequent crop yields is usually low. Mixtures of green manure crops are often more successful than sole crops, as they are less susceptible to pest attacks (see Box A4) and combine different characteristics needed for improved fallow, such as quick soil cover and deep rooting.

As legume growth depends on the presence of suitable *Rhizobium* strains, inoculation may be necessary. Plant growth and organic N_2-binding can be hindered by water stress, unfavourable pH, lack of other nutrients (particularly P, Ca, Mo, Zn) and/or Mn toxicity. Applying mineral or organic fertilisers (including rock phosphate, lime and ashes) can help improve legume establishment (Prinz 1986).

Also species in the natural vegetation should be considered for improved fallow, particularly those that are protected by local farmers, e.g. *Acioa barterii, Chlorophora excelsa, Alchornea cordifolia, Anthonota macrophylla* and *Dialium guineense* in southern Nigeria (Radwanski & Wickens 1981, Obi & Tuley 1973). Also tropical grasses such as *Pennisetum purpureum, Panicum maximum* or *Tripsacum laxum* can produce much biomass and accumulate phosphorus and potassium more quickly than most legumes (Prinz 1986).

Table A2   Criteria for selection of plants for improved fallow

Criteria	Effects
High biomass production	Mobilisation of nutrients from soil into vegetation; suppression of weeds
Deep rooting system	Pumping up of weathered and/or leached nutrients from soil layers not occupied by roots of main crop
Fast initial growth	Quick soil cover for effective soil protection; suppression of weeds
More leaf than wood (low C/N ratio)	Easy decomposition of organic matter leading to enhanced availability of nutrients for succeeding crops; easy to handle during cutting and/or incorporation into the soil
Nitrogen fixing	Increase of nitrogen availability
Good affinity with mycorrhiza	Mobilisation of phosphorus leading to improved availability for crops
Efficient water use	Possibility to grow after main cropping season on residual soil moisture or with less rainfall
Nonhost for crop-related pests and diseases	Decrease in pest and disease populations
No rhizomes	Controllable growth
Easy and abundant seed formation	Propagation in farmers' fields
Useful 'by-products' (e.g. fodder, wood)	Integration of animal husbandry and forestry

Source: Copijn (1985).

Farmer acceptance of green manuring. If green manure crops are not associated with a direct increase in income, farmers are not likely to be interested in them. It is therefore important that green manuring raises the farmer's income not only indirectly by improving soil fertility but also directly, e.g. by yielding by-products of economic importance such as fuel, stakes for climbing plants, food, fodder and local medicines. All forms of sown fallow demand a great deal of labour. Even more important can be the point in time when this labour is needed. If this coincides with other farm activities that cannot be delayed, improved fallow is not likely to be accepted by the farmers (Prinz 1986). Where forms of alley cropping are practised, farmers often prefer to plant the green manure crops in a looser configuration than the recommended model.

Two leguminous plants which show great promise as green manure are velvet bean *(Mucuna pruriens)* and sunnhemp *(Crotalaria juncea, C. ochroleuca)*. Sunnhemp is also effective in controlling insects (Stoll 1986). An example of farmers' use of velvet bean in Honduras is given in Box A5, and farmers' experience with sunnhemp in Tanzania is described in Box A6.

The use of green manures to improve animal feed is dealt with under 'Integrated farm systems' in Section A5.

Box A4
Diversified alley cropping in northern Thailand

Attempts to promote alley cropping and soil conservation among farmers in northern Thailand who formerly practised swidden (slash-and-burn) agriculture were focused mainly on *Leucaena leucocephala*. This grows quickly, bears coppicing well and produces leaves with high protein content which can be used as both soil conditioner and fodder. However, the insect *Heterophylla incisa*, which devastated *Leucaena* stands in Fiji, the Philippines and Indonesia, has now arrived in Thailand.

Fortunately, a large number of other leguminous tree species can be used in alley cropping. In northern Thailand, we are diversifying the hedgerows with a mixture of leguminous trees in an attempt to interrupt the 'attack formation' of the insect. Our mixture includes *Azadirachta indica* (neem), *Cajanus cajan* (pigeon pea), *Calliandra calothyrsus, Crotalaria juncea* (sunnhemp), *Gliricidia sepium, Sesbania roxburghii* and *S. rostrata*, in addition to *Leucaena*.

Planting sole stands of *Leucaena* was a mistake, but the crucial element of the alley cropping technology – the basic planting and pruning design – has proven effective for erosion control and soil amelioration. Now is the time to diversify the benefits of the technology to include nutritional and economic ones. Perhaps in the not-so-distant future, Thai farmers will walk their hedgerows for a handful of *Leucaena* for the pig, a spray of *Gliricidia* flowers to mix with a *Cajanus* stew to go with their sticky rice, and a sprig of *Azadirachta* to keep the bugs out of the rice (Taylor 1987).

Contact: Farming Systems Research Institute, Northern Agricultural Development Centre, Chiang Mai 50 000, Thailand.

Box A5
Improving smallholder farming with velvet bean

Hundreds of farmers in small villages on the northern coast of Honduras are using velvet bean in association with maize. They are obtaining good results in terms of higher yields, erosion control, and reduction in weeding and land preparation costs. In this humid tropical region, temperatures (mean of 28°C) and precipitation (over 3000 mm/year) are high, and altitudes vary from sea level to high mountains. There are two cropping seasons: January – June and July – December. Most farmers grow only one maize crop per year, during the first season.

Farmers using velvet bean for the first time sow it 1 – 2 months after sowing maize. When the maize is harvested, its stalks are bent over and left on the fields. Velvet bean starts covering the stalks and soon takes over the field. By December, the large quantities of legume foliage (50 – 70 t/ha) begin to dry out and cover the soil with a layer up to 20 cm thick. The next maize crop is planted directly through this layer, which suppresses weeds and allows adequate establishment of the corn. In the second year, velvet bean seeds volunteer from the year before and the cycle continues with the sowing of new maize. The farmers obtain corn yields of 2700 – 3250 kg/ha, more than double the national average, without using chemical fertiliser.

The continuous use of legumes is bringing about changes in the entire farming system. For example, ploughing is being replaced by no-tillage. Migratory farming is slowly disappearing.

Farmers have found a simple and cheap way to make their land more productive.

Velvet bean has been fairly rapidly adopted in the region without promotion by private or government agencies. Because the farmers are familiar with growing maize, they quickly recognise the benefits of the innovation. They gain most of their income from maize. Low yield means low income, which was the situation before velvet bean was introduced. Moreover, using velvet bean has next to no financial cost. The seed is passed on from farmer to farmer. Cultivation practices for velvet bean fit into the normal farming practices for maize, resulting in more efficient use of available labour and resources (Milton 1989, Bunch 1990).

Contact: CIDICCO, Apdo 278-c, Tegucigalpa DC, Honduras.

**Box A6
Experience with sunnhemp
in Tanzania**

The origin of cultivating sunnhemp
(Crotalaria ochroleuca) in
Tanzania was to help weak
farmers suffering from the
effects of leprosy to control
weeds and other pests in their
crops. In the first year of maize
after a sole crop of sunnhemp,
no weeding at all may be
needed. In the second and third
year, one weeding is needed
instead of two or three. In a rice
crop after sunnhemp, only one
weeding is required. Sunnhemp
can also be sown as an intercrop, usually at first weeding.
No second weeding is required,
as sunnhemp suppresses the
weeds. Near Tabora, farmers
are successfully using sunnhemp
to fight the weed Striga.

The first effect of sunnhemp
noticed by the farmers is the
strong initial growth of the
crops: generally only half the
usual amount of fertiliser is
needed in the first year after
sunnhemp. If intercropping with
sunnhemp is practised, there is
no need for fallowing.

Sunnhemp attracts certain
insects which infest vegetables
(e.g. cabbage), coffee, citrus
trees, flowers etc. The insects
sometimes nearly destroy the
sunnhemp, but the plants usually
recover again, even developing
a sturdier growth. The insects
do not move back to the other
crops as long as sunnhemp
flowers are available. According
to the Utengule Tea and Coffee
Estates in Mbeya, all insects
moved from their 20 ha medicinal flowerbeds to the sunnhemp beds placed between
them. It is not yet known
whether sunnhemp has any
negative effect on crops or livestock, but sunnhemp seed
should not be stored in a room
where people are working, if
fresh air cannot enter easily.

Growing sunnhemp does not
work miracles and does not
exclude other means or methods
such as crop rotation. For a few
kinds of weed, herbicides may
still have to be applied. But
chemical fertilisers and herbicides
are only for cases where natural
means, like sunnhemp, have failed
to work (Rupper 1987).
Contact: Sunnhemp Seed Bank,
Box 1, Peramiho, Tanzania.

Use of mineral fertiliser

Many tropical soils are very poor in nutrients or have specific nutrient
deficits which constrain crop growth. Nutrients are removed with the
harvested parts of the plants. When increased yields are sought, external
inputs of nutrients are needed to replace these 'lost' nutrients and to
increase the stock of available nutrients. Like plants, soil microbes and
soil animals also need mineral nutrients. Mineral fertiliser normally
increases the availability of biomass for organic fertiliser and may
enhance soil life when applied moderately. Soil life, in turn, increases
the efficiency of mineral fertilisers.

Applying mineral fertilisers in low to moderate amounts and in
balanced combination with organic fertilisers and possibly also
micronutrients (e.g. by seed dressing) can greatly enhance soil balance,
nutrient availability and, hence, the level and sustainability of crop
production and crop health. The efficiency and recycling of fertiliser
can be further increased by controlling weeds, pests, diseases, erosion
and leaching; by rotating shallow- and deep-rooting crops; by
synchronising fertiliser application (e.g. split doses of nitrogen) and by
applying it below the soil surface and near the root zone.

Although external inputs of mineral fertilisers (artificial fertiliser,
bone meal, rock phosphate etc.) can greatly increase crop production,
they can also have harmful effects. Concentrated and continuous use
of easily soluble mineral fertiliser may disturb soil life and lead to
acidification, micronutrient depletion, soil degradation, poor crop
health and lower crop yields. For example, ammonium sulphate is a
very strong biocide which hinders nitrogen fixation and kills nematodes
and earthworms. Superphosphate has a negative effect on free-living

Box A7
Rock phosphate for mixed smallholdings in Sri Lanka

In tropical soils, nitrogen and/or phosphate are usually the nutrients most limiting to crop growth. Also subsistence-oriented smallholders in Sri Lanka have to sell some products to obtain cash. These products contain certain amounts of nutrients essential for plant growth. Nitrogen thus exported from the farm can be replenished by atmospheric nitrogen fixation in legumes grown on the farm, but other nutrients exported via the sold products have to be reintroduced through artificial fertilisers. These are expensive, especially when they have to be imported from abroad.

Sri Lanka has a large, but little used reserve of alkaline rock phosphates. Legumes have a special ability to mobilise and utilise rock phosphates. When phosphate is the primary limiting factor in integrated crop-livestock systems, applying rock phosphate to legumes can raise the quantity and quality of fodder, leading to better animal health and higher productivity. The phosphates in the forage derived from applied rock phosphate can be recovered in the manure of stall-kept animals. After biogas production, the residue can be used to improve the fertility of arable land. The organic manure not only supplies nutrients but also improves soil physical characteristics.

In crop-livestock systems involving alley cropping, *Leucaena* and *Gliricidia* are often used as woody leguminous species which fix atmospheric nitrogen. Thought could be given to growing less permanent leguminous forages, such as *Pueraria* and *Mucuna* species, as both are very vigorous growers capable of utilising sparsely soluble rock phosphates. With such nonwoody legumes, crop rotation could be practised as a means of controlling soil-borne diseases and improving crop health and, thus, crop productivity (van Diest 1988).

nitrogen-fixing bacteria, which may be favoured by 'mild' fertilisers such as Thomas slag, thermophosphate or bone-meal when added to stubble mulch or straw (Primavesi 1990). Among mineral fertilisers, calcium ammonium nitrate is less harmful than urea, which is still to be preferred over ammonium sulphate. This is because urea to a degree — and ammonium sulphate even more so — acidifies quickly, leading to high levels of toxic soluble aluminium in the soil (Smaling 1990).

Phosphate, which is often in short supply, is an important element for plant growth. Rock phosphate may be locally available (see Box A7). Mineralisation of rock phosphate can be accelerated in acid conditions. Farmers can enhance acidulation by composting rock phosphate together with manure and plant residues. Rock phosphate is best applied to legumes to enhance nitrogen fixation and to improve animal rations and, hence, manure quality.

Table A3   Effect of different phosphate sources on yield of pigeon pea

Treatment	Yield (100 kg/ha) Grain	Straw	P uptake* (kg/ha)
Control	14.7	51.0	4.37
Mussoorie rock	15.7	54.0	5.59
Compost alone	16.9	55.0	5.90
Superphosphate	16.4	56.5	6.67
P-enriched compost	18.3	61.5	7.12
Least significant difference	1.4	6.8	0.53

$(p \leqslant 0.05)$

*P was added at the rate of 17.3 kg/ha

Source: Mishra & Bangar (1986).

Box A8
Composting rock phosphate

The agronomic effectiveness of several types of rock phosphate is highest for acid soils with low P and Ca levels, but rock phosphate has not been an effective source of P in neutral or alkaline soils. Compared to water-soluble P fertilisers, rock phosphates are generally considered less effective, presumably because their limited solubility does not maintain a sufficiently high level of P in the soil solution.

A variety of micro-organisms, such as fungi, bacteria and actinomycetes, produce substances which can dissolve inorganic phosphates in the soil. Concentrated microbial activity during composting produces such dissolving substances. To test the potential of this microbial activity to solubise rock phosphate, plant material, cattle dung, old compost and rock phosphate were mixed in a ratio of 6:1:0.5:2.5 on a dry weight basis. The rock phosphate (Mussoorie rock) was a carbonate apatite of sedimentary origin, containing about 67% apatite. The mixed material was allowed to decompose for three months in a pit.

The water-soluble phosphorus decreased with the addition of rock phosphate to the composed material. However, the phosphorus soluble in 2% citric acid increased dramatically after 90 days and accounted for over 40% of the total phosphorus. The effect of the P-enriched compost on pigeon pea *(Cajanus cajan)* grown in neutral to slightly alkaline (pH 7.6 – 7.8) soils was superior to rock phosphate or compost alone and superior also to superphosphate (see Table A3). The phosphorus uptake was highest with the P-enriched compost. It is believed that the higher efficiency of P-enriched compost was due to the slower release of the phosphorus. With single superphosphate, although there was more water–soluble P, a large part became fixed upon addition to the soil.

Adding rock phosphate to compost is one method farmers might use to better utilise the limited phosphorus available to them. Responses will depend on the composition of the rock phosphates and soil types (Mishra & Bangar 1986).

Contact: Department of Microbiology, Haryana Agricultural University, Hissar 125004, India.

The choice of mineral fertilisers will depend on soil type and other variables. When available and affordable, soil testing by an independent soil analyst experienced in ecological farming can help farmers find the right kind and amounts of mineral and organic fertilisers. The availability and price of mineral fertilisers may limit their use. Appropriate and balanced levels of application can improve the economics of using mineral fertilisers.

Appendix A2

Managing flows of solar radiation, air and water

There is a great deal of overlap between techniques of microclimate management, water management and erosion control. Here, several techniques are discussed which, under certain conditions, contribute to creating favourable conditions for plant and animal life, conserving water and soil, and reducing climatic risks. Techniques that improve water availability to crops, especially in drought-prone areas, play an important role in increasing biomass production and water availability for humans and animals.

Mulching

Mulch can be defined as a shallow layer at the soil/air interface with properties that differ from the original soil surface layer (Stigter 1984). Mulching is an important technique for improving soil microclimate; enhancing soil life, structure and fertility; conserving soil moisture; reducing weed growth; preventing damage by impact from solar radiation and rainfall (erosion control); and reducing the need for tillage. Widely used traditional mulches include layers of dry grass; crop residues (straw, leaves etc.); fresh organic material from trees, bushes, grasses and weeds; household refuse and live plants (cover crops, green manures) (see Box 3.3).

The effects of mulch depend on its composition and colour, the amount applied, the timing of application and the rate at which the mulch decomposes. This rate depends, in turn, on the form and timing of application and the meteorological conditions in the soil and air. As many aspects are involved, considerable experimentation is required to find the best way to apply mulch.

Mulch application on the soil surface can replace seedbed preparation in zero-tillage systems. Seeds are sown with minimum soil disturbance by opening a small slit in the mulch layer or by punching holes in the soil for seed placement. Weeds are controlled chemically or manually. Benefits ascribed to zero tillage, in addition to water and soil conservation, include reduced labour and energy requirements, equal or higher crop yields, and greater net return as compared with conventional tillage. Under LEIA conditions, zero tillage is often recommended in combination with alley cropping (see 'Green manuring').

Zero tillage is most suitable for soils with low susceptibility to compaction and crusting, good internal drainage, high biological activity, friable consistency over a wide range of water contents and a coarse surface or a surface of self-mulching clay with high initial infiltration. It is less suitable for severely degraded soils or those that undergo severe hardening during the dry season. On such soils, tillage is required to create a favourable zone for water infiltration, crop establishment and· root penetration.

Possible constraints to mulching are insufficient availability of mulch material, unsuitable soils, soil compaction and pest problems (rodents, insects, fungi, persistent weeds). Further development of mulching/no-tillage systems is needed, particularly for those areas where conservation of soil and water is important to maintain the crop production capability of the land (Unger 1987).

Windbreaks

To improve the microclimate or decrease wind erosion, windbreaks can be formed by living hedges – narrow bands of closely planted woody species – generally planted around fields, gardens or farm compounds. Stone walls or scattered trees can serve a similar function. Besides influencing the microclimate, hedges can be useful in keeping animals out of fields and/or producing fruits, herbs, fodder, mulch, thatching material or fuel. They also play a role in balancing pest populations.

The improvement in microclimate as a result of windbreaks can lead to an increase in crop yield and may compensate for the land lost to planting hedges or scattered trees. The total biomass production (i.e. of both crops and hedges) may be raised considerably. Hedge management techniques will depend on the function they are meant to perform.

Constraints to planting windbreaks may be excessive competition for light and root space with the adjacent crop, and pest transmission from plants in the hedge to certain crops. Establishment may be difficult

Box A9
Windbreaks in the Maggia valley

In response to farmers' request for assistance in protecting their land from wind erosion, CARE has worked since 1975 with the Niger Forestry Service to plant windbreaks in the Maggia Valley. Across the width of the 1500 km² valley, windbreaks about 2 km long are planted 100 m apart. In recently established ones, neem *(Azadirachta indica)* is combined with *Acacia nilotica*, thus making the windbreaks two-tiered and more aerodynamic (the lower growing acacia on the windward side and the taller neem on the lee side) than those planted in the first years, when only neem trees were used. With support from the local communities, over 400 km of windbreaks have been planted.

In a survey made in 1985 of 420 valley residents in 17 villages, the respondents overwhelmingly noted an increase in crop yields in plots protected by windbreaks. Only 4% claimed a decrease. On account of shading effects, 17% of the area was removed from cultivation in the windbreak zone. Per unit of cultivated land, the farms inside the windbreaks yielded about 15% more grain. Overall biomass inside the windbreak zone increased by 68%. This is important since crop residues are used for fodder, fuel and thatching.

In effect, the land protected with windbreaks produced about the same yields as land outside the windbreak zone, and brought added benefits in the form of by-products and possibly also increased organic matter content in the soils. The windbreaks of the Maggia Valley show that wind erosion can be reduced at a specific site through tree planting. Satisfactory results can be achieved in a relatively short time, and the techniques are neither complicated nor inappropriate for the farming communities (Steinberg 1988).
Contact: CARE International, Agriculture and Natural Resources Program, 660 First Avenue, New York, NY 10016, USA.

because of long dry periods and free-roaming animals. Allowing natural regrowth of trees on the borders of fields or contour lines can be a viable alternative to deliberately planting trees.

Water harvesting

In rainfed agriculture, good management of water is of great importance. Rainfall may be too low or too high or too irregular, creating high risks of yield losses and/or unfavourable growth conditions or damage by erosion. Techniques are needed to conserve the available water and/or to guide excess water safely from the field. Where there is not enough rain to grow a crop or where rainfall is very irregular, water-harvesting techniques can be used to concentrate rain or flood water in such a way that it can be used for crop growth.

Water harvesting not only secures and increases crop production in regions where rainfall is normally insufficient; it can also serve to control soil erosion and to recharge aquifers tapped for irrigation. An additional benefit is the improvement in soil fertility. Silt, manure and other organic matter are 'harvested' together with the water. The soil profile stays moist longer, stimulating soil life and improving humus formation, nutrient availability and the soil's capacity to hold water. Water harvesting is a resource-enhancing technique which, due to synergetic effects, reaches its full potential when used in combination with other techniques, such as improved seed and application of organic and inorganic fertilisers.

A classic example of water and nutrient harvesting was in the Nile Valley before the Aswan Dam was built. The annual flooding of cropland permitted permanent agriculture in this arid region. After the construction of the dam, this was no longer possible. Now farmers have advantages of greater water availability, but great problems in keeping the land fertile. Small water-harvesting systems with an external catchment area function more or less like the traditional flooding system along the Nile. Water-harvesting systems with a 'within-field' catchment harvest less nutrients but control erosion on site.

Many variations of water harvesting are possible. Water-harvesting systems can be classified as follows (Reijntjes 1986b):

1 Systems with an external catchment area for collecting run-off water or flood water from small watersheds:

   i) Agricultural use, without any special arrangements, of natural depressions where run-off water or flood water is concentrated temporarily and water infiltration is relatively high (traditional in, for example, West and East Africa).

  ii) Simple techniques for water spreading and infiltration by means of low, permeable bunds (ridges) of stones, bundled sticks, crop residues or fences of living plants along contour lines (traditional in, for example, Mali).

 iii) Water pockets or pits: holes for seeding, collecting run-off and managing organic matter (*zai* in Burkina Faso, *covas* in Cape Verde, both traditional).

 iv) Half-circular or V-shaped ridges used mainly for tree planting and rangeland improvement (new).

  v) Water collection: graded bunds or furrows divert run-off from cropland, village land and wasteland to tanks located at a lower level; the water is used for supplementary irrigation in dry periods or as full irrigation (traditional in India; new in, for example, West Africa).

 vi) Run-off farming: run-off water from treated (e.g. by spraying chemicals or clearing gravel stones to increase run-off) or untreated catchment areas is diverted to lower-lying cropland (traditional in, for example, Israel and Tunisia, *meskat*, Figure A3).

Figure A3
Meskat water-harvesting system in Tunisia: run-off from the catchment area (*meskat*) infiltrates into the culti-vated terraces (*mankaa*). (After El Amami 1983)

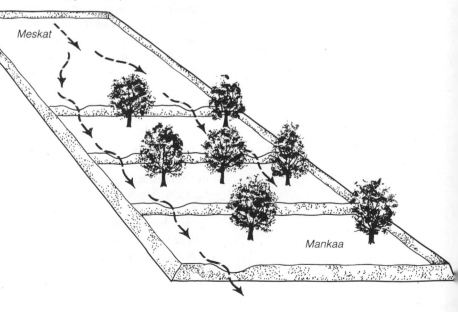

Figure A4
View and profile of *jessour* water-
harvesting system in Tunisia: run-off
water and sediments from the hills are
captured behind the dams (*tabia*) in the
valley bottom. (After El Amami 1983)

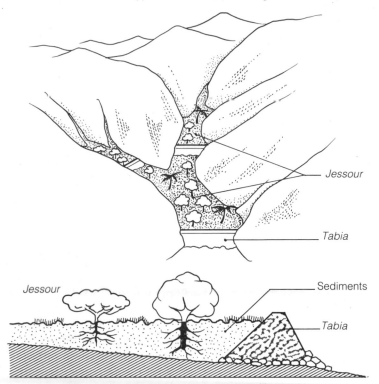

Box A10
Water harvesting by
nomads

Technically, it is possible to
harvest water for reafforestation
and rangeland improvement.
Especially for tree planting in
dry areas, water-harvesting —
ypes 1iv) and 3ii) in the
classification — can be very useful. However, the probability
that nomadic people will accept
these techniques is not high.

In Africa south of the Sahara,
water-harvesting projects for
impoverished nomads have not
been successful for several
reasons:

- cropping does not fit well into
  the nomadic strategy for survival, in which mobility plays
  an important role;
- working the soil is not highly
  esteemed among nomads;
- land use is often communal,
  which complicates issues of
  individual or group investment
  in rangeland improvement or
  tree growing and how the
  benefits will be shared;
- because of the high climatic
  variability, especially in drier
  areas with red soils, the reliability of water-harvesting
  systems is low.

Where simple water-harvesting
techniques for cropping are
already known by the local
people, the interest in and the
probability of success with
improved water-harvesting
techniques may be higher
(Reijntjes 1986a).

vii) Run-on farming: run-off water and silt from small watersheds
is captured by dams in seasonal stream beds or diverted to
cropland. In front of these dams, the silt forms terraces which
are used for farming. The infiltrated water makes cropping
possible (traditional in, for example, Israel and Tunisia, *jessour*,
Figure A4).

2 Systems for storage and agricultural use of flood water (flood water
farming): making use of the run-off concentrated by natural
watersheds in seasonal or permanent streams. The flood water is
diverted from its natural channel by dams or barrages and led to
the cropland where the water is kept impounded by earthen dams
around the fields. The infiltrated water is used for farming
(traditional in, for example, the Nile Delta before the Aswan Dam;
north India, *ahars* and *khadirs*; south Pakistan, *sailabas* and
kurkabas).

3 Systems with a 'within-field' catchment area called *'in situ'* water
harvesting or 'microcatchments':
   i) *Negarim:* run-off from a small plot (micro- or within-catchment)
   is captured at one side, where it infiltrates the soil and directly
   contributes to the available moisture in the rooted profile of
   an individual productive tree or shrub (traditional in, for
   example, Morocco; new in, for example, Israel);
   ii) Contour ridges or bunds: the same system as 3i), but instead
   of small plots, strips are used. Crops can be seeded in front
   of the bunds where water infiltration is concentrated (relatively
   new in, for example, India and Africa);

iii) Contour beds: the same system as 3ii), but the beds are W-shaped, with alternating wide and narrow ridges. The wide ridges serve as a catchment zone, the narrow ridges as a planting zone and the furrows can serve as drainage or irrigation channels. Mechanisation can be used (new in, for example, Brazil).

The choice of water-harvesting technique depends on numerous factors, such as climate and soil characteristics, availability of stones and labour, previous experience of farmers with water harvesting, the degree of social organisation and other socioeconomic factors.

Although water harvesting – as a resource-enhancing technique – is of particular importance in the semiarid zone, it can also be applied in the subhumid zone, for example, to gain maximum profit from the first rains for the production of early seedlings or an early crop.

The more complicated or extended water-harvesting systems (e.g. flood-water farming) involve many risks and require considerable labour to construct and maintain the structures. A high degree of social organisation is therefore needed. 'Within-field' systems, which are less risky and labour-intensive, may have a greater chance of being adopted, especially when sufficient (preferably organic) fertiliser is available to maintain soil fertility and, thus, allow continued productive use. As in the case of all improvements in farming systems, water harvesting must raise yields sufficiently to justify the investment (labour, capital, land) by the farmers.

Tied ridging

Tied ridges alternating with furrows can be constructed by inter-tying main ridges ploughed along the contour lines with smaller perpendicular cross ridges every few metres. The cross ridges are somewhat lower than the main ridges to prevent their erosion. Seeds or tubers are placed either near the top of the ridge to avoid waterlogging, or towards the bottom of the basin when moisture is limited. In some traditional systems, crop residues are left in the furrow and covered by the soil of the old ridge, thus forming the new ridge in the place of the old furrow and composting the crop residues on the spot.

Tied ridging has led to striking increases in crop yields on the alfisols of the Sudano-Sahelian tropics (Hulugalle 1989a). The bunds increase water infiltration, improve soil physical properties and decrease run-off and erosion.

Tied ridging can be used only where rainfall does not exceed the storage capacity of the furrows; otherwise, severe erosion may result. Tied ridging is more successful on coarser soils, mainly because waterlogging in very wet years could negate the results. Vertisols give better overall production with broad bed and furrow techniques (IBSRAM 1987, Jutzi et al. 1987).

Strip cropping

A valuable technology for farmers who have to grow their crops on slopes is strip cropping (also known as 'in-row tillage'), which can increase crop production and prevent soil erosion at the same time. The

Box A11
Tied ridges and legumes boost yields in Burkina Faso

Stylosanthes hamata is a potentially valuable forage crop for the Sudano – Sahelian region, but information on cereal/forage – legume cropping systems is sparse. In the predominantly cereal-based farming systems, stover is an important source of dry-season fodder. Nitrogen-fixing legumes undersown or intercropped with a cereal can improve stover quality as well as soil physical and chemical properties.

At the Kamboinse Research Station in Ouagadougou, a 3-year study was made of the effects of tied ridging and undersowing *S. hamata* on soil properties and maize yield. The experimental treatments included tied and open ridges, both of which were sown to either sole maize or maize undersown with *S. hamata*.

Profile water content was greater with tied than open ridges. With tied ridges, under-sowing with *S. hamata* resulted in drier soil profile in 2 of the 3 years, compared with sole maize. The furrows of tied-ridge plots were higher in clay content, soil organic matter, total cation exchange capacity and exchangeable Ca, Mg and K. Relative leaf water content of maize was increased and subsoil root densities were greater with tied ridges. The plots undersown with *S. hamata* had deeper root systems.

Tied ridging significantly increased grain and dry matter production of maize but did not affect that of *S. hamata*. Grain and dry matter yield of maize in tied-ridge plots was reduced by undersowing with *S. hamata* when drought occurred during reproductive and late vegetative growth. In open-ridge plots, dry matter and maize yield were not affected by the cropping system. Highest total dry matter production was observed in undersown plots with tied ridging (Hulugalle 1989b).
Contact: NR Hulugalle, 117/1 Pieris Avenue, Kalubowila, Dehiwela, Sri Lanka.

---

crop is sown in narrow, tilled rows along contours on a hillside. The strips of land between the rows, which are left untilled in natural grasses, slow the flow of rainwater down the slope and prevent it from washing away the topsoil. More water penetrates into the soil and provides moisture for the crop. The grass strips also provide a natural habitat for insects, many of which will prefer this to the crop. Organic matter and fertiliser are concentrated in furrows where the seed is planted.

By tilling only the strips to be planted, the farmer saves labour in comparison with cultivation of the entire field. In subsequent seasons, the rows are replanted and the strips of grass cut back periodically, but they are never dug up. The steps recommended by World Neighbors for in-row tillage are presented in Box A12.

Box A12
Farming by the in-row tillage method

1 Cut back the grass and dig rows about one and a half to two hands wide along the contours, determined by using, for example, an A-frame.
2 Break up large clods of soils and remove roots, stones and other obstacles in the dug rows.
3 Prune back the grass growing between the rows but do not remove the roots, as these 'tie' the soil together.
4 Make a furrow along the length of each row, using a hoe or other pointed implement.
5 Fill the furrow with organic matter, such as poultry manure and, if available, a small amount of chemical fertiliser. Cover this with soil and mix together.
6 Plant the seed with the appropriate spacing for that particular crop or crops and cover the seed with soil.

World Neighbors has prepared a film strip showing how this is done in Honduras (World Neighbors 1988).
Contact: World Neighbors, 5116 North Portland Avenue, Oklahoma City, OK 73112, USA.

Permeable contour-line barriers

Constructing permeable ridges of stones, stalks, branches, trunks or other organic material or planting hedges of grasses or shrubs/trees along contour lines at regular vertical intervals to conserve water and soil may bring considerable increases in yield. The ridges or hedges do not completely stop the run-off but slow it down and spread the water over the field, thus enhancing water infiltration and reducing soil erosion. Silt trapped on the higher side of the barrier forms natural terraces. Compared with impermeable earthen bunds, permeable contour-line barriers have the advantages that they spread rainwater more evenly over the field and present less risk of erosion when damaged. Water conservation can be improved still further by constructing ditches on the lower side of the barrier, as is traditionally practised in Mexico (Mountjoy & Gliessman 1988, see Figure A5).

An important advantage of using hedges of grasses or shrubs as contour-line barriers is that the land reserved for the barrier is productively used, yielding fodder, mulch, fuelwood etc. (see also 'Contour-line farming'). A particularly promising plant for vegetative barriers is vetiver grass (see Boxes A13 and A14).

Box A13
Moisture conservation with vetiver grass

A World Bank-supported project in India (Pilot Project for Watershed Development in Rainfed Areas) promotes a system of soil and moisture conservation based on stabilising soil with vegetative contour barriers of vetiver grass *(Vetiveria zizanioides)*. This is native to India and exhibits a wide range of adaptability: from over 2000 m in the Himalayas, where it is covered with snow in winter, to the deserts of Rajasthan, the swamps near Delhi and the wastelands of Andhra Pradesh.

In the wet tropics, vetiver hedges can be established in five months; in arid areas, it takes three seasons. Once established they completely stop sheet erosion (erosion of the top layer of soil). Rather than concentrating run-off water into streams and so making it more erosive, hedges slow run-off, spread it out and filter out the silt, while letting the water seep through the entire length of the hedge. Silt trapped behind the grass barrier spreads back across the field. Vetiver grass grows through the silt, forming a natural terrace over the years.

Vetiver-based moisture conservation systems can be found in Sri Lanka, Nigeria, Somalia, Indonesia, the Philippines, Burma, Thailand, Nepal, China and the southern USA. The system is cheap, replicable and sustainable. The farmer can do all the planting and maintenance without assistance. It costs less than 1/10 the cost of engineered soil conservation systems, and extends the cropping range. With engineered systems, arbitrary limits are set for growing food crops: 12% is the maximum 'safe' slope. With this system of contour ploughing and planting between the stabilising hedges, food crops have been produced safely on 100% (45°) slopes. Vast areas of land, hitherto classified as unstable, can now be safely used for production, so long as the hedges are maintained.

Some examples of positive experiences with vetiver in other parts of the tropics are:
- Vetiver has satisfactorily established on the almost 'bauxite' borrow pit of a large dam in the Kandy Hills, Sri Lanka, where it was planted using crowbars to make the planting holes.
- On the island of St Vincent in the West Indies, vetiver grass has been stopping erosion on slopes of up to 100% for over 50 years and, in some areas, has resulted in the build-up of natural terraces to a height of 4 m.
- In Trinidad, stabilising road embankments with vetiver prevents erosion on 100% slopes of 'scree', shale and red yellow podsolic soil in areas with 2000 mm rainfall (Greenfield 1988).

Contact: Vetiver Information Network, Attn. Mr RG Grimshaw, World Bank, 1818 H St. NW, Washington DC 20433, USA.

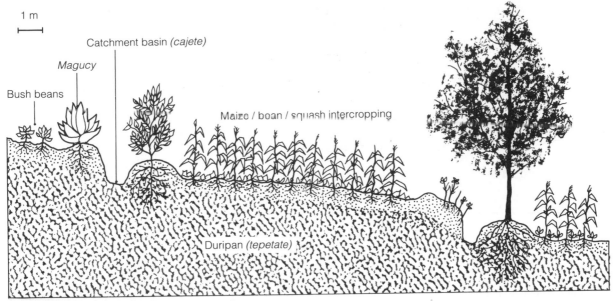

1 m

Bush beans

Magucy

Catchment basin *(cajete)*

Maize / bean / squash intercropping

Duripan *(tepetate)*

Figure A5 Cross-section of field surface in Tlaxcala, Mexico, showing terrace/catchment complex. (*Source:* Mountjoy & Gliessman 1988)

Box A14
Indian peasants have long used vetiver grass (khus)

As implementing officers of a khus-based vegetative barrier system for soil conservation, we were initially skeptical, as the technology had not been tested by researchers. We asked: What if the grass spreads like a weed? What if it gets diseased? What if it is browsed? Can it endure for many years?

Vetiver grass grows wild in many parts of Karnataka State. In July 1988, we came upon farmers in some villages of Gundlupet Taluka of Mysore District using khus for soil conservation. Even the old farmers (over 80 years) say they used it in their fields since they were young, just as their fathers did. Where irrigation and intensive land shaping were adopted, khus appeared less important for soil and water conservation but it is still used in the drylands. It has been planted in all vulnerable areas where rills would otherwise have formed.

Even on almost flat fields, some farmers plant khus to mark boundary lines, as it is a perennial plant. These lines have remained for several decades. The farmers also use khus to protect waste-weirs and to stabilise drop structures.

Farmers regard the fodder value of khus as an added merit. They said that 3 – 4 cuttings can be obtained at 45-day intervals, mainly during and shortly after the monsoon.

The farmers have developed their own ways of multiplying and propagating khus. On sloped land, they form small section bunds across the slope and plant 2 – 3 slips per rill 20 – 30 cm apart on the upstream side. In flat fields, the slips are simply planted in the plough furrow. In either case, they chop off the top of the plant and avoid planting inflorescence axles. Khus establishes well if planted after the first monsoon shower. Even without irrigation, hedges form in about a year. The slips for further planting are taken from 3-year-old bunds. When waste-weirs or drop structures are to

be treated, clumps of khus are taken and placed at appropriate locations.

During field visits, we noticed a sole case of diseased khus. The plants had been affected by *Ustilago raysiae*, a smut disease without serious consequence. None of the farmers regarded khus as a weed or as a host for pests and diseases. A few farmers in Tumkur District said that growing khus prevented the occurrence of striga, a root parasite.

Most scientists are unaware that khus has long been used by Indian farmers. The knowledge that these farmers have gained in dealing with khus-based soil conservation systems needs to be documented and the other uses of khus, e.g. for fodder, should be studied (Subramanya & Sastry 1990).
Contact: S Subramanya, Government of Karnataka, Vidhanasaudha Building, Bangalore 560 001; and KN Ranganatha Sastry, Visvesvaraya Centre, Dr Ambedkar Road, Bangalore 560 001, India.

Besides vetiver grass, many other grasses/shrubs or combinations of these, with similar growth characteristics, can be used. For soil conservation in the Himalayas, for example, indigenous shrubs that have their crowns (offsets) beneath the surface (so the shrubs will not be killed by browsing animals) are being transplanted into hedges. These shrubs act like vetiver grass, and supply essential fuelwood in a 3-year cycle. So far, more than 100 different indigenous plants for this purpose have been identified (Greenfield 1988).

Water ponds

Small ponds or dams are traditionally used in many parts of the world to store water for livestock and domestic purposes as well as for

---

Box A15
Wetland use in a small farm system in Zimbabwe

Agricultural development in Zimbabwe was based on assumptions that shifting cultivation had been the precolonial system and that fertilising upland was the only way to maintain production levels under today's population pressure. However, using mineral fertiliser in dry areas is barely economic and cattle numbers cannot be kept high enough to provide sufficient manure.

In Zvishavane (ca. 500 mm annual rainfall) local farmers knew that intensive wetland farming had been practised in precolonial times but had been largely abandoned for various reasons. Those interested in reintensifying their farming by developing wetlands attended meetings initiated by two British students. The main advantage of using wetlands is that it stabilises production by maintaining more constant water supply. Also, the wetlands are usually rich in clay, and the soils tend to have more organic matter than the sandy uplands. However, if heavy rains fall, land preparation is difficult. Water-logging can damage crops, and occasional surface flows can cause soil erosion.

One of the farmers, Phiri Maseko, had been experimenting

for many years with controlling hydrology in a patch of wetland within his 4-ha plot. He had dug a 30 x 15 m pond on the upper margin of the wetland, where water seeps out of the ground on encountering layers of clay. This pond catches and stores water from heavy rains, and prevents the damaging effects of surplus water in the fields. With a series of wells, water is then circulated within the wetland to irrigate crops during dry spells in the wet season and during the dry season. A third component of managing hydrology was identifying areas where water would flow during storms and turning the areas over to Kikuyu grass.

Using this water supply, an intensive integrated system was developed. Within banana groves established in areas too wet for cultivation, bees are kept. Fish are farmed in the pond, and reeds suitable for making baskets are grown for sale. Several hundred fruit trees, mainly citrus and mango, are planted within and around the fields, where they make use of the water supply and provide a valuable cash income as well as a food crop. In the diverse cropping system, all the major cereals, including rice, are intercropped with legumes. Vegetables are also grown, especially in the dry season. The cattle are fed abundant crop residues, banana

leaves and grass cuttings.

Water management in this semiarid area thus permits a level of intensification which improves the economic position of the dryland holding and reduces risks. The excitement of farmers visiting the farm is rooted also in their own awareness of the historical significance of wetland use. Many had wanted to make similar innovations but felt constrained by the administrative banning of wetland use, in effect since colonisation.

The meetings of farmer groups were directed to understanding the hydrology and soils of local wetlands. A model was thus developed for discussing potential land-use innovations. Working in groups permits cross-checking of information and ideas. Farmers often disputed issues with each other. This deepened their understanding of the complexity of the system. They learned from each other and gained confidence in their own abilities. The meetings revealed new development options provided by local knowledge, stimulated farmers to implement their own projects and promoted diffusion of ideas (Maseko et al. 1988).
Contact: Phiri Maseko, Zvishavane Water Resources Project (ZWRP), PO Box 118, Zvishavane, Zimbabwe.

Box A16
Upland soil conservation in the Philippines

Social and biological scientists from IRRI and the Philippine Department of Agriculture are conducting research in the municipality of Claveria. The area receives ca. 2200 mm annual rainfall and has moderately well-drained, clay, acidic soils. Over half the land has over 15% slope. Farms are less than 3 ha in size. Upland rice – fallow rotation and cassava are the main cropping patterns at 400 – 500 m elevation; maize – maize and maize – fallow rotation at 500 – 650 m; and maize, vegetables and perennials at 650 – 950 m.

A random sample of 55 farmers were interviewed using open-ended guide questions formulated after exploratory research had revealed some key issues facing farmers. Fields were visited, and farmers discussed crop and resource management. Possibilities of limiting soil erosion, by planting perennials, digging diversion canals and planting bananas in gullies to trap soil and eroded nutrients, were mentioned by about half the farmers, but only 20% had actually made diversion canals along upper plot borders. Several had left weedy strips between plots. Weeds and crop residues were commonly piled across and in erosion channels, and bananas planted in them to benefit from deposited nutrients, to decrease water flow rate and destructiveness, and to trap soil. While a few landowners had planted perennials on slopes, none had planted trees in strips or on upper slopes. Farmers with sloping land reported declining yields due to nutrient losses from soil erosion and depletion from continuous cropping.

World Neighbors, working with upland farmers on the nearby island of Cebu, had introduced contour ditches, bunds and legume tree/grass hedgerows which led to natural terraces between strips. This was identified as a possible way to control soil erosion in Clavaria. A group of farmers were trained by farmers in Cebu in using the A-frame to lay out contours, bunding-ditching to establish strips, and hedgerow planting of fodder grass (Napier grass, *Pennisetum purpureum*) and legume trees (*Gliricidia sepium*). On their return, members of the group established contour bunds, ditches and hedgerows on their parcels. Local sources of *P. purpureum* and *G. sepium* were located and planted. The group established almost 7000 m of contour bunds on 10 parcels of 0.8 ha mean size. Depending on soil compaction and ground cover, each worker planted 17 – 57 m of strips per day.

The farmers tested establishment methods and various hedgerow species and combinations including other grasses (e.g. *Panicum maximum*), trees (indigenous *Cassia spectabilis*), cash perennials (coffee, cacao, fruit), wild sunflower (*Helianthus annuus*) and even weeds that are otherwise serious crop pests e.g. *Digitaria longiflora* and *Paspalum conjugatum*. They observed that terracing was faster and more effective using weed-grasses, compared to trees, on the bunds. They saw that *G. sepium* on the down-slope side of the bund suffered from competition with Napier grass planted above. A farmer's test of trees placed above the grass looks promising; this is now being tested by researchers.

G. sepium cuttings suffered more than seedlings from termites and poor rooting. Farmers and researchers are working together on a small nursery to supply *G. sepium* and *C. spectabilis* seedlings.

Some researchers are impatient for farmers to adopt all improved procedures immediately, such as modifying bunding methods to save labour, growing legume trees and incorporating biomass, removing grassy weeds, and cutting back Napier grass to reduce competition with *G. sepium* and the alley crop. But the farmers want to proceed gradually and modify contoured areas in their own step-by-step manner.

Farmer-to-farmer technology dissemination continued. Cebuano farmer-trainers visited Claveria after the contours and a first crop had been established. Fields and crops were observed and discussed, and further ideas about hedgerow spacing, bund species and spacings, ditching and soil traps, cash crops, rice cultivars, local markets, a live-stock (goat) component and group vs individual labour were shared.

In summary, our approach to on-farm research and technology transfer consisted of understanding farmer practice, perception and technical knowledge; using this and farmer experiments to help identify technical possibilities; back-up research on alternatives that integrate farmer and researcher concerns and contributions; and technology transfer from adaptor-adopters to farmers who want solutions to problems addressed by the technologies (Fujisaka 1989).
Contact: Sam Fujisaka, IRRI, PO Box 933, Manila, Philippines.

small-scale irrigation. Strategically applied water from these ponds not only improves water availability where rainfall in unreliable but also permits extension of the growing season for earlier planting or a second crop. In drought-prone areas, water ponds can play an important role in securing and increasing farm productivity (see Box A15).

Farmer-centred development of water and soil conservation techniques

Numerous low-cost technical options for water and soil conservation adapted to LEIA conditions are available to farmers. The techniques mentioned in Section A1 also help in some way to regulate water infiltration and decrease erosion. The choice of plants and animals used by the farmer and the planting configurations in terms of time and space also affect water infiltration and soil erosion. Finding the combination of techniques which fits each specific farm situation is a task that can be accomplished only by the farmers themselves. PTD approaches can help farmers find ways of making better use of the available rainfall and protecting their farm system against soil erosion. An example of such a PTD process is presented in Box A16.

Appendix A3

Pest and disease management

In ecologically-sound agriculture, ways of controlling a pest outbreak are limited. As natural pesticides are less effective than chemical ones, ecological pest management is based on understanding the life cycles of pests and preventing the build-up of excessive pest populations. The Agriculture, Man and Ecology (AME) Programme has developed a checklist which can be used in a PTD process, by farmers and extensionists together, as a tool for developing ecological techniques of pest management. This checklist (Table A4) consists of three parts:

1 **Problem definition**. First it must be understood what the problem is, what pest causes it and what its life cycle is. In most cases, this information can be obtained from conventional agricultural literature. Information can be collected on damage done, the period when the crop is sensitive to the pest, and the economic threshold level.

2 **Preventive measures**. Then the cultivation practices are analysed, step by step. With the knowledge about the pest gained during problem definition, one can see whether cultivation practices can be changed to limit build-up of the pest population. The primary sources of information are farmers cultivating the affected crop. Further information can be found in traditional farming practices and in literature on modern Integrated Pest Management (IPM) and ecological farming.

3 **Control measures**. Specific control measures are considered, starting from the weaker ones, and going towards the stronger ones which have more environmental side-effects. In normal years, preventive measures should suffice for pest management. Where problems with

Table A4   Tackling a pest problem: a checklist

Problem definition
Affected crops
Type of pest and damage done
Economic threshold
Pest's active period, population development

Preventive measures
Crop rotation
Multiple cropping systems
● companion planting
● repellent planting
● trap planting
Conservational biological control
● natural enemies and damage done
● population development of natural enemy
Tillage
Fertiliser application
Water
Sowing and planting
● timing
● depth
● plant density
● synchronisation
Harvesting
Storage
Sanitation
● removing crop residues
● removing other host plants
Resistant varieties

Control measures
Biological control
● classic
● augmentative
Mechanical control
● hand-picking
● barriers
● scaring
Natural pesticides
● attractant
● repellent
● antifeedant
● crop-strengthening extracts
Rock powders
Botanical pesticides
● application method, timing and effect
● (phyto-)toxity
● effects on pest's natural enemies
Synthetic pesticides
● see botanical pesticides
Others

Source: van der Werf (1985).

Box A17
Dealing with maize stemborers

Pest
Maize stemborer (*Busseola fusca*), found in Africa.
Spotted stemborer (*Chilo partellus*), found in East Africa and Indian subcontinent.
Pink stemborer (*Sesamia calamites, Ostrinia furnacalis*), found in Africa.

Problem definition
1 Affected crops: Maize, sorghum, millet, rice and sugarcane.
2 Type of pest and damage done: Moth lays eggs on young seedlings (3 – 4 weeks); eggs hatch into caterpillars, which feed on young leaves. Caterpillars enter deeper into the plant as they grow and finally feed on the growing centre. A wilting top is a clear sign of infestation.
3 Economic threshold: 5 – 10%.
4 Pest's active period: Most active when crop is knee-high, 1 month after sowing. More active after a dry period and during full moon.

Preventive measures
1 Crop rotation: Can break the pest's life cycle.
2 Multiple cropping systems:
   ● Companion planting. Inter-cropping maize with legumes (e.g. field lablab, peanut, cowpea) or potatoes will reduce egg-laying by moths, as they recognise maize by the stalk silhouette, which is less clear with an intercrop. The change in microclimate can increase the population of *Trichogramma chilonis*, which is parasitic on the eggs of *Ostrinia furnacalis*, and the activity of predatory spider (*Lycosa* spp) against stemborer.
   ● Trap cropping. A susceptible maize hybrid can be planted along the field edges as a trap crop, and can be used as fodder.
3 Conservational biological control: see 2.
4 Tillage: Creating the best
5 Fertiliser: conditions will
6 Water: stimulate healthy
7 Sowing and planting: and strong growth.
   ● Time. As stemborer moths are more active during full moon and attracted by a crop 3 – 4 weeks old, planting is best done between two full moons. Early planting is also important.
   ● Plant density. Increased plant density will increase percentage infestation but sowing 10 – 15% more than the desired final population can be beneficial. The diseased plants can be taken out to be used as fodder (or burned/buried).
   ● Synchronisation. Simultaneous planting in a large area reduces the time period when mass reproduction of the pest can coincide with the susceptible stage of the host plant.
8 Harvesting: –
9 Storage: –
10 Sanitation: As the pest survives in stubble, all crop residues should be cleared by burning or grazing to destroy the larvae or pupae. Some farmers who grow maize every year make ridges on top of crop residues. Covering crop residues with a thick layer of soil hampers development of the stemborer moth. The new crop is sown between the ridges. Thus, maize is sown on the same spot only every 2nd year.
11 Other host plants: Sorghum and millet.
12 Resistant varieties: Maize varieties resistant to *Busseola fusca* and *Sesamia calamites* are known.

Control measures
1 Biological control: –
2 Mechanical control: Hand picking and killing caterpillars. Barrier of wood ash or soil in the funnel of the growing plant. Moth can be attracted by light traps, which should be set up before eggs are laid (usually before maize flowering). Moths are attracted best during full moon from 7:00 to 10:00 p.m.
3 Natural pesticide: Cow urine is collected and allowed to stand for 2 weeks, then diluted with 6 parts water and sprayed.
4 Rock powders: –
5 Botanical pesticides:
   ● Neem (*Azadirachta indica*). 250 – 500 g dried seeds are pitted and broken. A cloth is filled and hung overnight over a 10 l container. Extract is applied during the evening.
   ● Infusion of tobacco poured into the funnel of the growing plant one month after seeding and repeated one week later will kill most of the young caterpillars.
   ● *Ryania speciosa*. Dried roots, leaves or stems are pounded and diluted with talcum or clay dust (40% plant material, 60% dust). This is applied in the funnel of the growing crop at 44 kg/ha 1 week after the moths start to fly (use light traps for indication of activity). This is effective for 5 – 9 days against maize stemborer as well as maize smut.
   ● *Tephrosia vogelii*. Infusion made from the leaves is an effective insecticide against stemborer. Effective treatment depends on timing. The most vulnerable time in the stemborer's life cycle is when the young larvae rest on the leaves and in the sheaths (before boring into the stalk). It is important to ensure that enough spray enters the sheaths (van der Werf 1985).

pests require control measures regularly, cultivation of that crop under the given agroclimatic conditions should be reconsidered (van der Werf 1985).

In AME, three cases of pest management (maize stemborers, rice stemborers and brown planthoppers) have been worked out. The first one is presented in Box A17 as an example to illustrate that there are many low-cost technical options for pest and disease management. Some of these options are discussed in this section.

Intercropping

Intercropping generally appears to have positive effects in terms of reducing the occurrence of insect pests, diseases and weeds.

Effects on insects. A survey of research involving 198 kinds of insects revealed that intercropping led to a reduction in insect population in 53% of all insect/crop combinations, an increase in 18% of the combinations, no effect in 9% and a variable response in 20%. When more than one of the intercrops is a host plant for the insect, the chance of decrease in number was much smaller (Andow 1983).

Natural enemies of insect pests tend to be more abundant in intercrops than in monocrops, as they find better conditions, such as better temporal and spatial distribution of nectar and pollen sources and more microhabitats for their special requirements (e.g. ground cover for nocturnal insect predators). It is also more likely that insect pests land on hosts growing in dense or pure stands. An insect has more difficulty locating host plants when these are less concentrated, as the visual and chemical stimuli from the host are not so strong and aromatic odours of other plants can disrupt host-finding behaviour, e.g. grass borders repel leafhoppers in beans. A host may be protected from insects by other overlapping plants (e.g. standing rice stubble can camouflage bean seedlings and protect from bean fly). Intercropping can also interfere with the population development and survival of insect pests, because companion crops block their dispersal across the field and it may be more difficult for them to locate and remain in microhabitats which favour their rapid development (Altieri 1987). Such mechanisms can be used to change or improve existing intercropping systems so as to reduce insect populations.

Effect on diseases. With few exceptions, intercrops suffer less diseases than pure crops with the same overall density (Steiner 1984). Because the density of susceptible plants is lower, the amount of potential inoculum is also lower. The nonsusceptible crop acts as a barrier to spread of the disease. However, the overall higher plant density of intercrops as compared with sole cropping induces a change in microclimate (usually a rise in relative humidity), making it more favourable for fungus and bacterial diseases. The spatial arrangement of the combination must then be adjusted to minimise the negative impact of these diseases. Provided the pathogen/host/environment relationship is understood in a cropping system, the use of intercropping shows great possibilities for reducing disease (Altieri & Liebman 1986).

Effect on weeds. The time spent weeding is often the main factor that limits farm size. Most crop combinations suppress weed growth by providing an early ground cover, either because plant population is high or because a component crop (e.g. melon) grows quickly. In many intercropping systems, only one weeding is required as compared with 2 – 3 in sole crops. The weeding is often combined with planting another intercrop, thus reducing the time spent solely on weeding (Steiner 1984).

Trap and decoy crops

Pests can be strongly attracted by certain plants. When these are sown in the field or alongside it, insects will gather on them and can thus be easily controlled. For example, the cotton bollworm *(Heliothis zea)* prefers maize and lays its eggs on cotton only when this grows as a sole crop. When a few rows of maize are sown in the cotton field, the eggs will be laid on these plants and can then be destroyed. The example of using sunnhemp as a trap crop was already given in Box A6.

This mechanism may also occur with intercropping. One crop may attract insects but not be damaged by them, while few insect pests may be found on the crop that is susceptible to them. Some weeds may also play the role of trap crop. However, using the trap-crop system can be risky, as insects from outside the system may also be attracted.

Trap crops can also be used in nematode control. The trap crops are then host plants for the nematodes and attract them, but are harvested or destroyed before the nematodes have multiplied. For example, in pineapple plantations, tomatoes are planted and destroyed again before root-knot nematodes can produce eggs.

When a crop activates nematodes but is not a suitable host plant for them, it is called a decoy crop. Activated nematode larvae find no food and die. Examples of decoy crops can be found in Table A5.

Constructed traps

Various kinds of trap can be constructed to catch insects, rodents or other creatures which threaten crops or livestock. The most common is the light trap, set up to catch night-flying insects. The highest catches are obtained by ultraviolet lamps, but electric and kerosene lamps can

Table A5   Decoy crops used to reduce nematode populations

Crop	Nematode species	Decoy crop
Eggplant	*Meloidogyne incognita,* *M. javanica*	*Tagetes patula,* *Sesamum orientale*
Tomato	*M. incognita* *Pratylenchus alleri*	*T. patula,* castorbean, chrysanthemum
Okra	*Meloidogyne* sp.	*T. patula*
Soybean	*Rotenlenchus* sp. *Pratylenchus* sp.	*T. minuto,* *Crotalaria spectabilis*
Various	*Pratylenchus penetrans*	*T. patula,* hybrids of *Gaillardia* and *Hellenium*
Various	*P. neglectus*	Oil radish
Oats	*Heterodera avenae*	Maize
Various	*Trichodorus* sp.	Asparagus

Source: Altieri & Liebmann (1986).

also be used. A wooden framework is anchored firmly in the ground, a light source is mounted on the frame and a shallow bowl of water is placed immediately underneath, e.g. a kerosene lantern can be suspended from a simple bamboo tripod. A couple of spoonfuls of oil can be added to the water so that, when moths fall in after being attracted by the light, the oil sticks to their wings and they cannot fly away again. The lamp can be set at different heights above the ground. If the light intensity is varied, a greater variety of insects or more insects will be attracted. The trap must be sturdily built so that it cannot be blown down or knocked over by animals.

Species which can be caught with light traps include: American bollworm *(Heliothis armigera)*, army worms *(Spodoptera* spp*)*, brown rice planthopper *(Nilaparvata lugena)*, cutworms *(Agrotis* spp*)*, green rice leafhopper *(Nephotettix nigropictus)*, rice gall midge *(Orseolia oryzae)* and tomato hornworm *(Manduca quinquemaculata)*. The optimum timing for placing the traps depends on the life cycle of the insect and the development stage of the crop. The best time is soon after the adult moths have emerged but before they have laid their eggs (Stoll 1986).

Light traps can also provide useful information about population dynamics and development of pest populations (for monitoring purposes).

Box A18
A trap, a fish poison and pest control

Moru and Morokodo farmers trying to grow citrus trees in Southern Sudan encountered the problem that trees weakened by water stress during the long dry season were easily attacked and killed by termites. An investigation of traditional pest control revealed no example (except wood ash) of using local preparations on crops. However, there were several examples of using poisons for fishing and hunting, some of which appeared suitable to adapt for use in farming.

One of the most promising of these was used by the Morokodo for termite control. To catch game, the Morokodo make a technically sophisticated spring trap from a wooden bow sprung with a piece of hide. It has numerous wooden and leather parts which are vulnerable to termite damage. To repel termites, the Morokodo use the bulb and fruit of *Catunaregan spinosa*, which they pound together with water into a concentrated pulp and pour over the traps.

The Moru do not have this kind of trap, so they do not know about using *C. spinosa* against termites. However, they use it as a poison to stun fish in pools and slow-moving streams. They therefore know where to find the plants, and how to collect and prepare them.

Having 'discovered' this local preparation, we tried it on some trees in our fruit tree nursery. Some senior staff, who were familiar with using the poison for traps, prepared it and poured it around the foot of randomly selected trees in the early dry season. The preparation proved to be very successful in both preventing and curing termite attacks and, thus, helping the trees survive the dry season.

Since water stress renders the trees most vulnerable to termites, the need for pest control is seasonal. *C. spinosa* forms fruit in the dry season, when it is most needed. Also at this time, there are no heavy rains which could wash the mixture away from the tree base. If a young tree is treated for the first couple of years, it is then normally strong enough to resist the termites.

The recommendation to the farmers that resulted was this: ''Make a mixture of bulbs and fruits as you would to protect a game trap (for the Morokodo) or to use as a fish poison (for the Moru) and pour it over the base of the fruit trees in the first couple of dry seasons of the tree's life, to protect the tree from termites''. This was a recommendation firmly rooted in the indigenous knowledge, but adapted for use in a different situation (Sharland 1990).
Contact: Roger Sharland, 1 York Road, Reading RG1 8DX, UK.

Box A19
Protecting fruit trees from stray animals

A common reason why farmers do not plant useful trees on bunds or in fields is because the young trees are grazed by stray cattle. But there are simple ways of preventing this. For example, droppings of cattle, goats and/or sheep can be collected and mixed in an equal amount of water. The slurry is swabbed on the young tree's leaves with one tablespoon of soap powder per litre of slurry. This repels stray animals from the leaves.

Another method of protecting young trees is to erect a 'wall' of morning glory (*Ipomoea cornea*) plants. This is done by taking 5-foot long stalks about one inch in diameter and planting them around the trees. The cuttings root quickly and create a solid fortress (Rao 1991).
Contact: Prabhanjan Rao, Panchajanya 70, New Santosh Nagar Colony, Hyderabad 500 659, India.

Repellents

The opposite of attracting pests is to repel them, as illustrated by the case of using plant preparations to keep termites away from fruit trees (Box A18). Repellents can also be used to keep animal pests away from crops and economic trees (Box A19).

Biological control

In biological control, pests are suppressed by their natural enemies, such as birds, spiders, mites, fungi, bacteria, viruses or plants (e.g. cover crops to control weeds). In traditional farming systems, structures and practices have evolved that enhance biological pest control (see Box 5.9), although farmers may not be conscious of this effect. On the basis of recent research in pest ecology, ways of using natural enemies for pest control are being developed.

Biological control can be cheap, efficient, selective and ecologically sound, but there are also disadvantages. Biological control, like chemical control, is sensitive to external factors. It does not always work fast enough to avoid damage. Many different factors may be important for biological control to be successful (e.g. climate, type of crop, size of the plot, intensity of breeding measures). Therefore, heavy demands are made on research and extension services when biological control programmes are introduced (Meerman et al. 1989).

However, some biological control measures can be applied by smallholders without outside support, e.g. the conservation approach of promoting natural enemies already in the area. A varied agroecosystem is favourable for natural enemies, as it offers alternative food sources and hiding possibilities (Altieri & Letourneau 1982). This variation may be created by intercropping, allowing certain weeds to grow, planting or retaining hedges and patches with wild vegetation, cover-cropping, mulching and composting (for natural enemies that live in the soil). However, creating variation can also favour some pests. Depending on the specific ecosystem and combination of pests, the best way of stimulating the occurrence of natural enemies must be sought.

A specific form of biological pest control involves the use of bacteria, fungi, protozoa and viruses. Mixed with water or another fluid, these micro-organisms are applied like chemical pesticides. Microbial pesticides have many advantages over chemicals: their effect is generally selective, they do not damage useful organisms and man, and resistance to them is not likely to develop. However, they must be applied more often, because they disintegrate more rapidly.

The best known and most widely used micro-organism is the bacterium *Bacillus thuringiensis (B.t.)*. During sporulation it produces a protein which is toxic to most caterpillars. Symptoms of poisoning show only minutes after a caterpillar starts eating a sprayed plant. The pesticide is sold to control caterpillar pests in various crops. *B.t.* is not known to be harmful to aquatic organisms, wildlife, livestock, beneficial insects (including bees) or man. Until now, no susceptible species have become resistant to *B.t.* (Gips 1987, Kumar 1984).

The microbial pesticides produced by Western companies with the aid of high-technology equipment are usually too expensive for smallholders. However, there are also examples of pesticide production

Box A20
Bakulo virus

Some caterpillars are highly susceptible to the bakulo virus disease. To control them, the disease can be promoted with a spray prepared from virus-infected caterpillars found in the field. They can easily be recognised by the practised eye. For example, when first infected, caterpillars of cabbage loopers (*Trichoplusia ni*) become white and inactive and tend to move to the upper parts of the plant, where they hang from the underside of the leaves. In the last stages, they turn black and erupt into liquid. Of the infected whitish caterpillars, 8 – 10 are homogenised in a mixer with water and diluted to a volume sufficient to spray over 0.5 ha. After 3 – 4 days, the caterpillars become infected and die. It is crucial to apply the spray as early as possible so that the caterpillars cannot cause great damage before they die. This requires close observation in endangered regions and immediate action.

The bakulo virus has been used successfully to control the American bollworm (*Heliothis armigera*), armyworm (*Spodoptera litura*), cabbage looper (*Trichoplusia ni*), corn earworm (*Heliothis zea*), cabbage worm (*Pieris rapae*), large cabbage worm (*Pieris brassicae*) and lucerne butterfly (*Colias philodice*) (Stoll 1986).

Contact: Gaby Stoll, Bühlengasse 2, D-7609 Hohberg 1, Germany.

in developing countries using cheap raw materials and simple, locally available equipment. To control the sap-sucking bug *Mahanarva posticata* in Brazil, thousands of hectares are treated with *Metarnizium anisopilae*, a fungus produced locally in simple laboratories, using bottles of sterilised rice as a substrate. To control the stemboring caterpillar *Ostrinia nubilalis* in China, 0.4 million hectares of maize are treated with locally prepared *Bauveria bassiana*, a fungus produced on steamed but not sterilised rice, bran or corn stalks (Prior 1989).

Use of pesticides

Where curative pesticides or drugs may still be necessary, they can sometimes be prepared from local plants or other materials (e.g. urine, ashes, minerals). In this way, dependency on chemical pesticides can be avoided or reduced. Plant-derived and natural pesticides and drugs are still poorly researched. Little is known about their effectiveness and their side effects on health and environment (which may not be less than in the case of chemical pesticides). Most work on 'alternative' pesticides is being done by NGOs and farmers and has seldom been scientifically evaluated (see ILEIA 1989). Where suitable plants and/or knowledge about these practices are not available, the use of chemical pesticides or drugs may be necessary.

Chemical pesticides. Generally speaking, chemical pesticides work quickly and effectively. They can be used in diverse ecological conditions. However, their use can also have serious disadvantages, not only for the farm but also for its surroundings. Moreover, they are often expensive and difficult to obtain. As most smallholders have little choice in pesticides, they cannot choose ones that work selectively. This constrains the application of IPM and exposes the farmers to toxic substances. For many of the toxic pesticides, there are alternative control methods: Gips (1987) lists alternatives for 12 of the most dangerous pesticides. However, in some emergencies, such as severe locust attacks, chemicals may have to be used.

To avoid damaging effects when choosing and applying pesticides, many precautions must be taken. However, labels of commercial pesticides are often incomprehensible to smallholders (or completely missing after repackaging), and economic damage thresholds are seldom known. Much work must still be done in the fields of regulation, infrastructure, research and extension before smallholders can use pesticides in an effective and responsible way.

Plant-derived pesticides. Using plant-derived pesticides is an age-old control method. Unfortunately, this knowledge is rapidly being lost, particularly where chemical pesticides have been introduced. Using natural pesticides is cheap and does not make farmers dependent on external inputs. However, some natural pesticides can also be toxic for humans and animals. Extreme care should therefore be taken in applying toxic materials, such as nicotine from tobacco. Numerous plants have defensive or lethal effects on vertebrates, insects, mites, nematodes, fungi or bacteria. Active components can be extracted from various plant parts: seeds, leaves, stems, fruit or roots. There are also many ways of extracting these components and dispersing them over

Box A21
Neem for pest control

Scientific interest in neem is based on its traditional use for pest control. In many Indo-Pakistani households, for example, neem leaves are placed between folds of clothing and pages of books to protect against termites, silverfish and other household pests. For the same reason, neem leaves are mixed with stored wheat, rice, maize, sorghum and other grains.

The use of neem cake (residue after oil is extracted from neem seed) for controlling nematodes and other soil-borne pests is increasing, even among affluent growers of cardamom, citrus, vegetables and sugarcane in India. In the Cardamom Hills of Kerala State, an informal marketing mechanism has evolved, and 3000 tons of this product is now sold annually – mostly by pesticide dealers. The product is transported from Karnataka and Andhra Pradesh (more than 250 km away). Farmers are willing to invest Rs 100/ha or more, as neem cake appears to be the most effective product for nematode control (Ahmed 1990).

Contact: Saleem Ahmed, Resource Systems Institute, 1777 East-West Road, Honolulu, Hawaii 96848, USA.

the crop (Stoll 1986). The lethal or defensive effect is usually the result not of one component, but of a mixture of active components, so that the chance for the pest organism to develop resistance is small (Stoll 1986, Project Consult 1986).

Box A22
Natural crop protection in Cameroon

Seminars with farmers and extensionists were organised by the Rural Training Centre Mfonta and INADES Formation Bamenda, and much information on indigenous methods and plants traditionally used to protect crops was gathered. Based on this information and literature, trials were designed and conducted at Mfonta and by farmers on their fields.

Good results against leaf-eating caterpillars, aphids and garden bugs were obtained with the following sprays:

- **Jimson weed** (*Datura stramonium*). One bucket (1 kg) of fresh leaves, stems, flowers and seeds are shredded and soaked in 10 l water, together with 2 tablespoons kerosene and soap (one handful or 50 g). It is left to stand for at least 3 hours (or overnight) before it is sifted and sprayed.

- **Castor oil** (*Ricinus communis*). Four glasses (0.5 kg) of shelled or five glasses (0.75 kg) of fresh unshelled seeds are mashed or ground and heated for 10 minutes with 2 l water, 2 teaspoons kerosene and some soap. This decoction is sifted and diluted to 10 l water and sprayed immediately.

- **God's tobacco** (*Lobelia columnaris*). One bucket (1 kg) fresh leaves are shredded and soaked in 10 l water with 2 teaspoons kerosene. The leaves are squeezed and washed like bitterleaf. It is left to stand for at least 3 hours (or overnight) and is then sifted and sprayed.

- **Papaya** (*Carica papaya*). One bucket (1 kg) fresh leaves are shredded and soaked in 10 l water, together with 2 tablespoons kerosene and some soap. It is left standing for at least 2 hours (or overnight), sifted and sprayed.

The effectiveness of these botanical sprays depends on the plant material used. The insecticidal power varies from plant to plant with growing conditions, plant age and variety, e.g., the castor oil variety with red stems is more effective than the one with green stems. It is also important that the sprays are not exposed to sunlight.

Striking results were obtained in maize storage. Using cow dung ash, maize can be stored free of weevils for more than 9 months. After harvesting and shelling, the maize is mixed well with the ash: about 5 kg (15 l bucket) per 100 kg maize (one big jute bag). Before the maize is eaten, the ash is removed by sifting and the maize is washed. When used for animal feed, the maize and ash are ground together (Schrimpf & Dziekan 1989).

Contact: Berthold Schrimpf, In der Gasse 1, D-3410 Bühle, Germany; or Irene Dziekan, INADES, PO Box 252, Bamenda, Cameroon.

Box A23
Innovative health care of water buffalo in the Philippines

Two farmer-innovators shared with us their experience in caring for the health of their water buffalo. In Pamahawan, Leyte, Ramon Pelisco is a tenant farmer who has one water buffalo with calf, some chickens and pigs. He uses the herbaceous plant *'albahaka'* *(Hyptis suaveolens)*, which grows abundantly in marginal areas, to treat buffalo for severe diarrhoea. The procedure is as follows:

- take 3 fresh rootstocks of *albahaka*,
- wash thoroughly with water,
- boil in 1 litre water until about 375 ml is left,
- allow it to cool,
- put it into a 375 ml bottle and drench it to the animal,
- give fresh solution every morning and afternoon for 4 consecutive days.

This herbal plant is also used to treat diarrhoea in humans, and is placed in chickens' nests to minimise lice infestation.

In Altavista, Leyte, Tito Pael owns and farms 7 ha of sloping land. Besides growing coconut, coffee and maize, he raises water buffaloes, goats, pigs and chickens. To deworm the calves, he mixes about 40 ml pure coconut milk (extracted from finely ground coconut meat without adding water) with one egg of native chicken. The mixture is poured to the calf in the afternoon. Then the calf is allowed to wallow and is observed for the following day. If there is no sign of internal parasites coming out yet, the medication is repeated. The medication is usually effective after two drenchings (Bantugan et al. 1989).

Natural medicines in animal health care

As in plant protection, so also in protecting and treating livestock, farmers have a long tradition of using locally available natural resources – and innovative farmers are experimenting with both old and new techniques (see Box A23). Preparations of natural veterinary medicines are described in Matzigkeit (1990) and numerous other examples can be found in the annotated bibliography of ethnoveterinary medicine prepared by Mathias-Mundy and McCorkle (1989).

Farmer-centred development of pest control techniques

Many activities on the farm influence the life cycle of organisms that are potential pests or diseases when their numbers become too high. Knowledge of these organisms' life cycles and damage levels and insight into the impact of specific techniques are necessary for effective, ecologically-sound techniques of pest and disease control. These techniques are often very site-specific. The activities of pest and disease organisms do not stop at the farm border. Synchronised, collective action is often necessary to prevent or control specific pests. Improving the capacity of farmers to control pests and diseases in an ecologically sound way therefore demands a participatory process in which the whole community is involved (see Box A24).

Box A24
Catch a moth to kill a caterpillar

The red hairy caterpillar is a notorious pest that can destroy crops and ruin farmers in a matter of a few days. Its attack is usually severe and wide-spread, encompassing not just a few fields but a vast area with numerous villages. Janseva Mandal, a voluntary organisation, works with 'tribals' around Nandurbar, Maharashtra, India. One year, soon after the outbreak of the monsoons, the red hairy caterpillar was noticed in the fields. The pests grew rapidly in size and number, infecting the crops at an alarming rate. The organisation realised the uselessness of responding with chemical pesticides, when the caterpillar survived even after being immersed in the pesticide. It therefore sought other ways to control the pest, based on a study of its life cycle. The study revealed the following:

● With the first heavy showers, moths emerge from pupae hibernating in cocoons in the soil, on bunds and near tree roots.
● After emerging, the females lay yellowish-white eggs on nearby grass and tree leaves.

At night, when the moths flock near fluorescent lights, the females lay eggs near light sources.
● After egg-laying, the moths die. The eggs hatch within 2 – 4 days and thousands of caterpillars crawl out. They feed voraciously on the leaves on which they are laid, con-suming matter many times their own weight. They then descend to the fields to feed on the tender leaves and shoots of newly germinated crops, moving from field to field. Their red-brown hair camouflages them in the mud and protects them against chemical pesticides.
● Roughly 6 weeks after the eggs hatch, the caterpillars go down into the soil for pupation until the next monsoon. Thus, the pest is usually active between July and August, and there is generally one genera-tion a year.

During the study, photographs were taken and a slide show explaining the stages in the caterpillar's life cycle was pre-pared. This was then shown in several villages around Nandurbar. To show that the caterpillar and moth were the same creature at different stages in its life cycle, caterpillars were kept in bottles

until the moth emerged out of their cocoons. Actual observation convinced the farmers.

Meetings were then arranged in the villages to find ways to control the pest. Farmers came up with the idea of trapping moths by hanging broad-mouthed vessels filled with water and some kerosene near electric lights. They also decided that egg-laden leaves be collected by village children who were given sweets, as an incentive, according to the number of leaves col-lected. The leaves were then burnt. Finally, the farmers pro-posed that the still surviving caterpillars be physically picked from the roots of seedlings, col-lected and destroyed by burning – again a task for children.

The farmers realised the need to organise caterpillar collection at the community level, as the pests do not observe field and village boundaries. This way of destroying the red hairy cater-pillar proved to be effective. It required almost no cash input or special skills, but it depended on a high level of community participation (Sashi & D'Silva 1989).

Contact: S Sashi and Brian D'Silva, 12/271 c Liela Villa, 36th Road, Bandra, Bombay 400 050, India.

Appendix A4

Choosing, conserving and improving genetic resources

Before any initiative is taken to introduce new cultivars or breeds, a good picture of the demand side is needed, e.g. the wishes of the farm household (men and women), the ecological constraints, the local varieties and the local capacity to manage them (e.g. selection, supply). Strengthening community capacity to manage genetic resources (local supply, conserving local varieties and breeds, handling and storing seed) is of vital importance for LEISA. If scientists, government agencies and NGOs collaborate with farmers in trying to improve the selection, conservation and distribution of genetic resources, valuable indigenous knowledge about genetic resource management can be tapped and kept alive. The challenge is to create conditions which enable the rural community to improve local varieties (Juma 1989).

Underutilised plant and animal genetic resources

Valuable genetic resources have been conserved and improved by indigenous peoples throughout the world. Modern agricultural science

---

Box A25
The multipurpose tuna plant

The spineless cactus or tuna (Opuntia ficus-indica) has a wide area of distribution. In ancient Mexico, it played an important role in culture and religion. In the Middle Ages, tuna was planted on the African coast so that its fruit (desert figs) could be harvested and eaten by sailors to prevent scurvy during long journeys to India. Tuna is found in the Andes countries from Peru's desert coast to inland Bolivia.

It has two kinds of roots: thick spongy ones that can bind water for a long time and support the plant, and small ones that function only for one rainy period. Most of these die with each dry period and thus cause rapid formation of organic material in the soil. Two-year-old leaf discs can be used for new cultivation of the plant. Growth is good on soil with a high pH and on sloping land. Already six months after planting, up to four new leaf discs can form. Fruit can be harvested in three years. Even in dry years, fruit yields are 15 – 20 t/ha.

The tuna fruit has a thick rind with many prickly hairs, is full of seed and very tasty. It can be eaten at home or sold. The desert fig is valued as a table fruit, and prices are high. In Peru, tuna is processed into liquor, marmalades and dried fruit pulp (Turkish Delight).

Particularly in dry periods, tuna serves as fodder. A study has shown that cattle and goats kept their weight for three months on a sole diet of tuna. Tuna is low in protein, but contains much roughage and several minerals and vitamins.

A third product of tuna growing is cochinilla (Dactilopius coccus Costa), an insect that sucks the juice of tuna leaves and is used as raw material for carmine pigment. A large market for this has been forecast in food and pharmaceutical industries, as toxic chemical pigments are increasingly replaced by natural organic ones.

Tuna can also be used for reafforestation. Thus, erosion control in the long term can be combined with short-term productive uses. Tuna helps control erosion in three ways: its use as fodder reduces grazing pressure on natural range; it contributes to formation of organic matter which protects the soil; and it can be grown along infiltration ditches (30 cm × 30 cm × 5 m), preventing quick run-off of rain-water and benefiting from the water thus caught.

In the Cochabamba area of Bolivia, tuna is being reintroduced with the aim of improving the existing mixed farming system in ecological and economic terms. In this subtropical highland (2200 – 2800 m above sea level), average rainfall is 500 – 600 mm per year; the dry season lasts 8 months. The ground is mainly chalky with a

pH of 7.0 – 7.8. A thin topsoil lies above a rocky subsoil. The farmers grow maize and wheat, but the land is eroded and both production and income are low. Past development attempts to increase potato and grain production were without success. Nor has afforestation with eucalyptus succeeded, as many of the trees dried out.

The farmers are already familiar with tuna and are beginning to revalue the plant. It has multiple uses and does not compete with traditional food crops on scarce arable land, because it grows on chalky, rocky ground where these cannot be grown. The labour of planting and maintaining tuna does not coincide with the sowing and harvesting of food crops, nor does it seem to interfere with important religious festivities. Tuna growing competes primarily with the regular migration to the (illegal) coca production areas of the tropical lowlands.

Growing tuna on communal land was not very successful at first, as too little attention was paid to grazing cattle. The farmers now recognise the need for protective measures. As transport causes problems for remote villages, ways are being sought to reduce the weight of the planting material, e.g. by cutting tuna plants to pieces for multiplication. Further experimentation is still needed (Tekelenburg 1988).
Contact: Tonnie Tekelenburg, Casilla 2521, Cochabamba, Bolivia.

Box A26
Bitter cassava

Cassava was introduced to northeast Mozambique about 200 years ago. Now, various cultivars are grown, with more reliance on bitter ones in areas of low and irregular rainfall. The main harvest is in the dry season, when fresh roots are peeled, cut into pieces, sun-dried for several weeks and stored. They are eventually consumed throughout the year in the form of a paste, after pounding and boiling.

In 1980/81 there was hardly any rain, causing severe crop losses and hunger. But the bitter cassava variety *Gurue* continued to thrive. Because the traditional processing technique takes so long, tubers and leaves were eaten without adequate processing. People who ate *Gurue* experienced symptoms of cyanide intoxication, but they did not starve.

After the drought, many farmers were eager to obtain cuttings of *Gurue*, which they value not only for its drought resistance but also for its high yields. After 11 months, the tubers weigh twice as much as those of other cultivars. This means less work and less losses at peeling. Also because of the plague of rats after the drought, farmers sought *Gurue* to plant. They knew about its toxicity, but securing food availability meant fighting pests, and rats do not like *Gurue*.

Also dried pieces of *Gurue* reportedly withstand storage pests better than other cultivars do. This may be related to *Gurue*'s high cyanide content. Peasants say: ''By the time insects get into the dried cassava pieces, it is good enough for us to eat.'' They mean that, by then, it is no longer toxic.

Thus, cassava cultivars with high levels of cyanide may have important benefits for farmers. To prevent cyanide intoxication, scientists have focused on developing low cyanogenic cultivars. Instead, emphasis should be shifted to processing methods that ensure sufficient cyanide removal from the tubers and leaves. For emergency situations, suitable rapid processing methods may have to be developed (Essers 1988).

Contact: Sander Essers, Roghorst 138, NL-6708 KR Wageningen, Netherlands.

has given attention to only a small fraction of the genetic resources useful for sustaining human life. Many plant species that are cultivated or collected and many animal species that are tended or hunted in LEIA systems are not known to formal science or have been underestimated in their potential (see Boxes A25 and A26). Excellent overviews of the sporadic and scattered information available about some of these species have been made by BOSTID (Board on Science and Technology for International Development; see Appendix C). An example of simultaneously conserving and making productive use of indigenous trees and a traditionally hunted animal species is given in Box A27.

Producing and conserving genetic resources

Genetic resources are being collected by research centres, but are usually stored in high-technology gene banks either in developed countries or in international agricultural research centres. The shortcomings of these systems are well known and documented (e.g. Mooney 1983, Plucknett 1987). The stored genetic resources may be maintained, but they are not easily available to smallholders. The alternative to this approach is *in situ* conservation: collecting, evaluating, safeguarding, improving, multiplying and distributing indigenous genetic resources in their place of origin.

In situ conservation is gaining popularity, particularly among NGOs, as a farmer-oriented approach to seed supply. The many current activities are often linked to local-level smallholder support programmes or function independently, and involve establishing seed banks, seed

Box A27
Fitting iguanas and trees into Central American farms

The green iguana has been hunted to extinction in many areas of its range (Mexico to Paraguay) because it is highly prized by local people as a food source. Moreover, deforestation for agriculture is destroying the iguana's natural habitat.

The Smithsonian Tropical Research Institute studied the iguana's reproductive behaviour and designed techniques to increase reproduction and survival rates. As the iguana depends on trees for habitat and food, promoting iguana management for local food and sales can encourage the protection and planting of farm forests for the iguanas.

Since 1985 the Pro Iguana Verde Foundation is promoting iguana management in rural communities in the Peninsula de Azuero, the most deforested area of Panama. Farmers are enthusiastic about the iguana's return and are planting mainly native multipurpose trees to provide a habitat for it. These include live-fencepost species (Bursera simaruba, Diphysa robinioides, Erythrina poeppigiana, Gliricidia sepium, and Spondias mombin), fruit trees (Anacardium occidentale, Crescentia sp., Inga sp., Psidium guajava, and Tamarindus indica) and timber trees (Acacia mangium, Bombacopsis guinatum, Cordia alliodora, Cedrela odorata, and Leucaena leucocephala). They are planted as shelterbelts along waterways or along existing fence lines in strips 20 m wide.

The communities receive brood stock with enhanced production capabilities. Reproductive iguana colonies are established near farmhouses. Feeding stations are set up to provide supplementary feed. Egg-laying sites are installed from which eggs can be easily collected for incubation. Simple incubation chambers and rearing cages permit optimal hatching rates and numbers of young which can be released into forest patches. The survival rate of iguanas can thus be multiplied 45-fold compared to natural survival. Two years after release of the first generation at the age of 7 months, iguana harvesting can start.

As iguana management is further investigated in collaboration with the farmers, its environmental and economic feasibility becomes ever more evident. As iguanas have low energy requirements, they consume little purchased feed while in captivity. Although it is not a necessary component of iguana management, supplementary feeding after release permits a 10-fold increase in the carrying capacity compared with natural iguana density, resulting in 400 harvestable adults per year and hectare. Meat production from iguana can be three times that of cattle per hectare. The iguanas consume less than half the purchased feed needed to raise a chicken or rabbit to the same size, as the feed merely supplements their natural diet: tree leaves, fruits and flowers – a resource for which no domestic animal competes.

Iguana management brings numerous social, environmental and economic benefits. It conserves the species and provides a local source of protein. Iguana products, such as meat, eggs and skins, can be sold on local or international markets. The re-establishment of forests in farming areas provides farmers with tree products (fuel, fruit, timber) and simultaneously protects soil and water resources (Werner 1989).

Contact: Fundacion Pro Iguana Verde, Apdo 1501, 3000 Heredia, Costa Rica.

exchange and breeding. In the long run, however, the most decisive factor in reversing genetic erosion will be strengthening community-based systems of conserving local genetic resources and local knowledge about them. Examples of how farmers' abilities to conserve plant genetic resources can be strengthened are given in Boxes A28 and A29.

A low-cost approach to genetic resource management would be local seed supply units: small farms established to produce sufficient improved seed to satisfy local needs. Such farms could be managed privately, by an NGO or under community control. The decentralised seed farms could be supplied annually with foundation seed from a research station or the seed industry, and could concentrate on multiplying seed and selling it locally. The village-based seed farms would need extension advice from decentralised seed inspection units.

Box A28
Saving seeds at village level

NGOs and other campaigners against genetic erosion who promote seed conservation at village level find that seed saving is more complicated than it first appeared. Categorisation, planting out, characterisation, storage and quality control are essential but are time-consuming and require financial resources.

Planting out collected seeds is probably the most expensive step. Seeds must be observed for 2 – 3 seasons so that characteristics, such as planting season, flowering habits and pest susceptibility, can be recorded. Many local varieties are bound to a certain season and give poor results in other seasons. Others do badly if fertilised and do well only under low fertility conditions. Seed collection, drying and cleaning are labour-intensive. However, all these activities can be easily learned and require systematic efforts rather than expertise.

Most problems occur in seed storage and maintaining viability. Here, guidance from a seed technologist is important. Seeds can be dried too much or not enough; they can be infested with fungus while in storage, affecting their viability. Success is determined primarily by the seed moisture content but also by the temperature and humidity under which the seed is stored.

The importance of labelling seed packets with place of origin, local name and collection date cannot be overemphasised. Any observations from farmers will help in future characterisation of accessions. For example, some cultivars store better than others, an important characteristic of old bean varieties but lost in the new ones. If this information is not recorded, one might not know it by merely observing plants in the field. Another attribute that farmers know well and may use to describe their 'collections' is versatility, e.g. legumes that can be sold as green vegetables but are equally good as dried grains.

Institutional efforts to collect seeds are justified (as opposed to farmers collecting local seeds for *in situ* conservation) if local seed diversity is lacking. Then, seeds must be brought from other areas and exchanged in order to reinstate the required diversity. But these efforts are useful and relevant only if the seeds collected and multiplied are given to farmers for them to try out and conserve. The sooner the materials are moved out, the better. The ideal approach is to give farmers a diversity from within each crop (e.g. 6 kinds of mung beans instead of just one), so they can choose from a range. Some farmers retain what others will not.

Farmers save varieties for a host of reasons, the least of which may be purely for the sake of conservation. The idea of getting growers to continue raising varieties only in order to save them, as done in some Western countries, is unrealistic. The focus must be on seed accessions whose attributes will, in themselves, result in their being conserved (Gonsalves 1990).

Contact: International Institute -of Rural Reconstruction (IIRR), Silang, Cavite, Philippines.

When they have become familiar with the techniques, the farmers could take over maintenance breeding of the improved composites and thus become self-sufficient with foundation seed. Crop processing, dressing seeds with fungicides and packing in bags can be done at village level, using already known, labour-intensive techniques. Improved composite maize seed could be produced and used in the local area, thus reducing the transport costs, which today make up the major part of present seed prices. At 1988 prices, the estimated production cost of improved composite maize by village-based seed farms is one-tenth of the consumer price of hybrid seed (Friis-Hansen 1989).

Box A29
Farmer-based rice
breeding in the Philippines

When high-yielding varieties (HYVs) of rice were first promoted in the Philippines, farmers thought these would deliver them from poverty. But then they realised that big harvests also mean big expenses. Production costs (seed, fertiliser, pesticide, labour, machine rental, irrigation) and land amortisation or rental fee took a large chunk of the harvest. Extensive use of pesticides caused disappearance of beneficial insects, while certain pests became resistant to commonly used pesticides. Farmers also noticed the disappearance of fishes, frogs and snails in the rice fields.

Thousands of rice varieties used to be grown countrywide, but now only 4 – 5 HYVs are widely planted in the country. The farmers have to buy certified seeds from seed producers. Loss of indigenous seed means erosion of genes that have been adapted to local environments for centuries.

A small multidisciplinary group of scientists at the University of the Philippines at Los Baños (UPLB) met with farmers and listened to their experiences with HYVs. The farmers wanted higher rice yields but not by accumulating unpaid loans for inputs and poisoning the streams and fields. They wanted a wide choice of rice varieties to plant.

A farmer-NGO-researcher partnership was formed within a project called MASIPAG. Field studies of problems identified by farmers are worked out by them together with NGOs and researchers. Similarly, farmer-based training begins with what farmers want to learn. Project components include:

- collection of rice cultivars, identification, multiplication, maintenance and evaluation
- breeding (hybridisation of farmers' selections)
- alternative pest management
- biofertiliser usage (crop residues, local organic resources, green manures, microbial inoculants)
- diversified farming (cropping systems, crop-livestock-poultry systems).

From 1986 to 1988, 140 rice cultivars were collected from various parts of the country. These are being purified and characterised in field plots. Cultivars evaluated by farmers also included 21 advanced lines from UPLB Department of Agronomy. As of 1988, farmers' top 11 selections in field plots include 4 traditional and 5 improved varieties and 2 advanced lines.

Some farmers learned and practised the art of crossbreeding. The parental materials were their own selections for specific characters and not necessarily from the 11 top selections. This is the first time in the history of Philippine agriculture that farmers have developed their own crosses. The farmer – scientist cooperation is also leading to good production without fertilisers and pesticides. In 1988 the top traditional rice varieties were already yielding as much or more than improved varieties from conventional breeding (Briones et al. 1989).

Contact: Angelina Briones, Dept of Soil Science, UPLB, Laguna, Philippines.

Appendix A5

Integrated farm systems

By selecting and adapting appropriate techniques and genetic resources, such as those mentioned above, farmers can create integrated LEISA systems for their specific biophysical and sociocultural setting. Such integrated systems can provide farm families with many of their daily needs: a variety of nutritious foods, wood for building and cooking, various other products for home use, and cash for things that the farm cannot provide. Combining different plant and animal species and applying a variety of techniques to create favourable conditions for them and to protect the environment also helps farmers maintain the productivity of their land and reduce farming risks, especially on sloping land and under unpredictable climatic conditions.

For orientation as to the general direction in which a farm system could be developed, case examples of integrated systems can be useful. Such cases are not intended as fixed models; they serve as a basis for discussion of technical options and must be adapted to the specific characteristics of each farm through a process of PTD. Some cases of integrated farm systems are presented here, including examples of participatory initiatives to create or improve integrated farm systems.

Bio-intensive gardening

Conventional promotion of home gardens has involved providing seeds (often imported), chemical fertiliser, pesticides and tools. The success of such gardens lasted only as long as the supply of external material inputs was guaranteed. When project funding ended, so did the gardens. It is within this context that an alternative strategy was developed, using locally available materials and knowledge-intensive methods: the bio-intensive gardening (BIG) approach. At the International Institute of Rural Reconstruction (IIRR) in the Philippines, ideas derived from John Jeavons of Ecology Action in California have been adapted for small-scale, family-centred food production in urban or rural settings.

BIG is a biological form of agriculture in which a small area is intensively cultivated, using natural ingredients to rebuild and maintain soil productivity. At the heart of the approach is the effort to improve the soil's capability to nurture soil and plant life.

Bed preparation. The initial deep digging (30 – 60 cm) of the narrow raised beds (Figure A6) for permanent, year-round use is labour-intensive, especially when double-dug (60 cm). Although this is more productive, it is less preferred by farmers who have not experienced its benefits. It is critical to keep the plot constantly covered with crops or, if no water is available during part of the year, to maintain a 10 – 15 cm thick layer of dry straw/grass or other mulch material. Minimum tillage can then be practised.

Manuring. The bio-intensive gardener tries to maximise the use of plant and animal wastes. Where no compost or manure is available, green-leaf manuring is recommended, as in the alley-cropping BIG option shown in Figure A7. Alternate rows of fast-growing leguminous trees are planted every 4 – 5 m with two beds between them. When the trees are at least one year old, they are cut 50 cm above ground level 3 – 4 times a year.

A major adaptation in IIRR's approach to BIG is the use of liquid fertiliser: 30 – 50 kg of weeds and dung is placed in a bag in a 50-gallon drum which is then filled with water. Three weeks later, one part of the solution mixed with four parts water is used to fertilise the soil around plants. Experiments showed that, if the leaves of *Leucaena leucocephala* or *Gliricidia sepium* (or both) are placed in the bag, the nutrient contribution is greater than with liquid fertiliser from animal dung. Liquid fertiliser is of critical importance in the wet season because of the constant leaching of natural and chemical nutrients from the bed. The impact is greatest on plants younger than 40 days. Liquid fertiliser is applied every 10 days, starting at transplanting.

Crop diversity. IIRR emphasises the importance of relying primarily (at least 7 out of 10 varieties used) on indigenous varieties. Imported cultivars are used only if they have been grown in the area for 10 years without the aid of heavy chemical inputs. Seeds that perform reasonably

Figure A6
Bed preparation for a bio-intensive garden. (*Source:* Gonsalves 1989)

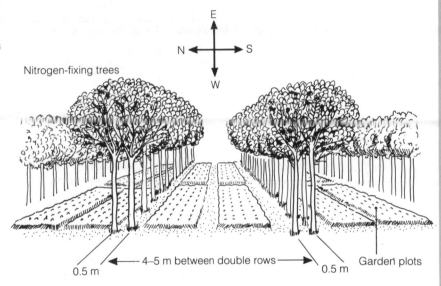

Figure A7
Integration of alley cropping and bio-intensive gardening. (*Source:* Gonsalves 1989)

Nitrogen-fixing trees

←— 4–5 m between double rows —→

0.5 m   0.5 m   Garden plots

well under adverse conditions (e.g. a lone cowpea or leafy amaranth from a previous season that continues to survive through the long summer) are of potential value. Other vegetables might not produce large, individual fruits but may be prolific and may yield over a longer period. Diversity is an important factor in reducing the insect threat, and is achieved through relay cropping, intercropping and other mixed cropping systems preferred by the gardeners. Crop rotation is critically important.

The beds are intensively sown so that, when the plants are fully grown, the soil is kept completely covered by the plant canopy, thereby eliminating weeds and reducing water evaporation from the soil surface. Where summer seasons are harsh, a drought-tolerant cover crop is sown into the stubble of the previous crop, without redigging the bed. In the humid conditions of the Philippines, *Dolichos lablab* (hyacinth bean) and rice bean have been found especially useful as a soil cover and source of green manure.

Locating contours

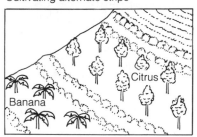

Planting double hedgerows

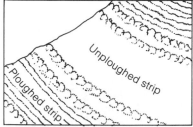

Cultivating alternate strips

Planting perennial crops

Sowing annuals between perennials

Figure A8
Contour farming according to the Sloping Agricultural Land Technology (SALT) scheme. (*Source:* Mindanao Baptist Rural Life Center)

Pest control. The use of predominantly indigenous or acclimatised vegetable varieties, crop diversity and good soil/water conditions result in few or no pest problems. However, virus in tomatoes, beetles on cucurbits and moths on cabbage are problems if some natural pesticides are not used. No soil-borne diseases such as nematodes or root rot have been experienced on the plots now used for four and a half years.

There are now over 20 000 bio-intensive gardens in the Philippines as a result of IIRR promotion. UNICEF and the Government of the Philippines are involved in what is popularly referred to as Family Food Production through BIG. The slogan is: 'Think BIG' (Gonsalves 1989). *Contact:* IIRR, Silang, Cavite, Philippines.

Contour farming

The contour farming scheme called Sloping Agricultural Land Technology (SALT), developed at the Mindanao Baptist Rural Life Center, is a way of turning a sloping piece of land into a productive upland farm. It enables farmers to stabilise and enrich the soil, conserves soil moisture, reduces pests and diseases and reduces the need for expensive inputs, such as chemical fertilisers. Moreover, it replaces an eroded hillside with a terraced, green landscape. Most important of all, the technology can increase the farmer's annual income almost threefold after only 5 years.

SALT is tailored for small family farms raising both annual food crops and permanent crops. It involves the following steps (Figure A8):

- locating the contour lines and cultivating the ground along them, 4 – 6 m apart on steep hills and 7 – 10 m apart on more gradual slopes;

- planting nitrogen-fixing shrubs and trees as double hedgerows in two furrows 50 cm apart along each contour line;

- cultivating and planting permanent crops (e.g. coffee, cocoa, citrus) in every third or fourth strip;

- cultivating alternate strips between the hedgerows before they are fully grown (thereafter, every strip is cultivated);

- planting short- and medium-term crops (e.g. maize, sorghum, upland rice, pineapple, sweet potato) between strips of permanent crops as sources of food and regular income;

- trimming the hedgerows down to 1 m above ground and using the trimmings as organic manure;

- rotating the nonpermanent crops to maintain productivity, fertility and good soil formation;

- building green terraces by piling stalks, leaves and stones at the base of the hedgerows to capture and enrich the soil (Tacio 1988).

This can be varied to include multipurpose trees in the hedgerows which can be cut for fodder and making a forage garden in the lower portion of the sloping farm. A goat barn can be built in the middle between the food and forage crops. Forage can be carried up to it and manure can be carried to the food crops above it (see Box A30).

Box A30
A contour farm in the Philippines

On 0.9 ha of hilly land near Villaba, Leyte, the Sulad family has planted double hedgerows of ipip-ipil *(Leucaena leucocephala)* along contours 5 – 6 m apart. Between the hedgerows, various mixtures of maize, upland rice, mungbean, peanut, onion, vegetables and sweet potato are grown at different times of the year, depending on weather conditions. The Salud family also raises chickens, water buffaloes, pigs and goats.

Vicente Sulad has innovated in the contour farm by incorporating certain crops which are normally planted only along field borders and around farmhouses. He had observed that, despite the hedges, some topsoil erosion continues. The formation of bench terraces behind the hedges results in relatively less fertile upper portions of the terrace. To maximise the use of these areas, he plants banana, yautia and cassava. On the other hand, because the ipip-ipil shades maize planted in the more fertile 'toe' of the terrace, he sows it 0.5 m upslope from the hedges and plants shade-tolerant ginger between the maize and the hedge. These additional crops reduce topsoil erosion and provide alternative sources of food and cash, especially during lean months (de Pedro & Mercado 1989).
Contact: Mindanao Baptist Rural
Life Center, Kinuskusan, Bansalan, Davao del Sur, Philippines.

Integrated crop – livestock – fish farming

Integrating crops, livestock and fish in smallholder farming systems has many ecological and economic advantages. Such systems are conducive to nature conservation as they promote habitat stability and diversity for wildlife living on the farms and in adjacent areas. As these integrated systems optimise the use of on-farm and adjacent resources, they encourage habitat conservation rather than destruction. Such systems are productive and profitable because they utilise waste as inputs in other enterprises within the farm, and because fish are a highly nutritious and valuable traditional food. They use microenvironments within a farm system which add to farm productivity and security (Chambers 1990), as already described in Box A15.

The following bioresource flow models, prepared by farmers in Vietnam and Malawi, illustrate how farm enterprises are integrated to exploit different land and water resources and to substitute for external inputs. Making these models helps farmers transform their farm systems by clarifying how existing enterprises can be better integrated and how new enterprises can be better fitted into the farm system.

Figure A9
Bioresource flow model of integrated farm in Vietnam. (*Source:* Lightfoot & Minnick 1990)

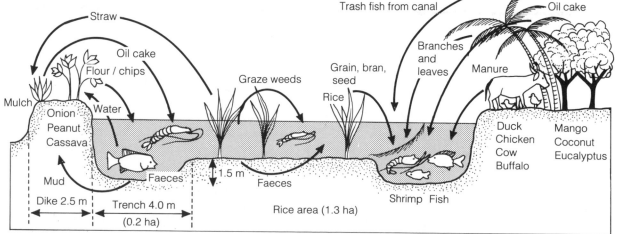

In Vietnam, the bioresource flow models show how water from the trenches is used to irrigate vegetables, grown in a mulch of rice straw, on dikes made from trench mud (see Figure A9). Immediately after trenching, chicken and cattle manure is put into the rice-field trenches to promote phytoplankton blooms for the fish and shrimp to feed on. Although the shrimps' diet is primarily natural, they are also fed during their first two months with farm-grown by-products, such as germinated rice grain, cassava flour, rice bran, coconut and groundnut oilcake, and trash fish from the irrigation canals. Mango and eucalyptus branches are also put in the trenches to keep out cattle and poachers, and to provide an undisturbed habitat which shrimps need. Such integration reduces the need for external inputs. As the shrimps and fish eat rice weeds, weeding expenses can be reduced by one third. Moreover, chemical fertiliser input can be reduced by 30% with no detrimental effect on rice production, because animal manure and fish faeces fertilise the paddy.

In Malawi, bioresource flow models show how fish ponds are integrated with other farm enterprises (see Figure A10). The pond water is fertilised with rotten fruits of guava, papaya and avocado, and manure from cattle, sheep and goats. Leaves of leucaena, pumpkin, wild vegetables and cowpea are also put into the pond to feed the fish, and coarse maize bran is used as a fish food when available and affordable. Apart from fish, other outputs from the pond are fertile sediments for the vegetable gardens and water for irrigation. The main purchased input on these farms – maize bran – can be supplemented and

Figure A10
Bioresource flow model of fish-pond integration into farm in Malawi.
(*Source:* Lightfoot & Minnick 1990)

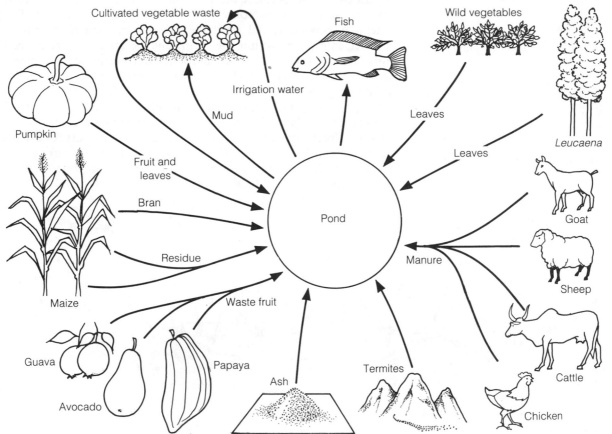

partly substituted with cocoyam leaves and Napier grass.
Contact: Clive Lightfoot or Roger Pullin, ICLARM, MC PO Box 1501, Makati, Metro Manila 1299, Philippines.

Integrated forage production

Various ways of integrating forage production into mixed farming and agropastoral systems are being developed by farmers and scientists. Using contour hedges to feed goats has been mentioned in the SALT example above. Further examples of integrated forage production are:

- an intensive zero-grazing system based on Napier grass and fodder hedges, practised by dairy smallholders in the highlands of Kenya (van Gelder & Voskuil 1990);
- alley farming, i.e. the diversification of alley cropping with multi-purpose woody species to provide forage which can be cut for small ruminants, as practised in the humid zone of Nigeria (Reynolds 1989);

Box A31
Fodderbanks for
agropastoral systems

In the subhumid zone of Nigeria, an ILCA research team worked together with Fulani agro-pastoralists to find ways of improving cattle nutrition. The Fulani, who have settled close to crop farmers with traditional land rights, usually obtain land for cropping on loan from the farmers in return for token payments. Grazing areas are communal and comprise uncropped areas, fallow and harvested fields. Annual rainfall is 1200 mm within a 6-month wet season.

The Fulani and scientists cooperated in on-farm trials with 'fodder banks': improved legume/grass pastures to be grazed 2 – 3 hours daily when existing grazing resources are particularly scarce or low in quality. ILCA initially recommended that the fodderbanks be grazed by pregnant and lactating cows throughout the dry season, in order to increase milk production. For the first few trial pastures, ILCA paid the establishment costs (fencing, *Stylosanthes* seed, superphosphate fertiliser). Later participants in the trials covered the costs themselves. The scientists observed how the Fulani used their fodderbanks, and discovered that they:

- saved the fodderbank until the late dry season, preferring to take advantage of crop residue grazing in the first half;
- preferred to allow very weak animals or (at the end of the dry season) the entire herd to graze the fodderbank.

Further studies by ILCA revealed that, in the early dry season, crop residues offer a much higher quality diet than the scientists had assumed, as the cattle select particularly nutritious plant parts, e.g. immature seed heads, green leaves, upper stalks. Using the fodderbanks for survival feeding in the late dry season halved the normal cow and calf mortality. Moreover, household studies revealed that much more cash was gained from animal than from milk sales, and the milk income was controlled by women. The men bought the production inputs out of the proceeds from cattle sales. Therefore, feeding cows to increase milk yield was not of high priority to the men.

Not only the Fulani reaction to ILCA's recommendations but also new ideas introduced by the Fulani influenced the design of the fodderbank. For example, some Fulani countered the problem of termite damage to fenceposts by adapting an indigenous form of live fencing (mainly *Ficus* spp). This considerably reduced fencing costs, which – in the original fodderbank concept – represented three-quarters of total establishment costs. Other Fulani tried different ways of preparing land for sowing stylo, based on their traditional techniques of using cattle to manure and trample the ground in the early wet season for broadcast sowing of rice and *iburu (Digitaria iburua)*. Some Fulani started growing cereals in part of the fodderbank, taking advantage of the soil fertility improvement via the legumes and the protection the fence offers from crop damage by animals. This could lead to a form of ley farming suited to the circumstances of agropastoralists in the subhumid zone (Waters-Bayer & Bayer 1987).
Contact: ILCA Subhumid Research Site, PMB 2248, Kaduna, Nigeria.

● fodderbanks, i.e. small improved pastures for strategic use in agropastoral systems in subhumid areas of West Africa (see Box A31).

Integrated resource management in the semiarid zone

As the development of LEISA systems is particularly challenging in semiarid areas, where low and irregular rainfall severely limits biomass production, we close this selection of techniques with an encouraging example of integrated management of water, land, tree and livestock resources in the semiarid zone of West Africa.

In 1979, Oxfam commenced an agroforestry project (PAF) in an area with 400 – 600 mm annual rainfall in Yatenga Province, Burkina Faso. The original idea was to promote the use of microcatchments to enable farmers to grow trees for wood. The farmers showed little enthusiasm for planting trees, but were in favour of microcatchments for growing food crops. Fortunately, the project had the flexibility to change its direction, and began to promote rock bunds as a means of increasing crop production. These have long been used by Dogon farmers in Mali, who build them in straight lines across the slope. The project introduced the idea of building them along the contours, using a water level made of a hosepipe and two sticks. Rock bunds require a high initial investment of time and labour to collect and transport the rocks, but once the bund has been established, little maintenance is needed.

PAF also promotes various other improved agricultural practices. One of these is the *zai* method of tillage, a traditional practice of digging a 20×20 cm hole 10 cm deep during the dry season and filling it with mulch such as crop residue. This leads to increased termite activity which, in turn, increases the rate of water penetration when the rains come. Millet is planted in the individual holes, which also help protect the seedlings from wind damage.

Rather than relying on food-for-work as an incentive, as is commonly done in the region, PAF did not pay villagers to build the rock bunds. The project did, however, provide information and training, as well as materials and equipment, such as water-levels, pick-axes and carts, either free or on loan. Some food loans were also given, most of which were paid back after the next harvest. Despite the lack of material incentives, the rock bunds have been widely adopted. Local people are convinced they repay the input of time and labour by bringing higher yields and increased yield security. The bunds also extend the area under crops by enabling land that had previously been considered useless to be brought into production.

Measurements made by PAF showed that the average grain yields in plots with bunds were consistently higher than in those without bunds, ranging from 12% in 1982 to 91% in 1984. The increased food production is obviously of crucial importance to the Yatenga farmers. Contour rock bunds are now found throughout the province even where they have not been directly promoted by PAF or any other project and the technique is also spreading to other parts of the country.

PAF has been building up its extension programme since 1983. Initially, it relied on the use of the flannelboard and other techniques to raise awareness. It then moved on to providing training in rock bund construction. Since it is too costly to provide individual instruction to all farmers wishing to construct bunds, PAF asked villages to nominate

Box A32
Village-based land-use management in Burkina Faso

In Longa village of Yatenga Province, the use of rock bunds along the contours has been integrated into the activities of a project which concentrates on land-use management. The villagers constructed rock bunds on 70 ha of agricultural land. They also agreed to impose severe restrictions on movement of animals. All livestock are now held in shaded enclosures, and fodder is collected from the cultivated areas. Trees have been widely planted, and natural regeneration is also evident. Neighbouring villages are now copying Longa's example. Seedlings are supplied by the Forestry Department, and guidance in stall feeding is provided by extensionists from the Ministry of Agriculture. The costs of stall construction and other inputs are borne by the villagers themselves (Kerkhof 1990).

representatives who could learn the technique and then pass on the knowledge to others. In all, several thousand people have now been trained, covering about 500 villages. Initially, the focus of training and extension work was on men farmers but in 1985 the project began training women farmers in rock bund construction, and women are now included in the extension team.

PAF staff have found that attitudes to tree growing tend to change once the initial activities lead to higher food production. In some villages, farmers are now prepared to plant trees along the bunds if they are provided with seedlings. They are also willing to provide protection for the trees against grazing animals during the dry season (see Box A32).

One of the principle lessons of the project is the importance of mobilising the community. Although PAF found that the quality of rock bunds on individual land was better than on communal land, it gradually became clear that much of the work can be done effectively only if there is community consensus. If, for example, rocks are not available nearby, their collection needs to be organised on a community basis. Similarly, if land is to be protected from grazing, the animals of all farmers must be controlled. The community orientation also helps ensure that the poorer farmers also benefit from the project.

Because of the progress which has been made, there is no longer any need for measures to raise the general level of awareness. People see the innovations in neighbouring villages and discuss them as part of their normal day-to-day contacts. In areas where the innovations have not yet been adopted, PAF promotes exchange and communication through excursions. The villagers are then left to reflect on what they have seen, discuss the issues and organise themselves when they feel ready for action. The farmers know where to find the PAF office. In the final analysis, it is up to them to get together and undertake development activities for their own benefit.

PAF also recognises that problems remain. There is, for example, the question of uneven distribution of costs and benefits. Rock collection and bund construction make heavy demands on the available labour, particularly on women. Rich farmers are more likely to be able to mobilise and provide food for communal groups to build bunds on their land. The sustainability of the increased yields obtained when using the bunds is another cause for concern. Because higher crop production means greater mineral extraction from the land, there is a danger of long-term depletion unless methods of increasing inputs of organic matter and fertilisers can be put into operation. Thus, recovery of manure from stall-fed animals and encouragement of composting assume critical importance in the longer-term perspective.

The project accepts the need to identify and resolve such issues, but is fully aware that the proposed solutions must be acceptable to the local people. Project staff stress that the reasons for its success are not primarily technical. The techniques work because they fit into the local context and meet the farmers' need for low-risk and low-cost strategies. More important than its ability to develop techniques is the fact that the project has won the confidence of the local people and that they see that it is concerned with their needs and priorities (Kerkhof 1990).
Contact: Mathieu Ouedraogo, Projet Agro-Forestier (PAF), BP 200, Ouahigouya, Yatenga Province, Burkina Faso.

Appendix B   Glossary of key terms

Adaptability   The capability of adjusting a farm system to cope with changing conditions.

Agroecology   The holistic study of agroecosystems, including all environmental and human elements, their interrelationships and the processes in which they are involved, e.g. symbiosis, competition, successional change.

Agroecosystem   An ecological system modified by people to produce food, fibre, fuel and other products desired for human use.

Agroforestry   The deliberate use of woody perennials (trees, shrubs, palms, bamboo) on the same land-management unit as arable crops, pastures and/or animals, either in a mixed spatial arrangement in the same place at the same time, or in a sequence over time.

Agropastoralism   Land-use system in which arable cropping and the keeping of grazing livestock are combined.

Agropisciculture   Combining cropping and the controlled breeding, hatching and rearing of fish within a farm.

Agrosilviculture   Land-use system in which herbaceous crops and trees or shrubs are combined.

Agrosilvopastoralism   Land-use system in which arable cropping, use of woody vegetation and the keeping of grazing livestock are combined.

Allelopathy   The release by a plant of a chemical that influences the growth of other plants.

Alley cropping   Growing annual crops in spaces between rows of trees or shrubs, often leguminous ones that tolerate heavy and regular coppicing. The leafy and woody material of the trees and shrubs is used as mulch in the crop(s) and also often as fodder, timber, fuel etc. Also referred to as avenue cropping and hedgerow cropping.

Aquaculture   The deliberate production for human use of plant and animal organisms living in water.

Arid   A climate in which potential evaporation exceeds rainfall in all months of the year, so that cropping is possible only with the support of water harvesting or irrigation; refers here to an area with an average of less than about 200 mm annual rainfall.

Bacteria   Microscopic one-celled organisms, many of which play an important role in the soil by breaking down organic matter and in animal nutrition by breaking down cellulose into digestible nutrients.

Biodynamic farming   A holistic system of agriculture devised by Rudolph Steiner that seeks to connect nature with cosmic creative forces. An attempt is made to create a whole-farm organism in harmony with its habitat. Compost and special preparations (e.g. plant-derived sprays) are used. Synthetic fertilisers and pesticides are avoided.

Biofertiliser   Fertiliser containing minerals combined with micro-organisms, the action of which renders the minerals available to plants.

Biological control   The use of natural enemies to control pests, including both control with imported natural enemies and augmentation and conservation of natural enemies through manipulation of the pest host, the environment and/or the enemies themselves (IASA 1990).

Biomass   The weight of material produced by a living organism or collection of organisms, plant or animal; here, annual biomass refers to annual plant species and perennial biomass to perennial plant species.

Biotechnology   The application of scientific and engineering principles to the processing of materials by biological agents to provide goods and services (Bull et al. 1982).

Botanical pesticide   A plant-derived pesticide.

Bund   A ridge of earth placed in a line along the contour of a slope to control water run-off and soil erosion.

Carrying capacity   The maximum numbers of users that can be sustained by a given set of land resources at a particular level of technology.

Climax vegetation   The final stage of plant succession in which the plant community reaches a steady state in balance with the soil, climate and fauna of a given area.

Common land   Land collectively owned and managed by a defined group of users, ideally governed by a common property regime (system of rights and duties) which prevents overexploitation; to be differentiated from open-access land, which has no barriers to its use (IIED 1988).

Complementarity   A state in which one element, in combination with one or more other elements, completes the whole.

Compost   The fertiliser resulting from the decomposition of residues from plants and animals.

Continuity   The capability to conserve the natural, financial and human resources of a farm system so as to ensure its persistence.

Contour   An imaginary line on a field joining all places at the same height above sea level.

Contour farming   Growing crops between contour lines stabilised by, e.g. earth bunds, stones ridges or hedges which conserve both soil and water; the system may also include a livestock component.

Cover crop   Annual crop sown to create a favourable soil micro-climate, decrease evaporation and protect soil from erosion. Cover crops also produce biomass which can be used for soil fertility management.

Crop   Annual or perennial plants cultivated to yield products desired for human consumption or processing, e.g. grain, vegetables (edible roots, stems or leaves), flowers, fruit, fibre, fuel.

Desertification   Process of continued decline in the biological productivity of arid/semiarid land, resulting in skeletal soil that is difficult to revitalise; refers here also to land degradation, i.e. reduction in the capability of land to satisfy a particular use.

Diversity   The quantity of different types of organisms, species, cultivars and/or physical elements per unit area.

Ecological agriculture   Farming practices that enhance or, at least, do not harm the environment and are aimed at minimising the use of chemical inputs, rather than completely avoiding them as in organic farming. Also known as ecofarming.

Ecology   The science of the relationships between organisms and their environment.

Ecosystem   The communities of plants and animals (including humans) living in a given area and their physical and chemical environment (e.g. air, water, soil), including the interactions between them and with their environment.

Environment   Surroundings, including water, air, soil and living organisms and their interrelationships.

Erosion   The gradual displacement or disappearance of parts of a system under the influence of external factors, e.g. erosion of soil by water or wind, erosion of indigenous knowledge, genetic erosion.

Ethnoveterinary medicine   The indigenous knowledge, skills, methods and practices pertaining to the health care of animals.

Extension   Refers here to agricultural extension: activities that disseminate research findings and advice to farmers on agricultural practices and improve farmers' analytical capacity and communication so as to help them in their decision-making related to farming.

External inputs   Inputs that originate from outside the system (farm, village, region, country). Artificial external inputs are inputs that need high quantities of fossil fuel to be produced or distributed, e.g. synthetic fertilisers, pesticides, pumped irrigation water.

Fallow   Land left uncultivated for one or more growing seasons; it is often colonised by natural vegetation and may be grazed.

Farm system   All the components within a given farm boundary which interact as a system, including people, crops, livestock, other vegetation, wildlife, and the social, economic and ecological interactions between them and with the environment.

Farming system   A unique and reasonably stable arrangement of farming enterprises managed according to well-defined practices in response to the physical, biological and socioeconomic environments and in accordance with the households' goals, preferences and resources. Refers to several farm systems which show more commonality with each other than with farms in another farming system (Shaner et al. 1982).

Farming Systems Research (FSR)   An applied approach to agricultural research conducted by multidisciplinary teams who assess the scope for, and potential impact of, technology change within a holistic farming systems framework. The major steps consist of identifying relatively homogenous groups of farmers within specific agroclimatic zones; identifying problems and opportunities of these clients; designing new technologies to suit their conditions; testing these innovations in on-farm trials, evaluating the results and recommending successful innovations for wider dissemination (Shaner et al. 1982).

Functional diversity   The quantity of different organisms, species or cultivars that contribute to increasing the stability, productivity or continuity of an agroecosystem.

Fungi   Plural of fungus; any of a large group of plants (including moulds, mildews, mushrooms, rusts and smuts) which are parasites on living organisms or feed on dead organic material, lack chlorophyll, roots, stem and leaves, and reproduce by means of spores.

Genetic erosion   Disappearance of genetic resources.

Genetic resources   Plant and animal stock with distinct inheritable characteristics of (potential) use within an agroecosystem.

Green manure   Green plant biomass used as fertiliser.

Green revolution   The use of a package of inputs, including modern varieties, pesticides, fertilisers and frequently also irrigation, in an attempt to increase farm yields in developing countries.

Habitat   The environment in which a plant or animal lives and which responds to its specific needs.

HEIA   High-external-input agriculture; depends on significant levels of natural or, more commonly, artificial inputs, such as fertilisers, pesticides and fossil energy, which originate from outside the system (farm, village, region, country) and generally have to be purchased.

Herbicide   A class of pesticides that destroys or reduces the negative effects of weeds.

Holism   An approach that considers all components and aspects of a system; particularly referring to approaches in which material and non-material aspects are considered across disciplinary boundaries.

Horticulture   The science or art of gardening, i.e. of cultivating vegetables, fruit and/or flowers.

Humid   A climate in which rainfall exceeds potential evaporation during at least 9 months of the year; refers here to tropical areas which receive more than about 1500 mm annual rainfall.

Humus   End product of the degrading process of organic matter that improves soil structure, provides nutrients for plants and increases the capacity of the soil to store nutrients and water.

Hybrid seed   Seed produced by crossing genetically dissimilar plants, i.e. of different varieties or species; the yield potential is superior to that of the parent lines but cannot be maintained in succeeding generations; therefore, the seed must normally be purchased each year.

Identity   The collective aspect of a set of characteristics by which something is recognisable or known, referring here to the characteristics of a farm system which are recognised by its users as being in harmony with their culture, their social relations and their relations to crops, animals and nature in general.

Indigenous   Occurring or living naturally in a specific area, such as native plants or animals (opposite of exotic); to be differentiated from 'endogenous', which means having its origin within a specific area (opposite of exogenous).

Indigenous knowledge (IK)   Knowledge of the people living in a certain area, generated by their own and their ancestors' experience and including knowledge originating from elsewhere which has been internalised by the local people. Also referred to as local knowledge.

Infrastructure   The supportive features of an economy often provided by government (but sometimes by private industry), e.g. water, transportation, communications, state organisations (Shaner et al. 1982).

Inputs   Refers here to farm inputs, i.e. elements that farmers add to farm resources in order to influence productivity, stability or continuity. The most elementary farm inputs are water, energy, nutrients and information; examples of other inputs are seeds, agrochemicals and equipment.

Integrated Pest Management (IPM)   A strategy which, in the context of the farm's environment and the population dynamics of the pest species, uses all suitable measures (biological, genetic, mechanical and chemical) in the most compatible manner possible so as to maintain pest populations at levels below those causing economic injury.

Integrated plant nutrition   Strategy to maintain and possibly increase soil fertility for sustaining crop productivity through optimising all possible sources, organic and inorganic, of plant nutrients needed for crop growth and quality in an integrated way, appropriate to each crop system and specific ecological and socioeconomic situation (FAO 1990).

Intercropping   Growing two or more crops at the same time in the same field. Cropping is intensified in terms of both time and space.

Internal inputs   Inputs which originate from inside the system (farm, village, region, country).

Legume   Any of a family (Leguminosae) of trees, shrubs and herbs (e.g. beans and peas), many of which have the ability to live in symbiotic relationship with rhizobia that can fix atmospheric nitrogen.

LEIA   Low-external-input agriculture; based primarily on the use of local inputs (i.e. from the farm and immediate surroundings); makes little use of farm supplies obtained through exchange or purchase.

LEISA   Low-external-input and sustainable agriculture, in which most of the inputs used originate from the own farm, village, region or country, and deliberate action is taken to ensure sustainability.

Ley farming   Alternation of food crops and pasture on the same piece of land. After several years of cultivation, pasture plants are sown or establish themselves and are used for grazing for several more years. The land is then used again for cropping.

Litter   Uppermost layer of organic material on the soil surface, including leaves, twigs and flowers, freshly fallen or slightly decomposed.

Livelihood system   A combination of people, resources and environment in which the stocks and flows of food and cash are used to meet the basic needs of the people. The livelihood system of a rural household may include cropping, tree growing, animal keeping, fishing, hunting, gathering, processing, trading, paid employment and a wide variety of other nonfarm activities.

Manuring   Application of animal dung, compost or other organic material used to fertilise the soil.

Microbial insecticide   A class of insecticides produced from naturally-occurring micro-organisms, e.g. *Bacillus thuringiensis*.

Microclimate   The temperature, sunlight, humidity and other climatic conditions in a small localised area, e.g. in a field, in a stand of trees, in the vicinity of a given plant, or in the topsoil.

Minimum tillage   Soil management practices which seek to minimise labour inputs and soil erosion, to maintain soil moisture and to reduce soil disturbance and exposure. Crop stubble is left or mulch is applied to protect the soil. Also known as conservation tillage or reduced tillage. In its most extreme form (zero- or no-tillage), seeds are drilled directly into the otherwise undisturbed soil.

Monocropping   Repeated growing of the same sole crop on the same land.

Mulch   Protective covering of the soil surface by various substances such as green or dry organic matter, sand or stones applied to prevent evaporation of moisture, regulate temperature and control weeds.

Multiple cropping   Growing two or more crops in the same field in a year, at the same time, or one after the other, or a combination of both.

Multistorey cropping   Growing tall crops (often perennials) and shorter crops (often biennials or annuals) simultaneously.

Mycorrhiza   Symbiotic associations of the threadlike filaments of a fungus with the roots of higher plants, which can increase the plants' capacity to absorb nutrients from the soil.

Natural farming   A system of agriculture devised by Masanobu Fukuoka that seeks to follow Nature by minimising human interference: no mechanical cultivation, no synthetic fertilisers or prepared compost, no weeding by tillage or herbicides, no dependence on chemicals.

Nematodes   Eelworms, found in great quantities in moist topsoil, many of which are parasitic on plants and animals.

Networking   Establishing and strengthening links between individuals, groups and organisations with similar interests and objectives.

NGO   Nongovernmental organisation; nonprofit, voluntary group engaged in relief and/or development activities.

Niche   A space in the ecosystem which, because of the specific local ecological, physical and/or social characteristics, is suited to a particular plant or animal species or a particular activity.

Nitrogen-fixing   The ability of organisms (bacteria, actinomycetes or algae) to convert atmospheric nitrogen into a form which can be used by higher plants. These organisms may be free-living or live in a symbiotic relationship with the plants.

Nonrenewable resource   Resources such as oil, coal and mineral ores which cannot be naturally regenerated on a time scale that is relevant to human exploitation (IIED 1988).

Nutrient cycling   The recurrent flow of nutrients through a farm or larger agroecosystem such that the major part of the mobile nutrients are kept within the system and reused.

Nutrient harvesting   Deliberate activities to capture nutrients from outside the farm system or from other parts within the farm system and to concentrate them in particular areas in the farm.

Optimum   The best or most favourable condition, degree or amount for a particular situation.

Organic   Any chemical compound containing carbon or derived from living organisms.

Organic farming   A system of agriculture that encourages healthy soils and crops through such practices as nutrient recycling of organic matter (such as compost and crop residue), crop rotations, proper tillage and the avoidance of synthetic fertilisers and pesticides (IASA 1990).

Output   Refers here to farm outputs, i.e. products or functions which are obtained by means of farming activities and which are consumed by the farm household, re-invested in farming (used as internal inputs) or externalised (exchanged or sold).

Parasite   An organism that lives in or on another organism (the host), from which it obtains its food.

Participatory Technology Development (PTD)   The process of combining the indigenous knowlege and research capacities of the local farming communities with that of research and development institutions in an interactive way, in order to identify, generate, test and apply new techniques and practices and to strengthen the existing experimental and technology management capacities of the farmers. Also referred to as People-centred Technology Development.

Pastoralism   The rearing of livestock which graze primarily natural pasture. Nomadic pastoralism refers to a more or less constantly wandering mode of livestock-keeping. In transhumant pastoralism the herds are moved seasonally or periodically between two regions of differing climate regimes (e.g. mountain/valley); the pastoralists occupy a permanent residence in at least one of these regions. In sedentary pastoralism the animals are kept year-round near a permanent residence.

Pathogen   Any micro-organism or virus that lives and feeds (parasitically) on or in a larger host organism and thereby injures it.

Perennial   A plant that lives for three or more years and which normally flowers and fruits at least in its second and subsequent years.

Permaculture   A consciously designed, integrated system of perennial or self-perpetuating species of crops, trees and animals.

Pest   An organism (insect, mite, weed, fungus, disease, animal etc.) that humans wish to control or eliminate for any of various reasons, including possible harm to crops, animals or structures (IASA 1990).

Pest management   Manipulation of pest or potential-pest populations so as to diminish their injury or render them harmless (IASA 1990).

Pesticide   Any substance for destroying or controlling any pest, includes insecticides, herbicides, fungicides, acaracides etc.

Productivity   The relationship between the quantity of goods or services produced and the factors used to produce it; farm productivity can be expressed as output per unit of land, capital, labour time, energy, water, nutrients etc.

Relay cropping   Growing two or more crops simultaneously during part of the life cycle of each crop. The second crop is planted after the first crop has reached its reproductive phase but before it is ready for harvest.

Resistance   Ability of a living organism to survive the disruption of life processes caused by pesticides, disease, drought etc., which would normally cause the death of other similar organisms (IASA 1990).

Resource-enhancing   Improving the quality of existing resources, referring here to all raw materials, energy sources and human capabilities which can be used for farming.

Resurgence   The increase in a pest population after it is freed from natural controls, most commonly following the application of a pesticide that destroys its natural enemies (IASA 1990).

Ridging   Making long, parallel, raised strips of earth, into which seeds are sown. Ridges are usually made at right angles to the slope. They can serve to increase water retention, reduce soil erosion, bury and compost weeds, and create a favourable environment for the seed by concentrating topsoil and raising the root-growing area of the young plants above the water table.

Rotation   Repeated cultivation of a succession of crops (as sole or mixed crops), possibly in combination with fallow, on the same land. One cycle often takes several years to complete.

Run-off   Rainfall or other water that flows across the soil surface and does not infiltrate into the soil.

Semiarid   A climate with average annual rainfall of about 200 – 900 mm with high variability of rainfall.

Sequential cropping   Growing two or more crops after each other in the same field per year. The succeeding crop is planted after the preceding crop has been harvested.

Shifting cultivation   A form of agriculture in which soil fertility is maintained by rotating fields rather than crops. A piece of land is cropped until the soil shows signs of exhaustion or is overrun by weeds, when the land is left to regenerate naturally while cultivation in done elsewhere. New sites are usually cleared by firing (slash-and-burn). Also known as swidden agriculture (IIED 1988).

Silvihorticulture   Care and cultivation of small plots (gardens) devoted primarily to vegetables and fruits, including woody species.

Silvopastoralism   Agroforestry system in which livestock browse or are fed the forage produced by trees and shrubs, and graze shorter plants such as grasses and herbs.

Sole cropping   Growing one crop variety alone in a pure stand.

Stability   Collective aspect of systems characteristics that minimise the negative effects of abrupt, unexpected change on the farm system.

Subhumid   In the tropics, a climate with average annual rainfall of roughly 900 – 1500 mm.

Subsistence agriculture   Farming systems in which a large part of final yield is consumed by the producer. Most subsistence systems involve production of some crops or animals for sale, but the ratio of subsistence to cash production may vary greatly from year to year.

Succession   An orderly process of change in a community (of plants, animals, soil microbes etc.) that results from modification of the environment by organisms and culminates in a system attaining a steady state, or climax (Richards 1974).

Sustainable agriculture   Management of resources for agriculture to satisfy changing human needs, while maintaining or enhancing the quality of the environment and conserving the natural resources.

Symbiosis   The relationship of two or more different organisms in a close association that is beneficial to each organism.

Synergy   The action of two or more substances, organs or organisms to achieve an effect of which each is individually incapable.

Synthetic   Produced by a chemical or artificial process rather than of natural origin.

Systems approach   An approach for studying a system as an entity made up of all its components and their interrelationships, together with relationships between the system and its environment.

Technology   The combination of knowledge, inputs and management practices which are deployed together with productive resources to produce a desired output.

Tenure   The right to property, granted by custom and/or law, which may include land, trees and other plants, animals and water.

Tied ridging   Leaving or forming short ridges of earth at right angles to contour ridges so as to retain rainwater in the field.

Traditional agriculture   Farming systems which are based on indigenous knowledge and practices, and have evolved over many generations.

Transition   The process of changing from one form to another; refers here to the change of a farm system from either HEIA or LEIA to LEISA.

Usufruct   The right to use and enjoy the yield of resources (land, vegetation, livestock etc.) which belong to someone else.

Viability   The capability of living and developing in a given environment or, e.g. in the case of a technology, of being practised in the long term.

Water harvesting   Collection and storage (in a tank or in the soil) of water, either runoff or stream flow, for securing and improving water availability for crop growth and/or animal and human consumption.
Weed   A plant in a place where it is not wanted by humans.

Appendix C Useful contacts and sources of further information

Appendix C1

Further reading

The following list of selected literature is subdivided according to the foregoing chapters and the subsections of Appendix A. Key words are printed in italics; only those are listed that do not already appear in the title. Organisations and publishers are indicated by their acronyms; full names and addresses are given in Appendix C4.

Chapter 1 Agriculture and sustainability

Blaikie, P. 1985. *The political economy of soil erosion in developing countries.* Harlow: Longman. 188 pp.
soil conservation / erosion / government policy

Conway, G.R. and Barbier, E.B. 1990. *After the green revolution: sustainable agriculture for development.* London: Earthscan. 205 pp.
economic development / livelihoods / performance indicators / policy issues

Fowler, C. and Mooney, P. 1990. *Shattering: food, politics, and the loss of genetic diversity.* Tucson: University of Arizona Press. 278 pp.
bioengineering / food policy / genetic conservation / genetic erosion

George, S. 1984. *Ill fares the land: essays on food, hunger and power.* Washington DC: Institute for Policy Studies. 102 pp.
agricultural research / ecological food systems / food policy / food supply / industrial crops / primary health care / terms of trade

Gradwohl, J. and Greenberg, R. 1988. *Saving the tropical forests.* London: Earthscan. 207 pp.
agroforestry / ecology / environment / forestry / reforestation / sustainable agriculture

Kotschi, J., Waters-Bayer, A., Adelhelm, R. and Hoesle, U. 1989. *Ecofarming in agricultural development.* Weikersheim: Margraf/GTZ. 132 pp.
agroforestry / aquaculture / extension / IK / mulching / multiple cropping / participatory research / soil fertility / technical cooperation

Shiva, V. 1988. *Staying alive: women, ecology and development.* London: Zed Books. 224 pp.
agriculture / economic development / food production / forestry / gender issues / India / natural resource management / soil conservation

Third World Network. 1990. *Return to the good earth: damaging effects of modern agriculture and the case for ecological farming.* Penang: Third World Network. 570 pp.
bioengineering / biological pest control / environment / genetic resources / green revolution / IK / pesticides / seed selection

Timberlake, L. 1985. *Africa in crisis: the causes, the cures of environmental bankruptcy.* London: Earthscan. 233 pp.
agroforestry / apartheid / climate / desertification / erosion control / farmers / foreign aid / fuelwood / oral therapy / politics / primary health care / renewable energy / shelterbelts

Wolf, E.C. 1986. *Beyond the green revolution: new approaches for Third World agriculture.* Washington DC: Worldwatch Institute. 46 pp.
agricultural development / bio-engineering / ecodevelopment / traditional agriculture

World Commission on Environment and Development. 1987. *Our common future.* Oxford University Press. 383 pp.
food security / ecosystems / economic development / energy resources / human ecology / human resources / industry / policy issues / population / sustainable agriculture / sustainable development / urban development

Chapter 2 Sustainability and farmers

Chambers, R., Saxena, N.C. and Shah, T. 1989. *To the hands of the poor: water and trees.* London: ITP. 273 pp.

extension / forest management / India / land use / lift irrigation / rural poverty / water management

Dankelman, I. and Davidson, J. 1988. *Women and environment in the Third World: alliance for the future.* London: Earthscan. 210 pp.
energy resources / environmental conservation / fodder / food security / fuelwood / water management / women's organisations

FAO. 1989. *Household food security and forestry: an analysis of socio-economic issues.* Rome: FAO. 147 pp.
forest resources / natural resource management / nutrition / small-scale enterprises

Harwood, R.H. 1979. *Small farm development: understanding and improving farming systems in the humid tropics.* Boulder: Westview. 160 pp.
farming systems analysis / interdisciplinary approach / methodology / mixed farming / on-farm research

IRRI. 1985. *Women in rice farming.* Aldershot: Gower. 531 pp.
agricultural engineering / farming systems / household production system / land use / rural development / women workers

Merrill-Sands, D. 1986. *The technology applications gap: overcoming constraints to small-farm development.* Rome: FAO. 144 pp.
decision-making / farmer-scientist interaction / economics / social organisation / technology adoption / technology development / annotated bibliography

Snyder, M. 1990. *Women: the key to ending hunger.* New York: Hunger Project. 37 pp.
family labour / food policy / food security

Chapter 3   Technology development by farmers

Brokensha, D., Warren, D.M. and Werner, O. (eds) 1980. *Indigenous knowledge systems and development.* Lanham: University Press of America. 460 pp.
ethnoscience / extension / farming systems / folk media / traditional agriculture

Marten, G.G. (ed.) 1986. *Traditional agriculture in Southeast Asia: a human ecology perspective.* Boulder: Westview. 358 pp.
agroecosystems / ethnoecology / multiple cropping / natural resource management / shifting cultivation / socioeconomic aspects / soil management

McCorkle, C.M., Brandstetter, R.H. and McClure, G.D. 1988. *A case study on farmer innovations and communication in Niger.* Washington DC: AED. 125 pp.
extension / farmer-scientist interaction / IK / millet / seed dressing / technology transfer

Richards, P. 1985. *Indigenous agricultural revolution: ecology and food production in West Africa.* London: Hutchinson. 192 pp.
agricultural extension / ecological balance / farming systems / multiple cropping / seed selection

Wilken, G.C. 1987. *Good farmers: traditional agricultural resource management in Mexico and Central America.* Berkeley: University of California Press. 302 pp.
erosion control / farming systems / field-surface management / IK / irrigation / slope management / soil management / terracing

Chapter 4   Low-external-input farming and agroecology

Abadilla, D.C. 1982. *Organic farming.* Quezon City: AFA Publications. 213 pp.
composting / crop rotation / earthworms / health / mulching / multiple cropping / nitrogen fixation / nutrition / pest control / plant protection / soil fertility / soil management

Altieri, M.A. 1987. *Agroecology: the scientific basis of alternative agriculture.* Boulder/ London: Westview/ITP. 227 pp.
agroecosystems / agroforestry / cover crops / crop rotation / FSR / minimum tillage / mulching / multiple cropping / organic farming / pest control / traditional agriculture / weed control

Bayliss-Smith, T.P. 1982. *The ecology of agricultural systems.* Cambridge: Dept of Geography, University of Cambridge. 112 pp.
agricultural economics / ecosystems / energy flow / environment / farming systems / temperate / tropical

Dover, M. and Talbot, L.M. 1987. *To feed the earth: agroecology for sustainable development.* Washington DC: World Resources Institute. 88 pp.
ecological principles / environmental constraints / policy issues / sustainable agriculture

Edwards, C.A. (ed.) 1990. *Sustainable agricultural systems.* Ankeny: SWCS. 696 pp.
bioengineering / crop rotation / ecological aspects / integrated farming / nutrient cycling / pasture management / pest control / policy issues / socioeconomic aspects / tillage / tropical / USA

Gliessman, S.R. (ed.) 1990. *Agroecology: researching the ecological basis for sustainable agriculture*. New York: Springer. 380 pp.
agroecosystems / agroforestry / allelopathy / crop diversification / cultivation systems / energy flow / insect-borne diseases / intercropping / nutrient mobility / pest control / quantification

Kotschi, J. (ed.) 1990. *Ecofarming practices for tropical smallholdings*. Weikersheim: Margraf/GTZ. 185 pp.
Africa / agroforestry / economics / fodder production / green manure / multiple cropping / PTD / rural development / soil conservation / soil fertility / technical cooperation / water conservation

Singh, R.P., Parr, J.F. and Stewart, B.A. 1990. *Dryland agriculture: strategies for sustainability*. New York: Springer. 373 pp.
climatology / crop residues / economic analysis / fertility / nutrient cycling / organic matter / simulation models / soil conservation / soil diseases / water erosion / water utilisation / wind erosion

Stonehouse, B. (ed.) 1981. *Biological husbandry: a scientific approach to organic farming*. London: Butterworth. 352 pp.
agricultural technology / cropping systems / energy conservation / farming systems / mulching / plant protection / soil analysis / soil fertility / temperate / tropical / waste recycling

Chapter 5   Basic ecological principles of LEISA

Barrow, C.J. 1987. *Water resources and agricultural development in the tropics*. Harlow: Longman. 356 pp.
dams / dew / floodwater / fog / groundwater / irrigation / mist / rainfall / reservoirs / runoff management / water management

Chambers, R. 1990. *Microenvironments unobserved*. Gatekeeper Series 22. London: IIED. 18 pp.
agricultural research / extension / microclimate management / nutrient harvesting / sustainable development / water harvesting

CTA. 1988. *Agroforestry: the efficiency of trees in African agrarian production and rural landscapes*. Wageningen: CTA/Terres et Vie/GTZ/ICRAF. 394 pp.
economics / environmental conservation / farming systems / IK / social participation / tree-crop interactions / women / practical experiences

Douglas, M.G. 1990. *Integrating conservation into the farming system: land use planning for smallholder farmers – concepts and procedures*. London: Commonwealth Secretariat. 137 pp.
integrated systems / land degradation / pastoralism / policy issues / social participation / socioeconomic aspects / soil conservation

Falloux, F. and Mukendi, A. (eds) 1988. *Desertification control and renewable resource management in the Sahelian and Sudanian zones of West Africa*. Washington: World Bank. 119 pp.
energy resources / erosion control / fuelwood consumption / land tenure / pastoralism / social participation / water management

FAO. 1983. *Integrating crops and livestock in West Africa*. Animal Production and Health Paper 41. Rome: FAO. 112 pp.
agropastoralism / agrosilvopastoralism / crop residues / land use / manure / mixed farming

Foley, G. and Barnard, G. 1984. *Farm and community forestry*. London: Earthscan. 236 pp.
agroforestry / deforestation / eucalyptus / fuelwood / IK / project design / project implementation

Francis, C.A. (ed.) 1986. *Multiple cropping systems*. New York: Macmillan. 383 pp.
agricultural research / breeding / cereals / ecology / economics / Leguminosae / pest control / plant interactions / risks / sociocultural factors

Hansen, M. 1988. *Escape from the pesticide treadmill: alternatives to pesticides in developing countries*. Mount Vernon: ICPR. 185 pp.
banana / cassava / coconut / cotton / IPM / pest control / pesticide use / rice / soybean / case studies

Juma, C. 1989. *Biological diversity and innovation: conserving and utilizing genetic resources in Kenya*. Nairobi: ACTS. 139 pp.
bio-engineering / environmental conservation / genetic erosion / germplasm conservation / policy issues / sustainable development

Lal, R. and Stewart, B.A. 1990. *Soil degradation*. New York: Springer. 345 pp.
biological soil degradation / chemical soil degradation / salinity / soil compaction / soil erosion / soil wetness

Lipton, M. 1989. *New seeds and poor people*. London: Unwin Hyman. 473 pp.
genetic resources / green revolution / modern varieties / physical properties / seed selection / socioeconomic aspects / small farms

Lynch, J.M. 1983. *Soil biotechnology: micro-biological factors in crop productivity*. Oxford: Blackwell. 191 pp.
cultivation practices / fungus diseases / organic farming / soil bacteria / soil management / soil science / viral diseases

Ohm, H.W. and Nagy, J.G. (eds) 1985. *Appropriate technologies for farmers in semi-arid West Africa*. West Lafayette: Purdue University. 359 pp.
agropastoralism / animal husbandry / animal traction / cotton / cowpea / maize / millet / multiple cropping / plant improvement / rice / soil fertility / soil management / sorghum / water management

Pacey, A. and Cullis, A. 1986. *Rainwater harvesting: the collection of rainfall and runoff in rural areas*. London: ITP. 216 pp.
drinking water / irrigation / runoff farming / water storage

Prescott, R. and Prescott A.C. 1983. *Genes from the wild: using wild genetic resources for food and raw materials*. London: Earthscan. 102 pp.
amylaceous crops / aquaculture / fibre crops / fodder / fruits / gene banks / genetic diversity / industrial crops / selection / vegetables

Steiner, K.G. 1984. *Intercropping in tropical smallholder agriculture with special reference to West Africa*. Eschborn: GTZ. 304 pp.
agricultural research / experimental design / farming systems / land equivalent ratio / pest control / plant competition / socioeconomic aspects / weed control / yield stability

Subba Rao, N.S. 1984. *Biofertilizers in agriculture*. New Delhi: Oxford/IBH Publishing Co. 186 pp.
algae / Azolla / composting / microbiology / nitrogen fixation / rhizobium / soil fertility / soil management

Ubels, J. (ed.) 1990. *Design for sustainable farmer-managed irrigation schemes in sub-Saharan Africa*. Wageningen: Department of Irrigation and Soil and Water Conservation, Agricultural University.
farming systems / gender issues / small farms / social participation / sustainable agriculture / practical experiences

Chapter 6   Development of LEISA systems

Agarwal, A. and Narain, S. 1989. *Towards green villages: a strategy for environmentally-sound and participatory rural development*. New Delhi: CSE. 52 pp.
conceptual framework / government policy / institutional framework / PTD / social participation / tree management / village

FAO. 1984. *Improved production systems as an alternative to shifting cultivation*. Soils Bulletin 53. Rome: FAO. 201 pp.
agroforestry / agroecology / IK / institutional framework / socioeconomic aspects

Gregersen, H., Draper, S. and Elz, D. (eds) 1989. *People and trees: the role of social forestry in sustainable development*. Washington DC: World Bank. 273 pp.
agricultural research / agroforestry / cooking stoves / education / employment / fuelwood / land tenure / monitoring / productivity / project implementation / social participation / training / tree tenure

Kerkhof, P. 1990. *Agroforestry in Africa: a survey of project experience*. London: Panos. 216 pp.
extension techniques / forest nurseries / project design / project implementation / seedling production / tree growing / case studies

McRobie, G. (ed.) 1990. *Tools for organic farming: a manual of appropriate equipment and treatment*. London: ITP. 77 pp.
cultivation practices / manufacturers' index / plant protection / storage / sustainable agriculture

Panos. 1987. *Towards sustainable development*. London: Panos.
environment / forest management / pest control / policy issue / soil management / water management / case studies

Pingali, P., Bigot, Y. and Binswanger, H.P. 1987. *Agricultural mechanization and the evolution of farming systems in sub-Saharan Africa*. Baltimore: John Hopkins University. 216 pp.
animal traction / government policy / labour / modern agriculture / ploughing / tractors / tree crops

Tull, K. 1987. *Experiences in success: case studies in growing enough food through regenerative agriculture*. Emmaus: Rodale. 53 pp.
agroforestry / gardening / organic farming / soil conservation / sustainable agriculture

Chapter 7   Actors and activities in developing LEISA technologies

Basant, R. and Subrahmamian, K.K. 1990. *Agromechanical diffusion in a backward region*. London: ITP. 94 pp.
agricultural equipment / autonomous development / social participation / socio-economic aspects / technology transfer

Chambers, R., Pacey, A. and Thrupp, L.A. (eds) 1989. *Farmer first: farmer innovation and agricultural research*. London: ITP. 218 pp.
agricultural technology / agroforestry / IK / on-farm research / PTD / social participation / case studies

Farrington, J. and Martin, A. 1988. *Farmer participation in agricultural research: a review of concepts and practices*. Agricultural Administration Unit Occasional Paper 9. London: ODI. 79 pp.
farmers / IK / social participation / theory / practical experiences

IFAP. 1990. *Sustainable farming and the role of farmers' organizations*. Paris: IFAP. 62 pp.
agricultural research / extension / farming systems / FSR / social participation / women

Kaimowitz, D. (ed.) 1990. *Making the link: agricultural research and technology transfer in developing countries*. Boulder: Westview/ISNAR. 278 pp.
extension / farmer – scientist interaction / on-farm research / participatory research

Merrill-Sands, D. and Kaimowitz, D. 1990. *The technology triangle: linking farmers, technology transfer agents and agricultural researchers*. The Hague: ISNAR. 118 pp.
agricultural research / extension / farmer-scientist interaction / human resources

Rhoades, R.E. 1984. *Breaking new ground: agricultural anthropology*. Lima: CIP. 71 pp.
ecological anthropology / ethnobotany / farmer-scientist interaction / FSR / IK / technology transformation / training methods

Röling, N. 1988. *Extension science: information systems in agricultural development*. Cambridge University Press. 233 pp.
research diffusion / target categories / voluntary change / practical experiences

Chapter 8   Participatory Technology Development in practice

Ashby, J. 1990. *Evaluating technology with farmers*. Cali: CIAT. 95 pp.
farmer – scientist interaction / methods / problem analysis / technology development / technology transfer / handbook

Bunch, R. 1985. *Two ears of corn: a guide to people-centered agricultural improvement*. 2nd ed. Oklahoma City: WN. 250 pp.
appropriate technology / extension / planning / PTD / small-scale agriculture

Davis Case, D. 1989. *Participatory assessment, monitoring and evaluation*. Community Forestry Note 2. Rome: FAO. 150 pp.
community forestry / rural communities / social participation

ETC. 1991. *Learning for people-centred technology development: a training guide*. Leusden: ETC.
farmer-to-farmer extension / farmers' experiments / methods / participatory diagnosis / PTD / rural communities / social participation

Hope, A. and Timmel, S. 1984. *Training for transformation: a handbook for community workers*. Mambo Press, Gweru, PO Box 779, Zimbabwe.
Vol. I, 147 pp.; Vol. II, 131 pp.; Vol. III, 182 pp.
methods / participatory diagnosis / participatory extension / people's awareness / rural development

McCracken, J.A., Pretty, J.N. and Conway, G.R. 1988. *An introduction to Rapid Rural Appraisal for agricultural development*. London: IIED. 96 pp.
concepts / farmer – scientist interaction / methodology / project design / project implementation

Oakley, P. 1991. *Projects with people: the practice of participation in rural development*. Geneva: ILO. 284 pp.
development aid / rural communities / social participation / technology transfer

Rugh, J. 1986. *Self-evaluation: ideas for participatory evaluation of rural community development projects*. Oklahoma City: WN. 42 pp.
development cooperation / rural development / social participation

Stephens, A. 1988. *Participatory monitoring and evaluation: handbook for field workers*. Bangkok: FAO-RAPA. 51 pp.
extension / rural development / PTD / social participation

Verhagen, K. 1987. *Self-help promotion: a challenge to the NGO community*. Amsterdam: KIT/CEBEMO. 152 pp.
methodology/ rural development / social groups / social participation

Appendix A1　Soil and nutrient management

BOSTID. 1979. *Tropical legumes: resources for the future.* Washington DC: NRC. 331 pp.
  fodder crops / fruits / green manures / leguminous crops / nuts / oil crops / plant descriptions / research contacts
BOSTID. 1981. *Food, fuel and fertilizer from organic wastes.* Washington DC: NRC. 153 pp.
  biogas / biological fertilisers / fish farming / mushrooms / organic fertilisers / sewage collection / waste disposal
BOSTID. 1984. *Leucaena: promising forage and tree crop for the tropics.* Washington DC: NRC. 100 pp.
  agroforestry / byproducts / erosion control / fuelwood / reforestation / shifting cultivation / soil management / terracing
Dalzell, H.W., Biddlestone, A.J., Gray, K.R. and Thurairajan, K. 1987. *Soil management: compost production and use in tropical and subtropical environments.* Soils Bulletin 56. Rome: FAO. 177 pp.
  environmental aspects / socioeconomic aspects / soil fertility / practical experiences
Evans, D.O. and Macklin, B. 1990. *Perennial* Sesbania: *production and use.* Waimanalo: NFTA. 41 pp.
  fodder / food / management / seed / soil improvement / wood
Gerold, F. 1989. *Sunnhemp 'marejea':* Crotalaria ochroleuca. Peramiho: Benedictine Publications. 96 pp.
  botanical pesticide / cultivation practices / organic fertiliser / sustainable agriculture
Gershuny, G. and Smillie, J. 1986. *The soul of soil: a guide to ecological soil management.* St Johnsbury: GAIA Services. 109 pp.
  crop rotation / green manure / organic farming / soil fertility / soil science / soil testing methods / tropical / USA
Parnes, R. 1990. *Fertile soil: a grower's guide to organic and inorganic fertilizers.* Davis: agAccess. 190 pp.
  compost / soil fertility / soil nutrients

Appendix A2　Managing flows of solar radiation, air and water

Chleq, J.F. and Dupriez, H. 1988. *Vanishing land and water.* London: Macmillan. 117 pp.
  erosion control / irrigation / soil and water conservation / water lifting / wells
Hudson, N.W. 1983. *Field engineering for agricultural development.* Oxford University Press. 240 pp.
  drainage / irrigation / land classification / soil and water conservation / water storage
Hurni, H. 1986. *Guidelines for development agents on soil conservation in Ethiopia.* Addis Ababa: Community Forests and Soil Conservation Development Department, Ministry of Agriculture. 100 pp.
  agroecozoning / agropastoralism / alley cropping / checkdams / controlled grazing / cutoff drains / forestry / grass strips / IK / level bunds / soil management / terracing / tree planting
Kuchelmeister, G. 1990. *Hedges for resource-poor land users in developing countries.* Eschborn: GTZ. 256 pp.
  agroforestry / alley cropping / appropriate species / cultivation practices / erosion control / fodder / fuelwood / green manure / mulching / natural resource management / nutrition
Murnyak, D. and Murnyak, M. 1990. *Raising fish in ponds: a farmer's guide to* Tilapia *culture.* Little Rock: Heifer. 75 pp.
  aquaculture / fish management / integrated farming / pond construction
Reij, C., Mulder, P. and Begemann, L. 1989. *Water harvesting for plant production.* Washington DC: World Bank. 123 pp.
  cost-benefit analysis / environmental factors / extension / sociological aspects / training
Stern, P. 1979. *Small-scale irrigation: a manual of low-cost water technology.* London: ITP. 152 pp.
  canals / erosion control / hydraulic structures / rivers / soil management / soil science / streams / terracing / water lifting
Tillman, G. 1981. *Environmentally sound small-scale water projects: guidelines for planning.* New York: Codel. 142 pp.
  environment / health / irrigation / sanitation / waste treatment / water management
Unger, P.W. 1984. *Tillage systems for soil and water conservation.* Soils Bulletin 54. Rome: FAO. 278 pp.
  agricultural equipment / cultivation systems / erosion control / land degradation / mulching / minimum tillage / shifting cultivation

World Bank. 1987. *Vetiver grass: the hedge against erosion*. Washington DC: World Bank. 78 pp.
erosion control / local names / management / vegetative contour hedges / water conservation

World Neighbors. 1985. *Introduction to soil and water conservation practices*. Oklahoma City: WN. 33 pp.
A-frame / contour lines / erosion control / soil fertility / soil management / water management

World Neighbors. 1986. *Leucaena-based farming*. Oklahoma City: WN. 29 pp.
A-frame / animal feeding / contour planting / cultivation practices / erosion control / multiple cropping / terracing / weed control

Appendix A3   Pest and disease management

van Alebeek, F.A.N. 1989. *Integrated pest management: a catalogue of training and extension materials*. Wageningen: Dept of Entomology, Wageningen Agricultural University/CTA. 305 pp.
extension / pest control / training / catalogue

Appert, J. 1987. *The storage of food grains and seeds*. London: Macmillan/CTA. 146 pp.
centralised storage / food processing / pest control / rural storage / storage hygiene / storage pests

Golob, P. and Webley, D.J. 1980. *The use of plants and minerals as traditional protectants of stored products*. London: Tropical Products Institute. 32 pp.
agricultural products / ashes / insecticidal plants / minerals / neem plant extracts

Malaret, L. 1985. *Safe pest control: an NGO action guide*. Nairobi: ELCI. 69 pp.
biological control / health / IPM / pesticide use / pesticides / side effects / addresses

Mathias-Mundy, E. and McCorkle, C.M. 1989. *Ethnoveterinary medicine: an annotated bibliography*. Ames: Technology and Social Change Program, Iowa State University. 199 pp.
veterinary anthropology / bibliography

Matzigkeit, U. 1990. *Natural veterinary medicine: ectoparasites in the tropics and subtropics*. Weikersheim: Margraf/AGRECOL. 183 pp.
flies / insecticidal plants / lice / mites / repellent plants / ticks / traditional medicine

van Schoubroeck, F.H.J., Herens, M., de Louw, W., Louwen, J.M. and Overtoom, T. 1990. *Managing pests and pesticides in small-scale agriculture*. Wageningen: CON. 204 pp.
agricultural research / biological pest control / code of conduct / cultivation practices / extension / handling / IPM / legislation / plant protection / storage / case studies / annotated bibliography / list of organisations

Stoll, G. 1986. *Natural crop protection based on local farm resources in the tropics and subtropics*. Langen: Margraf. 186 pp.
ashes / bait / biological control / disease control / insecticidal plants / IPM / pesticides / stored products / traps / vegetable oils

Appendix A4   Choosing, conserving and improving genetic resources

Arraudeau, M.A. and Vergara, B.S. 1988. *A farmer's primer on growing upland rice*. Manila: IRRI. 283 pp.
cultivation practices / cultivation systems / disease control / fertilisation / morphology / pest control / seed selection / weed control

BOSTID. 1975. *Underexploited tropical plants with promising economic value*. Washington DC: NRC. 189 pp.
amylaceous crops / food science / forestry / fruits / nuts / oil crops / transport / research contacts

BOSTID. 1981. *The winged bean: a high protein crop for the tropics*. Washington DC: NRS. 46 pp.
cultivation practices / nutritive value / plant protection

BOSTID. 1983. *Little known Asian animals with a promising economic future*. Washington DC: NRC. 124 pp.
babirusa / banteng / indigenous genetic resources / Javan warty pig / kouprey / madura / mithan / yak

BOSTID. 1989. *Lost crops of the Incas: little-known plants of the Andes with promise for worldwide cultivation*. Washington DC: NRC. 415 pp.
cultivation practices / fruits / genetic conservation / grain legumes / IK / nuts / root crops / species / uses / vegetables

BOSTID. 1990. *Saline agriculture: salt-tolerant plants for developing countries.* Washington DC: NRC. 143 pp.
  fibre crops / fodder / fuelwood / grain legumes / oil crops / resins / root crops / saline soils / traditional plants
Kay, D.E. 1979. *Food legumes.* London: Tropical Products Institute. 451 pp.
  growth requirements / harvesting / planting / processing / storage / trade
Kay, D.E. (revised by E.G.B. Gooding) 1987. *Root crops.* London: ODNRI. 380 pp.
  botany / cultivation practices / handling / harvesting / pests / plant diseases / trade
Martin, F.W. Campbell, C.W. and Ruberte, R.M. 1987. *Perennial edible fruits of the tropics: an inventory.* Washington DC: US Dept of Agriculture. 250 pp.
  major fruits / minor fruits / basic information
Terra, G.J.A. 1973. *Tropical vegetables: vegetables growing in the tropics and subtropics especially of indigenous vegetables.* Amsterdam: Royal Tropical Institute. 112 pp.
  garden vegetables / nutritional requirements / plant protection / soil treatment

Appendix A5   Integrated farm systems

Capistrano, L., Durno, J. and Moeliono, I. (eds) 1990. *Resource book on sustainable agriculture for the uplands.* Silang: IIRR. 199 pp.
  agroforestry / reforestation / seed production / soil and water conservation / upland farming
Critchley, W. (ed. O. Graham) 1991. *Looking after our land: new approaches to soil and water conservation in dryland Africa.* Oxford: Oxfam. 88 pp.
  case studies / social participation / soil management / agroforestry / handbook with video film
Dupriez, H. and de Leener, P. 1988. *Agriculture in African rural communities: crops and soils.* London: Macmillan. 294 pp.
  cultivation practices / extension / hedges / humus / land settlement / land use / plant growth / soil life / training / trees / handbook
Dupriez, H. and de Leener, P. 1990. *African gardens and orchards: growing vegetables and fruits.* London: Macmillan. 354 pp.
  home gardens /. horticulture / multiple cropping / plant protection / propagation / soil management / traditional species / handbook
Folliott, P.F. and Thames, J.L. 1983. *Environmentally sound small-scale forestry projects: guidelines for planning.* Arlington: VITA. 109 pp.
  agroforestry / fuelwood / shelterbelts / project design / handbook
Hegde, N.G. 1987. *Handbook of wastelands development.* Pune: BAIF. 102 pp.
  afforestation / cultivation practices / India / land use / species / water management
Hoskin, C.M. 1973. *The Samaka guide to homesite farming.* Manila: Samaka Service Centre. 173 pp.
  compost / construction / fertilisation / fish ponds / fruit trees / goats / home gardens / horticulture / marketing / mushrooms / pigs / poultry / rabbits / vegetables / water buffalo / wells
IIRR. 1988. *The bio-intensive approach to small-scale household food production.* Silang: IIRR. (separate sheets)
  bio-intensive gardening / diversity / local seeds / natural pesticides / nutrition / organic fertiliser / rotation / soil preparation
IIRR/UNICEF 1988. *Regenerative agricultural technologies: trainer's kit.* Silang: IIRR. (separate sheets)
  alley cropping / animal husbandry / aquaculture / contour farming / fruit trees / horticulture / natural pesticides / nutrition / organic fertiliser / small-scale farming / sustainable agriculture
IIRR. 1990. *Low-external-input rice production (LIRP): technology information kit.* Silang: IIRR. (separate sheets)
  byproducts / cropping patterns / farm implements / economics / green manure / integrated crop-livestock-fish systems / nutrient cycling / organic fertiliser / pest control / seed handling / traditional cultivars / transplanting / water management / weed control
Jacobs, L. 1986. *Environmentally sound small-scale livestock projects: guidelines for planning.* Alexandria: VITA. 149 pp.
  agricultural wastes / agroforestry / animal husbandry / farming systems / organic manure / nutrient cycling / soil fertility / sugar
Mollison, B. 1988. *Permaculture: a designers' manual.* Tyalgum: Tagari Publications. 574 pp.
  Australia / cultivation practices / farming systems / organic farming / project design / soil management / water management

Orev, Y. 1986. *A practical handbook on desert range improvement techniques*. Geneva: WHO. 176 pp.
fencing / natural resource management / transhumance routes / vegetative cover / water resources

Rocheleau, D., Weber, F. and Field-Juma, A. 1988. *Agroforestry in dryland Africa*. Nairobi: ICRAF. 311 pp.
alley cropping / contour vegetation / cultivation practices / fallow / living fences / microcatchments / multipurpose trees / social participation / species / terrace building / water management

Scheinman, D. and Mchome, C. 1986. *Caring for the land of the Usambaras: a guide to preserving the environment through agriculture, agroforestry and zero grazing*. Eschborn: GTZ. 287 pp.
animal husbandry / crops / cultivation practices / ecodevelopment / erosion control / nutrition / Tanzania / handbook

Sommers, P. 1983. *Low-cost farming in the humid tropics: an illustrated handbook*. Manila: Island Publishing House. 38 pp.
fertilisation / organic farming / pest control / seed preparation / seed selection / site preparation / storage / water management / weed control

Vukasin, H.L. 1988. *Environmentally sound small-scale agricultural projects: guidelines for planning*. Arlington: Codel/VITA. 103 pp.
agroforestry / crop protection / erosion control / evaluation / IK / integrated nutrient management / irrigation / minimum tillage / organic farming / rural development / social participation / water supply / handbook

Wijewardene, R. and Waidyanatha, P. 1984. *Conservation farming for small farmers in the humid tropics: systems, techniques and tools*. London: Commonwealth Secretariat. 39 pp.
*alley cropping / erosion control / farm equipment / fuelwood / **Gliricidia maculata** / herbicides / land clearing / **Leucaena leucocephala** / minimum tillage / mulching / shifting cultivation*

World Neighbors. 1989. *Integrated farm management*. Oklahoma City: WN.
animal husbandry / crop husbandry / diversification / fodder / PTD / soil and water conservation / trees

Appendix C2

Annotated bibliographies on sustainable agriculture

AGRECOL/ENDA/GEYSER/INADES. 1989. *Agriculture écologique en Afrique francophone*. Langenbruck: AGRECOL. 96 pp., 180 documents, 33 journals, 62 organisations.
agroecosystems / agroforestry / animal husbandry / crop protection / genetic resources / nutrition / self-help / soil fertility / storage

AGRECOL/GEYSER. 1990. *Agroecologia en America Latina: una guia*. Langenbruck: AGRECOL. 144 pp., 104 documents, 21 journals, 82 organisations.
animal integration / crop production / crop protection / genetic resources / health / multiple cropping / nutrition / small-scale agriculture / soil and water conservation / soil fertility / tillage

Amanor, K. 1990. *NGOs and agricultural technology development: a collection of abstracts*. London: ODI. 62 pp., 109 documents, 18 journals and directories.
evaluation / journals and directories / methodology / organisation of research / project descriptions

Carls, J. *Abstracts on sustainable agriculture*. Eschborn: GATE. Vol. 1, 1988, 294 pp., 250 titles; Vol. 2, 1989, 372 pp., 263 titles.
agroecology / agroforestry / agrometeorology / cropping systems / erosion control / FSR / homegardens / integrated systems / plant protection / potential crops for marginal lands/ seed production / soil fertility / traditional landuse systems / water management

FAO. Annually. *FAO publications catalogue*. Rome: FAO.
agriculture / animal production and health / economic and social development / fisheries / food and nutrition / forestry / land and water development / plant production and protection / statistics

IRRI. 1989. *Publications of the international agricultural research and development centers*. Manila: IRRI/CGIAR. 730 pp.
ACIAR / AVRDC / CIAT / CIMMYT / CIP / FFTC-ASPAC / GTZ / IBPGR / ICARDA / ICIMOD / ICIPE / ICLARM / ICRAF / ICRISAT / IFPRI / IIMI / IITA / ILCA / ILRAD / IRRI / ISNAR / WARDA / Winrock

KIT. 1988. *Tropical agriculture: selected handbooks*. Amsterdam: KIT/CTA. 119 pp., 405 titles.

agricultural production / agroforestry / animal husbandry / appropriate technology / aquaculture / cooperatives / credit / crop production / crop protection / economics and policy / education and training / employment / energy / environmental protection / extension / forage production / FSR / marketing and input supply / pasture and range management / planning and management / plant breeding / post-harvest operations / social participation / soil and water management

KIT. 1990. *Environmental management in the tropics: an annotated bibliography*. Amsterdam: KIT. 236 pp., 468 titles.

agroforestry / energy and fuelwood / environmental impact assessment / environmental planning and education / extension / forest management / institutionalisation and legislation / landuse planning and management / range management / rural environment and sustainable development / socioeconomic issues / strategies / soil conservation and erosion control / sustainability of farming systems / water management

Merrill-Sands, D. 1986. *The technology application gap: overcoming constraints to small-farm development*. Part B: Annotated bibliography of selected references pertinent to understanding the technology application gap. Rome: FAO. 49 pp., 45 titles.

appropriate technology development / decision-making / farming systems and development / household and production economics / organisation and management of production / social organisation of households

Nanda, M. 1990. *Planting the future: a resource guide to sustainable agriculture in the Third World*. Minneapolis: IASA. 309 pp.

addresses / group experiences / periodicals / visual aids

SATIS. 1986/1988/1990. *Catalog: a world of information from the member bookshops of SATIS*. Utrecht: SATIS.

agriculture / animal husbandry / aquaculture / building / cultivation practices / energy / environment / farm equipment / food preparation / forestry / health / nutrition / soil management / storage / water

Appendix C3

Some periodicals on sustainable agriculture in the tropics

ABSTRECO (Abstracts on Sustainable Agriculture). Dept of Ecological Agriculture, Wageningen Agricultural University, Haarweg 333, NL-6709 RZ Wageningen, Netherlands.

ecological agriculture / temperate / tropical / 8 times a year

African Diversity. African Committee for Plant Genetic Resource (ACPGR), available from RAFI, European Office, c/o Institute for International Cooperation (IIZ), Wipplingerstr. 32, A-1010 Vienna, Austria.

biological diversity / biotechnology / plant genetic resources / Africa / 3 times a year

African Farmer. The Hunger Project, 1 Madison Ave, New York, New York 10010, USA.

agricultural development / IK / self-reliant development / small-scale farming / Africa / English / French / quarterly

African Livestock Research. ILCA, PO Box 5689, Addis Ababa, Ethiopia.

livestock development / FSR / Africa / scientific / quarterly

Agricultural Information Development Bulletin. UNESCAP, United Nations Building, Rajadamnern Nok Ave, Bangkok 10200, Thailand.

agricultural development / small-scale farming / Asia / Pacific / quarterly

Agriculture Administration (Research and Extension) Network Papers. ODI, Regent's College, Inner Circle, Regent's Park, London NW1 4NS, UK.

participatory research-extension / research-extension methodology / networking / twice a year

Agriculture and Human Values. Agriculture, Food, and Human Values Society, 1001 McCarthy Hall, University of Florida, Gainesville, Florida 32611, USA.

alternative agriculture / interdisciplinary / policies / practices / quarterly

Agricultures Actualité: réussir l'alternative. GEYSER, Vacquières, F-34270 St Mathieu de Treviers, France.

ecological agriculture / sustainable agriculture / temperate / tropical / abstracts / information / French / quarterly

Agroforestry Systems. Kluwer Academic Publishers in cooperation with ICRAF, PO Box 17, NL-3300 AA Dordrecht, Netherlands.

agroforestry / scientific / quarterly

Agroforestry Today. ICRAF, PO Box 30677, Nairobi, Kenya.

agroforestry / tree growing / information / institutional / quarterly / English / French

American Journal of Alternative Agriculture. Institute of Alternative Agriculture, 9200 Edmonston Road, Suite 117, Greenbelt, Maryland 20770, USA.
agroecosystems / ecological agriculture / sustainable agriculture / policies / temperate / tropical / information update / quarterly

Appropriate Technology. ITP, 103-105 Southampton Row, London WC1B 4HH, UK.
farm implements / rural development / information update / quarterly

At ICRISAT. ICRISAT, Patancheru, Andhra Pradesh 502 324, India.
chickpea / groundnut / millet / pigeonpea / sorghum / semiarid / Asia / Africa / research information / institutional / quarterly

AT-Source. AT-Source, PO Box 41, NL-6700 AA Wageningen, Netherlands.
appropriate technology / rural development / small-scale farming / information / practical / English / French / quarterly

The BAIF Journal. BAIF, Senapati Bapat Road, Pune 411 016, India.
agroforestry / livestock development / rural development / tribals / regional / institutional / quarterly

Baobab. Arid Lands Unit, Oxfam, 274 Banbury Road, Oxford OX2 7DZ, UK.
rural development / small-scale farming / semiarid / Africa / networking / information update / English / French / quarterly

Bean Newsletter. CIAT, AA 6713, Cali, Colombia.
bean / research highlights / English / Spanish / quarterly

Biological Agriculture and Horticulture. AB Academic Publishers, PO Box 42, Bicester, Oxon OX6 7NW, UK.
ecological agriculture / ecological horticulture / temperate / tropical / scientific / quarterly

Biotechnology and Development Monitor. Department of International Relations and Public International Law, University of Amsterdam, Oudezijds Achterburgwal 237, NL-1012 DL Amsterdam, Netherlands.
biotechnology / development / bi-monthly

BOS NiEuWSLETTER. BOS Foundation, PO Box 23, NL-6700 AA Wageningen, Netherlands.
agroforestry / forestry / social forestry / occasional

Buffalo Bulletin. International Buffalo Information Center (IBIC), Kasetsart University, Bangkhen, Bangkok 10900, Thailand.
buffalo / Asia / quarterly

Cassava Newsletter. CIAT, AA 6713, Cali, Colombia.
cassava / research highlights / English / Spanish / quarterly

Ceres. FAO, Via delle Terme di Caracalla, 00100 Rome, Italy.
agricultural development / information update / English / French / Spanish / Arabic / bi-monthly

CIAT international. Publications Unit, CIAT, AA 6713, Cali, Colombia.
bean / cassava / FSR / research information / institutional / 3 times a year

CIKARD News. CIKARD, 318B Curtiss Hall, Iowa State University, Ames, Iowa 50011, USA.
agriculture / IK / institutional / networking / quarterly

CONTOUR. Asia Soil Conservation Network (ASOCON), Manggala Wanabakti, Blok IV Lt 8, Jl Gatot Subroto, PO Box 133 JKWB, Jakarta 10270, Indonesia.
soil and water conservation / Southeast Asia / networking / 3 times a year

Cover Crops News. CIDICCO, PO Box 278-C, Tegucigalpa DC, Honduras.
cover crops / small-scale farming / humid / subhumid / Latin America / quarterly

The Cultivar. Agroecology Program, University of California, Santa Cruz, California 95064, USA.
agroecology / sustainable agriculture / temperate / tropical / quarterly

DAP Project Bulletin. ACIAR Draught Animal Power Project, Graduate School of Tropical Veterinary Science and Agriculture, James Cook University, Townsville, Queensland 4811, Australia.
draught animals / draught farming systems / occasional

Drylander. Drylands Research Institute, University of California, Riverside, California 92521, USA.
agroecology / desertification / sustainable agriculture / semiarid / USA / international / quarterly

ECHO News. Educational Concerns for Hunger Organization, 17430 Durrance Road, North Fort Myers, Florida 33917, USA.
food production / indigenous crops and trees / seeds / small-scale farming / practical / quarterly

Eco Africa. African NGOs Environment Network (ANEN), PO Box 53844, Nairobi, Kenya.
environment / grass-roots initiatives / natural resource management / NGOs / sustainable development / Africa / networking / English / French / bi-monthly

Ecoforum. ELCI, PO Box 72461, Nairobi, Kenya.
  environment / natural resource management / NGOs / sustainable development / networking / English / French / Spanish / bi-monthly
The Ecologist. Ecosystems Ltd, 29A High St, New Malden, Surrey KT3 4BY, UK.
  development / environment / natural resource management / international / bi-monthly
Ecology and Farming. IFOAM. c/o Okozentrum Imsbach, D-6695 Tholey-Theley, Germany.
  agroecology / ecological agriculture / marketing / policies / temperate / tropical / networking / information update / English / quarterly
Farm Forestry News. Multipurpose Tree Species Research Network Forestry/Fuelwood Research and Development Project, Winrock International, 1611 N Kent St, Suite 600, Arlington, Virginia 22209, USA.
  agroforestry / multipurpose trees / Asia / networking / quarterly
Farming for Development. IFAP, 21 rue Chaptal, F-75009 Paris, France.
  agricultural development / farmer organisations / networking / English / French / quarterly
Farming Systems Newsletter. CIMMYT, PO Box MP 154, Mt Pleasant, Harare, Zimbabwe.
  maize / wheat / Africa / research highlights / networking / quarterly
Food Matters Worldwide. Farmers' World Network, Arthur Rank Centre, NAC, Stoneleigh, Warks CV8 2LZ, UK.
  agricultural development / food security / international / farmer-to-farmer networking / bi-monthly
FTPP Newsletter. Forests, Trees and People Program, International Research Development Centre (IRDC), Swedish University of Agricultural Sciences, S-75007 Uppsala, Sweden.
  community forestry / networking / information update / quarterly
Gatekeeper Series. Sustainable Agriculture Programme, IIED, 3 Endsleigh St, London WC1H ODD, UK.
  agricultural development / briefing papers / occasional
GATE: question, answers, information. GATE/GTZ, PO Box 5180, D-6236 Eschborn 1, Germany.
  appropriate technology / rural development / small-scale farming / sustainable agriculture / information update / quarterly
Global Pesticide Monitor. PAN, 965 Mission St, 514 San Francisco, California 94103, USA.
  pesticide use / sustainable pest control / networking / quarterly
Grassroots Development. Journal of the Inter-American Foundation, 1515 Wilson Blvd, Rosslyn, Virginia 22209, USA.
  self-help development / small enterprise development / small-scale farming / technical cooperation / urban / rural / 3 times a year
Haramata, Bulletin of the Drylands: People, Policies, Programmes. Dryland Networks Programme, IIED, 3 Endsleigh St, London WC1H 0DD, UK.
  natural resource management / small-scale farming / sustainable rural development / semiarid / Africa / networking / information update / English / French / quarterly
Heifer Project Exchange. Hiefer Project International, PO Box 808, Little Rock, Arkansas 72203, USA.
  appropriate livestock technology / livestock development / low-cost veterinary techniques / twice a year
IBSRAM Newsletter. IBSRAM, PO Box 9-109, Bangkhen, Bangkok 10900, Thailand.
  soil management / research highlights / institutional / quarterly
IDRC Reports. IDRC, PO Box 8500, Ottawa K1G 3H9, Canada.
  agricultural science / health science / research / social science / information update / communication / English / French / Spanish / Arabic / quarterly
ifda dossier. International Foundation for Development Alternatives, 4 place du marché, CH-1260 Nyon, Switzerland.
  human development / sustainable development / English / French / Spanish / quarterly
IITA Research Briefs. IITA, PMB 5320, Oyo Road, Ibadan, Nigeria.
  alley cropping / cassava / cocoyam / cowpea / FSR / soybean / sweet potato / yam / humid / subhumid / research highlights / institutional / quarterly
ILCA Newsletter. ILCA, PO Box 5689, Addis Ababa, Ethiopia.
  livestock development / Africa / networking / research highlights / institutional / quarterly
ILEIA Newsletter. ILEIA, PO Box 64, NL-3830 AB Leusden, Netherlands.
  agricultural development / low-external-input agriculture / sustainable agriculture / PTD / IK / networking / information update / quarterly
International Agricultural Development. International Agricultural Development, 19 Woodford Close, Caversham, Reading RG4 7HN, UK.
  agricultural development / small-scale farming / sustainable agriculture / information update / bi-monthly

International AG-Sieve. Rodale Institute, 222 Main St, Emmaus, Pennsylvania 18098, USA.
ecological agriculture / small-scale farming / sustainable agriculture / networking / information update / quarterly

International Sweet Potato Newsletter. Philippine Root Crops Information Service, Philippine Root Crops Research and Training Centre, Visayas State College of Agriculture, Baybay, Leyte, Philippines.
sweet potato / research highlights / twice a year

International Tree Crops News. International Tree Crops Institute (ITCI), PO Box 283, Caulfield South, Victoria 3162, Australia.
agroforestry systems / multipurpose trees / twice a year

IRETA's South Pacific Agricultural News. Institute for Research, Extension and Training in Agriculture, USP Alafua, PMB Apia, Western Samoa.
agricultural development / FSR / IK / small-scale farming / regional / quarterly

IRED Forum. Informations et Réseaux pour le Développement, 3 rue de Varembé, Case 116, CH-1211 Geneva 20, Switzerland.
grassroots groups / rural and urban development / networking / information update / English / French / Spanish / quarterly

Irrigation Management Network Papers. IIMI/ODI, Regent's College, Inner Circle, Regent's Park, London NW1 4NS, UK.
irrigation / networking / twice a year

ISNAR Newsletter. ISNAR, PO Box 93375, NL-2509 AJ The Hague, Netherlands.
agricultural research / research policy / networking / institutional / free

Journal of Farming Systems Research-Extension. Association of Farming Systems Research-Extension (AFSR/E), 845 North Park Ave, University of Arizona, Tucson, Arizona 85719, USA.
FSR / on-farm research / scientific / networking / occasional

Journal of Pesticide Reform. Northwest Coalition for Alternatives to Pesticides, PO Box 1393, Eugene, Oregon 97440, USA.
pesticide / sustainable pest control / pesticide regulation / USA / networking / information update / international / quarterly

Journal of Soil and Water Conservation. SWCS, 7515 NE Ankeny Road, Ankeny, Iowa 50021-9764, USA.
soil and water conservation / USA / international / networking / scientific / bi-monthly

Journal of Sustainable Agriculture. Food Product Press, The Haworth Press, 10 Alice St, Binghamton, New York 13904, USA.
agroecology / sustainable agriculture / USA / international / scientific / quarterly

Manna. IASA, University of Minnesota, Newman Center, 1701 University Ave SE, Rm 202, Minneapolis, Minnesota 55414, USA.
education / policies / research / sustainable agriculture / documentation / networking / USA / international / quarterly

NAGA: The ICLARM Quarterly. ICLARM, MC PO Box 1501, Makati, Metro Manila 1299, Philippines.
aquaculture / fisheries / integrated crop-tree-livestock-fish systems / institutional / quarterly

Newsletter for Beekeepers in Tropical and Subtropical Countries. International Bee Research Association (IBRA), 18 North Road, Cardiff CF1 3DY, UK.
low-cost sustainable beekeeping / institutional / quarterly

NFTA News. NFTA, PO Box 680, Waimanalo, Hawaii 96795, USA.
agroforestry / nitrogen-fixing trees / small-scale farming / networking / quarterly

Panoscope. Panos Institute, 9 White Lion St, London N1 9PD, UK.
environment / natural resource management / sustainable development / bi-monthly

Pastoral Development Network Papers. ODI, Regent's College, Inner Circle, Regent's Park, London NW1 4NS, UK.
pastoral development / networking / twice a year

The Permaculture Activist. The Permaculture Activist, PO Box 101, Davis, California 95617, USA.
ecological agriculture / information update / USA / international / quarterly

Permaculture Edge. Permaculture, PO Box 650, Nambour, Queensland 4560, Australia.
ecological agriculture / sustainable agriculture / information update / quarterly

Perspectives. Earthscan/IIED, 3 Endsleigh St, London WC1H ODD, UK.
development / environment / natural resource management / policies / information update / quarterly

Plant Genetic Resources Newsletter. IBPGR, c/o FAO, Via delle Terme di Caracalla, I-00100 Rome, Italy.
genetic resource management / research highlights / bi-monthly

RED. Mesoamerican Center for the Study of Appropriate Technology (CEMAT), 4a Av. 32-21, Zona 12, PO Box 1160, Guatemala 01012, Guatemala.

agricultural development / appropriate technology / small farmers / networking / regional / quarterly

Resources. KENGO, PO Box 48197, Nairobi, Kenya.
sustainable natural resource management / quarterly

Seedling. GRAIN, Apto 23398, E-08080 Barcelona, Spain.
biotechnology / genetic erosion / genetic resources / policies / small farmers / networking / information update / tropics / international / bi-monthly

Seed Sowers. Center for PVO/University Collaboration in Development, Bird Bldg, Western Carolina University, Cullowhee, North Carolina 28723, USA.
seed growing / seed harvesting / seed storage / Senegal / The Gambia / On-Farm Seed Project / quarterly

The Small Farm Newsletter. CUSO-Thailand, 17 Phaholyothin Golf Village, Phaholyothin Road, Bangkhen, Bangkok 10900, Thailand.
small-scale farming / sustainable agriculture / humid / subhumid / regional / networking / quarterly

Social Forestry Network Papers. ODI, Regent's College, Inner Circle, Regent's Park, London NW1 4NS, UK.
community forestry / farm forestry / networking / quarterly

SPORE. CTA, PO Box 380, NL-6700 AJ Wageningen, Netherlands.
agricultural development / small-scale farming / Africa / Caribbean / Pacific / information update / English / French / bi-monthly

Sustainable Agriculture Newsletter. Sustainable Agriculture Network, CUSO-Thailand, 17 Phaholyothin Golf Village, Phaholyothin Road, Bangkhen, Bangkok 10900, Thailand.
small-scale farming / networking / regional / quarterly

The Tribune: a women and development quarterly. International Women's Tribune Centre (IWTC), 777 United Nations Plaza, New York, New York 10017, USA.
women / urban / rural / networking / information update / quarterly

TRI NEWS. Tropical Resources Institute, Yale School of Forestry and Environmental Studies, 205 Prospect St, New Haven, Connecticut 06511, USA.
natural resource management / sustainable resource use / quarterly

Vetiver Newsletter. Vetiver Information Network, World Bank, 1818 H St NW, Washington DC 20433, USA.
fodder / soil and water conservation / vetiver grass / information update / quarterly

VITA News. VITA, 1815 North Lynn St, Suite 200, Arlington, Virginia 22209-2079, USA.
appropriate technology / rural development / small-scale farming / information update / practical / quarterly

Voices. DCFRN, 595 Bay St, 9th Floor, Toronto, Ontario M5G 2C3, Canada.
extension / low-cost farming / radio communication / sustainable agriculture / networking / practical / occasional

Women in Action. Isis International, Casilla 2067, Correo Central, Santiago, Chile.
development / ecology / urban / rural / networking / information update / international / quarterly

World Neighbors in Action. WN, 5116 North Portland Ave, Oklahoma City, Oklahoma 73112, USA.
rural development / small-scale farming / sustainable agriculture / practical / English / French / Spanish / quarterly

WorldWatch. Worldwatch Institute, 1776 Massachusetts Ave NW, Washington DC 20036, USA.
environment / policies / sustainable development / bi-monthly

Appendix C4

Addresses of organisations concerned with sustainable agriculture

The publishers of books listed in the references and under further readings (Appendix 1) but not readily available in bookshops are included in this list. Further addresses of organisations which issue periodicals can be found in Appendix C3.

ACDEP (Association of Church Development Programmes), c/o Tamale Archdiocesan Agricultural Programme, PO Box 42, Tamale, Ghana.

ACIAR (Australian Centre for International Agricultural Research), GPO Box 1571, Canberra, ACT 2601, Australia.

ACTS (African Centre for Technology Studies), PO Box 45917, Nairobi, Kenya.

AED (Academy for Educational Development), 1255 23rd St NW, Washington DC 20037, USA.

AGRECOL Development Information, c/o Ökozentrum Langenbruck, CH-4438 Langenbruck, Switzerland.

EULEISA (European Network of Agricultural Networks for Low-External- Input and Sustainable Agriculture), c/o AGRECOL.

AGROMISA Foundation, PO Box 41, NL-6700 AA Wageningen, Netherlands.

AIT (Asian Institute of Technology), GPO Box 2754, Bangkok 10501, Thailand.

AME (Agriculture Man and Ecology Programme). See ETC.

APICA (Association pour la Promotion des Initiatives Communautaires Africaines), BP 5946, Douala Akwa, Cameroon.

ATIP (Agricultural Technology Improvement Project), Dept of Agricultural Research, Ministry of Agriculture, P/Bag 0033, Gaborone, Botswana.

AVRDC (Asian Vegetable Research and Development Center), PO Box 42, Shanhua, Tainan 741, Taiwan, Republic of China.

BAIF (Bharatiya Agro-Industries Foundation), Senapati Bapat Marg, Pune 411 016, India.

BIRC (BAIF Information Resource Centre). See above.

BOS (Foundation for Dutch Forestry Development Cooperation), PO Box 23, NL-6700 AA Wageningen, Netherlands.

BOSTID (Board on Science and Technology for International Development), Publications and Information Services, Office of International Affairs, National Research Council, 2101 Constitution Ave NW, Washington DC 20418, USA.

CAB (Commonwealth Agricultural Bureaux) International, PO Box 100, Wallingford, Oxon OX10 8DF, UK.

CABO (Centre for Agrobiological Research), PO Box 14, 6700 AA Wageningen, Netherlands.

CATIE (Centro Agronómico Tropical de Investigación y Enseñanza), Apto Postal 13, Turrialba, Costa Rica.

CCTA (Comisin de Coordinacin de Tecnologa Andina), Apto 140426, Lima 14, Peru.

CDCS (Centre for Development Cooperation Services), Van der Boechorststraat 7, NL-1081 BT Amsterdam, Netherlands.

CDR (Centre for Development Research), Ny Kongensgade 9, DK-1472 Copenhagen K, Denmark.

CDTF (Community Development Trust Fund), PO Box 9421, Dar es Salaam, Tanzania.

CET (Centro de Educación y Tecnología), Casilla 16557, Correo 9, Santiago, Chile.

CGIAR (Consultative Group on International Agricultural Research), 1818 H St NW, Washington DC 20433, USA.

CIAT (Centro Internacional de Agricultura Tropical), AA 6713, Cali, Colombia.

CIDICCO (International Cover Crop Clearing House), PO Box 278-C, Tegucigalpa DC, Honduras.

CIKARD (Center for Indigenous Knowledge for Agricultural and Rural Development), 318B Curtiss Hall, Iowa State University, Ames, Iowa 50011, USA.

CIMMYT (Centro Internacional de Mejoramiento de Maíz y Trigo), Londres 40, Apto Postal 6-641, 06600 Mexico DF, Mexico.

CIP (Centro Internacional de la Papa), Apto Postal 5969, Lima, Peru.

CODEL Inc., Environment and Development Program, 475 Riverside Drive, Rm 1842, New York, NY 10115, USA.

Commonwealth Secretariat, Marlborough House, Pall Mall, London SW1, UK.

CON (Centre for Development Work), PO Box 211, NL-6700 AE Wageningen, Netherlands.

CSE (Centre for Science and Environment), 807 Vishal Bhawan, 95 Nehru Place, New Delhi 110 019, India.

CTA (Technical Centre for Agricultural and Rural Cooperation), PO Box 380, NL-6700 AJ Wageningen, Netherlands.

CUSO (Canadian University Services Overseas), 135 Rideau Street, Ottawa, Ontario K1N 9K7, Canada.

DCFRN (Developing Countries Farm Radio Network), 40 Dundas St W, Box 12, Suite 227, Toronto, Ontario M5G 2C2, Canada.

DGIS (Directorate General for International Cooperation), Ministry of Foreign Affairs, PO Box 20061, NL-2500 EB The Hague, Netherlands.

EAP (Ecological Agriculture Projects), PO Box 191, Macdonald College, Ste Anne de Bellevue, Québec H9X 1C0, Canada.

Earthscan Publications Ltd. See IIED.

ELCI (Environment Liaison Centre International), PO Box 72461, Nairobi, Kenya.

ENDA (Environnement et Développement du Tiers-Monde), PO Box 3370, Dakar, Senegal.

ETC Foundation, PO Box 64, NL-3830 AB Leusden, Netherlands.

FAO (Food and Agriculture Organization of the United Nations), Via delle Terme di Caracalla, I – 00100 Rome, Italy.

FAO-RAPA (Regional Office for Asia and the Pacific), Maliwan Mansion, Phra Atit Road, Bangkok, Thailand.

FASE (Federação de Orgãos para Assistência Social e Educacional), Projecto Technologias Alternativas, Rua Bento Lisboa, 58-Catete 22221, Rio de Janeiro, RJ, Brazil.

FFTC-ASPAC (Food and Fertilizer Technology Center for the Asian and Pacific Region), 5th Floor, 14 Wenchow St, Taipei 10616, Taiwan, Republic of China.

GAIA Services, PO Box 84, RFD, St Johnsbury, Vermont 05819, USA.

GATE (German Appropriate Technology Exchange). See GTZ.

GEYSER (Groupe d'Etudes et de Service pour l'Economie des Resources), Vacquieres, 34270 St Mathieu de Treviers, France.

GRAAP (Groupe de Recherche et d'Appui pour l'Autopromotion Paysanne), BP 305, Bobo Dioulasso, Burkina Faso.

GRAIN (Genetic Resources Action International), Apto 23398, E-08080 Barcelona, Spain.

GRET (Groupe de Recherche et d'Echanges Technologiques), 213 Rue La Fayette, F-75010 Paris, France.

GTZ (German Agency for Technical Cooperation), PO Box 5180, D-6236 Eschborn 1, Germany.

Heifer Project International, PO Box 808, Little Rock, Arkansas 72203, USA.

IASA (International Alliance for Sustainable Agriculture), Newman Centre, University of Minnesota, 1701 University Ave SE, Rm 202, Minneapolis, Minnesota 55414, USA.

IBPGR (International Board for Plant Genetic Resources), Crop Genetic Resources Centre, FAO. See FAO.

IBSRAM (International Board for Soil Research and Management), PO Box 9-109, Bangkhen, Bangkok 10900, Thailand.

ICARDA (International Center for Agricultural Research in the Dry Areas), PO Box 5466, Aleppo, Syria.

ICDA Seeds Campaign. See GRAIN.

ICIMOD (International Centre for Integrated Mountain Development), PO Box 3226, Kathmandu, Nepal.

ICIPE (International Centre of Insect Physiology and Ecology), PO Box 30772, Nairobi, Kenya.

ICLARM (International Centre for Living Aquatic Resources Management), MCC PO Box 1501, Makati, Metro Manila 1299, Philippines.

ICPR (Institute for Consumer Policy Research), c/o Consumers Union, 256 Washington St, Mt Vernon, New York 10553, USA.

ICRAF (International Council for Research in Agroforestry), PO Box 30677, Nairobi, Kenya.

ICRISAT (International Crops Research Institute for the Semi-Arid Tropics), Patancheru PO, Andhra Pradesh 502 324, India.

IDRC (International Development Research Centre), PO Box 8500, Ottawa K1G 3H9, Canada.

IDS (Institute of Development Studies), University of Sussex, Brighton BN1 9RE, UK.

IFAD (International Fund for Agricultural Development), 107 via del Serafico, I-00142 Rome, Italy.

IFOAM (International Federation of Organic Agricultural Movements), c/o Ökozentrum Imsbach, D-6695 Tholey-Theley, Germany.

IFAP (International Federation of Agricultural Producers), 21 rue Chaptal, F – 75009 Paris, France.

IFPRI (International Food Policy Research Institute), 1776 Massachusetts Ave NW, Washington DC 20036, USA.

IIED (International Institute for Environment and Development), 3 Endsleigh St, London WC1H ODD, UK.

IIMI (International Irrigation Management Institute), PO Box 2075, Colombo, Sri Lanka.

IIRR (International Institute of Rural Reconstruction), Silang, Cavite, Phillipines.

IITA (International Institute of Tropical Agriculture), PO Box 5320, Ibadan, Nigeria.

ILCA (International Livestock Centre for Africa), PO Box 5689, Addis Ababa, Ethiopia.

ILEIA (Information Centre for Low-External-Input and Sustainable Agriculture). See ETC.

ILO (International Labour Office), BP 500, CH-1211 Geneva 22, Switzerland.

ILRAD (International Laboratory for Research and Animal Diseases), PO Box 30709, Nairobi, Kenya.

INADES (Institut Africain Développement Economique), BP 8, Abidjan 08, Côte d'Ivoire.

IRRI (International Rice Research Institute), PO Box 933, Manila, Philippines.

ISNAR (International Service for National Agricultural Research), PO Box 93375, NL-2509 AJ The Hague, Netherlands.

ITDG (Intermediate Technology Development Group), Myson House, Railway Terrace, Rugby, Warwickshire CV21 3LF, UK.

ITP (Intermediate Technology Publications), 103-105 Southampton Row, London WC1B 4HH, UK.

IUCN (International Union for Conservation of Nature), Ave du Mont Blanc, CH-1196 Gland, Switzerland.

KENGO (Kenya Energy and Environment Organization), PO Box 48197, Nairobi, Kenya.

KIT (Royal Tropical Institute), Mauritskade 63, NL-1092 AD Amsterdam, Netherlands.

KWDP (Kenyan Woodfuel Development Project), c/o Kenyan Woodfuel Agroforestry Project, PO Box 56212, Nairobi, Kenya.

LBL (Landwirtschaftliche Beratungszentrale Lindau), CH-8315 Lindau, Switzerland.

NRC (National Research Council). See BOSTID.

Nature et Progrès, Commission Tiers Monde, 40 route de Rouen, F-80500 Montdidier, France.

NERAD (Northeast Rainfed Agricultural Development Project), Northeast Regional Office of Agriculture, Tha Phra, Khon Kaen 40206, Thailand.

NFTA (Nitrogen Fixing Tree Association), PO Box 680, Waimanalo, Hawaii 96795, USA.

NPSAS (Northern Plains Sustainable Agriculture Society), RR1, Box 73, Windsor, North Dakota 58424, USA.

NRI (Overseas Development Natural Resources Institute), Pembroke House, Central Ave, Chatham Maritime, Chatham ME4 4TB, UK.

ODI (Overseas Development Institute), Regent's College, Inner Circle, Regent's Park, London NW1 4NS, UK.

OTA (Office of Technology Assessment), US Congress. Publications available from: US Government Printing Office, Washington DC 20402-9325, USA.

Oxfam, 274 Banbury Road, Oxford OX2 7DZ, UK.

ORSTOM (Office de la Recherche Scientifique et Technique Outre-Mer), 27 quai de la Tournelli, F-75005 Paris, France.

PAF (Projet Agro-Forestier), BP 200, Ouahigouya, Yatenga Province, Burkina Faso.

PAN (Pesticides Action Network International), Bollandistenstraat 22, B-1040 Brussels, Belgium.

PPST (Patriotic and People-Oriented Science & Technology Foundation), 6 Sec ⋅ . Cross St, Karpagam Gardens, Adyar, Madras 600 020, India.

PRATEC (Proyecto Andino de Tecnologías Campesinas), Pumacahua 1364 ʌ1, Peru.

PRONAT (Protection Naturelle). See ENDA.

RAFI (Rural Advancement Fund International), RR 1 (Beresford), Brandon, Manitoba R7A 5Y1, Canada.

RESADOC (Réseau Sahélien d'Information et de Documentation Scientifique et Technique), Institut du Sahel, BP 1530, Bamako, Mali.

Rodale Institute, 222 Main St, Emmaus, Pensylvania 18098, USA.

SAN (Seeds Action Network). See members, e.g. ELCI, SIBAT, ICDA, RAFI.

SATIS (Socially Appropriate Technology International Information Services), PO Box 803, NL-3500 AV Utrecht, Netherlands.

SIBAT (Sibol Ng Aghman At Akmang Teknolohiya), Rm 421, Singson Bldg, Plaza Moraga, Manila, Philippines.

SIDA (Swedish International Development Authority), Birger Jarlsgatan, S-10525 Stockholm, Sweden.

SKAT (Swiss Centre for Appropriate Technology), Varnbelstrasse 14, CH-900 St Gallen, Switzerland.

Sustainable Agriculture Information Project, Agroecology Program, University of California, Santa Cruz, California 95064, USA.

SWCS (Soil and Water Conservation Society), 7515 NE Ankeny Road, Ankeny, Iowa 50021-9764, USA.

TAC (Technical Advisory Committee of Consultative Group on International Agricultural Research), c/o FAO. See FAO.

Third World Network, 87 Cantonment Road, 10250 Penang, Malaysia.

Terres et Vie, 113 rue Laurent Delvaux, F-1400 Nivelles, Belgium.

TOOL (Transfer of Technology for Development), Sarphatistraat 650, NL-1018 AV Amsterdam, Netherlands.

UNESCO (United Nations Educational, Scientific and Cultural Organization), 7 Place de Fontenoy, F-75700 Paris, France.

UNESCAP (United Nations Economic and Social Commission for Asia and the Pacific), Rajadamnern Nok Ave, Bangkok 10200, Thailand.

UPLB (University of the Philippines at Los Baños), College Lacuna, Philippines.

VITA (Volunteers in Technical Assistance), 1815 North Lynn St, Suite 200, Arlington, Virginia 22209-2079, USA.

WARDA (West African Rice Development Association), PO Box 1019, Monrovia, Liberia.

Winrock International Livestock Research Training Center, Route 3, Petit Jean Mountain, Morrilton, Arkansas 72110, USA.

WN (World Neighbors), 5116 North Portland Ave, Oklahoma City, Oklahoma 73112, USA.

World Bank, 1818 H Street NW, Washington DC 20433, USA.

World Resources Institute, 1735 New York Ave NW, Washington, DC 20006, USA.

Worldwatch Institute, 1776 Massachusetts Ave NW, Washington, DC 20036, USA.

References

Agarwal, A. and Narain, S. 1989. *Towards green villages: a strategy for environmentally-sound and participatory rural development*. New Delhi: CSE.

Agboola, A.A. 1980. Effect of different cropping systems on crop yield and soil fertility management in the humid tropics. In: *Organic recycling in Africa* (Rome: FAO), pp. 87–105.

Ahmed, S. 1990. Neem for pest control. *ILEIA Newsletter* 6 (1): 7.

Akobundu, I.O. 1983. No-tillage weed control in the tropics. In: *No-tillage crop production in the tropics* (Monrovia: International Plant Protection Centre), pp. 32–44.

Akobundu, I.O. and Poku, J.A. 1984. Control of *Imperata cylindrica*. In: *International Institute of Tropical Agriculture Annual Report* 1984 (Ibadan: IITA), pp. 175–6.

Alexandratos, N. (ed.) 1988. *World agriculture toward 2000: an FAO study*. Rome/London: FAO/Belhaven Press.

Allan, W. 1965. *The African husbandman*. London: Oliver and Boyd.

Altieri, M.A. 1987. *Agroecology: the scientific basis of alternative agriculture*. Boulder/London: Westview/ITP.

Altieri, M.A. and Letourneau, D.K. 1982. Vegetation management and biological control in agroecosystems. *Crop Protection* 4: 405–30.

Altieri, M.A. and Liebman, M.Z. 1986. Insect, weed and plant disease management in multiple cropping systems. In: Francis, C.A. (ed.), *Multiple cropping systems* (New York: Macmillan), pp. 183–218.

Andow, D. 1983. The extent of monoculture and its effects on insect pest populations with particular reference to wheat and cotton. *Agricultural Ecosystems and Environment* 9: 25–36.

Arnold, J.E.M. 1990. Tree components in farming systems. *Unasylva* 160: 35–42.

Arntzen, J.W. 1984. *Determinants of field locations in Kgatleng, Gabarone*. Gabarone: University of Botswana.

Balasubramainam, A. 1987. Microclimate and its utilization in Indian farming. *ILEIA Newsletter* 3 (3): 9.

van den Ban, A.W. and Hawkins, H.S. 1988. *Agricultural extension*. Harlow: Longman.

Bantugan, S.C., Singzon, S.B., Obusa, A.P. and Tabada, E.M.Jr. 1989. Indigenous health practices and breeding management of carabao. *Farm and Resource Management Institute Information Service Newsletter* 3 (1): 11–13.

Barker, D. 1979. Appropriate methodology: an example using a traditional African board game to measure farmers' attitudes and environmental images. *IDS Bulletin* 10 (2): 37–40.

Barrow, C.J. 1987. *Water resources and agricultural development in the tropics*. Harlow: Longman.

Basant, R. and Subrahmamian, K.K. 1990. *Agro-mechanical diffusion in a backward region*. London: ITP.

Bayer, W. 1986. Agropastoral herding practices and grazing behaviour of cattle in Nigeria's subhumid zone. *ILCA Bulletin* 24: 8–13.

Bayer, W. 1989. Low-demand animals for low-input systems. *ILEIA Newsletter* 5 (4): 14–15.

Bayer, W. 1990. Use of native browse by Fulani cattle in central Nigeria. *Agroforestry Systems* 12: 217–28.

Bayer, W. and Maina, J. 1984. Seasonal pattern of tick load in Bunaji cattle in the subhumid zone of Nigeria. *Veterinary Parasitology* 15: 301–7.

Bayer, W., Niamir, M. and Waters-Bayer, A. 1987. Is Holistic Resource Management the answer for African rangelands? *Pastoral Development Network Paper* 24e. London: ODI.

Beets, W.C. 1982. *Multiple cropping and tropical farming systems*. Gower: Westview.

Bertol, O. and Wagner, O. 1987. A knife roller or chopping roller. *ILEIA Newsletter* 3 (1): 10.

Biggs, S.D. 1984. Linkages between agricultural research, extension and farmers: linkage analysis. In: *Planning Training Manual: Training Programme for Agricultural Planning at the District Level in Nepal* (Kathmandu: FAO).

Biggs, S.D. 1989. A multiple source of innovation model of agricultural research and technology promotion. *Agricultural Administration (Research and Extension) Network Paper* 6. London: ODI.

Biggs, S.D. and Clay, E. 1981. Sources of innovation in agricultural technology. *World Development* 9 (4): 321 – 36.

Blaikie, P. 1985. *The political economy of soil erosion in developing countries.* Harlow: Longman.

Bourn, D. 1983. *Tsetse control, agricultural expansion and environmental change in Nigeria.* PhD thesis, Oxford University.

Box, L. 1987. Experimenting cultivators: a methodology for adaptive agricultural research. *Agricultural Administration (Research and Extension) Network Discussion Paper* 23. London: ODI.

Box, L. 1989. Virgilio's theorem: a method for adaptive agricultural research. In: Chambers, R., Pacey, A. and Thrupp, L.A. (eds), *Farmer first: farmer innovation and agricultural research* (London: ITP), pp. 61 – 7.

Brader, L. 1982. Recent trends of insect control in the tropics. *Ent. exp. and app. 31: 111 – 20.*

Breman, H. 1990. No sustainability without external inputs. In: Directorate General for International Cooperation, Project Group Africa (eds), *Beyond adjustment: sub-Saharan Africa* (The Hague: Ministry of Foreign Affairs), pp. 124 – 34.

Breman, H. and de Wit, C.T. 1983. Rangeland productivity and exploitation in the Sahel. *Science* 221: 1341 – 7.

Briones, A.M., Cayaban, E.B.Jr., Vicente, P.R. and Aspiras, R.B. 1989. Farmer-based research for sustainable rice farming. *ILEIA Newsletter* 5 (4): 24 – 5.

Brokensha, D., Warren, D.M. and Werner, O. (eds) 1980. *Indigenous knowledge systems and development.* Lanham: University Press of America.

Brown, L.R. 1988. *The changing world food prospect: the nineties and beyond.* Washington DC: Worldwatch Institute.

Brown, L.R. (ed.) 1989. *State of the world 1989: a Worldwatch Institute report on progress toward a sustainable society.* New York: Norton.

Buck, L. 1991. NGOs, government and agroforestry research methodology in Kenya. *Agricultural Administration (Research and Extension) Network Paper.* London: ODI.

Budelman, A. 1983. *Primary agricultural research: farmers perform field trials – experiences from the Lower Tana Basin, East Kenya.* Wageningen: Dept of Tropical Crop Science, Agricultural University.

Bull, A.T., Holt, G. and Lilly, M.D. 1982. *Biotechnology: international trends and perspectives.* Paris: OECD.

Bunch, R. 1985. *Two ears of corn: a guide to people-centered agricultural improvement.* 2nd ed. Oklahoma City: World Neighbors.

Bunch, R. 1990. *Low input soil restoration in Honduras: the Cantarranas farmer-to-farmer extension programme.* Gatekeeper Series 23. London: IIED.

Cardono, A. and Orozco, L.A. 1987. *Informe anual de programa indígena.* Quibdo: Proyecto DIAR.

CDTF. 1977. *Appropriate technology for grain storage: report of a pilot project.* Dar es Salaam: CDTF/Institute of Adult Education/ Economic Development Bureau.

CIMMYT. 1988. *Annual report.* Mexico City: CIMMYT.

CGIAR. 1978. *Farming Systems Research at the International Agricultural Research Centers.* Rome: TAC Secretariat, Agriculture Dept, FAO.

CGIAR. 1985. *Summary of IARCs: a study of achievements and potentials.* Washington DC: CGIAR.

Chambers, R. 1983. *Rural development: putting the last first.* London: Longman.

Chambers, R. 1989. *To the hands of the poor: water and trees.* London: ITP.

Chambers, R. 1990. *Microenvironments unobserved.* Gatekeeper Series 22. London: IIED.

Chambers, R. and Jiggins, J. 1986. *Agricultural research for resource-poor farmers: a parsimonious paradigm.* Discussion Paper 220. Brighton: IDS, University of Sussex.

Chambers, R., Pacey, A. and Thrupp, L.A. (eds) 1989. *Farmer first: farmer innovation and agricultural research.* London: ITP.

Chavangi, N.A., Engelhard, R.J. and Jones, V. 1985. *Culture as the basis for implementing self-sustaining woodfuel development programmes.* Nairobi: Beijer Institute.

Clarke, R. 1986. *Restoring the balance: women and forest resources.* Rome: FAO/SIDA.

Clawson, D.L. 1985. Harvest security and intraspecies diversity in traditional tropical agriculture. *Tropical Botany* 39 (1): 56 – 67.

Cole, J.L. 1987. Tractor versus oxen: letter to editors. *ILEIA Newsletter* 3 (4): 24.

Conklin, H.C. 1957. *Hanunoo agriculture: a report on an integral system of shifting cultivation in the Philippines.* Forestry Development Paper 12. Rome: FAO.

Connell, J. 1990. Farmers experiment with a new crop. *ILEIA Newsletter* 6 (1): 18 – 19.

Conway, G.R. 1987. The properties of agroecosystems. *Agricultural Systems* 24: 95 – 117.

Conway, G.R. and Pretty, J.N. 1988. *Fertiliser risks in the developing countries: a review.* London: IIED.

Conway, G.R., McCracken, J.A. and Pretty, J.N. 1987. *Training notes for Agroecosystem Analysis and Rapid Rural Appraisal.* 2nd ed. London: IIED.

Conway, G.R., Chambers, R., McCracken, J. and Pretty, J. (eds) 1988. *RRA Notes 1*. London: IIED.

Cooper, J.P. and Tainton, N.M. 1968. Light and temperature requirements for the growth of tropical and temperate grasses. *Herbage Abstracts* 38 (3): 167 – 76.

Copijn, A.N. 1985. *Bio-ecological soil regeneration methods*. Leusden: AME Programme, ETC Foundation.

Crouch, B. 1984. *Problem census: farmer-centred problem identification: training for agriculture and rural development*. Rome: FAO.

CTA. 1988. *Agroforestry: the efficiency of trees in African agrarian production and rural landscapes*. Wageningen: CTA/Terres et Vie/GTZ/ICRAF.

Czech, H.J. 1986. *The Truco concept*. Eschborn: GTZ.

Dalzell, H.W., Biddlestone, A.J., Gray, K.R. and Thurairajan, K. 1987. *Soil management: compost production and use in tropical and subtropical environments*. Soils Bulletin 56. Rome: FAO.

Dankelman, I. and Davidson, J. 1988. *Women and environment in the Third World: alliance for the future*. London: Earthscan.

Daxl, R. 1985. Integrierte Schädlingsbekämpfung in der Baumwollkultur Nicaraguas. In: Kranz. J. (ed.), *Integrierter Pflanzenschutz in den Tropen* (Giessen: Wissenschaftliches Zentrum Tropeninstitut), pp. 151 – 65.

De Datta, S.K. 1981. *Principles and practices of rice production*. New York: Wiley & Sons.

Dela Cruz, C.R., Lightfoot C., Costa-Pierce, B.A. and Carangal, V.R. (eds) (in press). *Rice-fish research and development in Asia*. ICLARM Conference Proceedings 24. Manila: ICLARM/IRRI/Central Luzon State University.

van Diest, A. 1988. Integrated farming strategies. *ILEIA Newsletter* 4 (4): 28.

Dogra, B. 1983. Traditional agriculture in India: high yields and no waste. *The Ecologist* 13 (2/3): 84 – 7.

Donkers, H. and Hoebink, P. 1989. Schatkamers voor de westerse landbouw. *NRC Handelsblad* 2/5/89.

Douglass, G.K. (ed.) 1984. *Agricultural sustainability in a changing world order*. Boulder: Westview.

Doutt, R.L. 1964. The historical development of biological control. In: DeBach, P. and Schlinger, E.I. (eds), *Biological control of insect pests and weeds* (New York: Reinhold), pp. 21 – 41.

Dover, M. and Talbot, L.M. 1987. *To feed the earth: agroecology for sustainable development*. Washington DC: World Resources Institute.

Duckham, A.M. and Masefield, G.B. 1970. *Farming systems of the world*. London: Chatto and Windus.

Dupre, G. (ed.) 1990. *Farming knowledge and development: proceedings of the 7th Section on Science, Diffusion of Technical Knowledge and Systems of Local Knowledge of the 7th World Congress of Rural Sociology, June 1988, Bologna, Italy*. Paris: ORSTOM.

Dupriez, H. and de Leener, P. 1988. *Agriculture in African rural communities: crops and soils*. London: Macmillan.

Edwards, C.A. 1987. The concept of integrated systems in lower-input, sustainable agriculture. *American Journal of Alternative Agriculture* 2 (4): 148 – 52.

Edwards, P., Pullin, R.S.V. and Gartner, J.A. 1988. *Research and education for the development of integrated crop-livestock-fish farming systems in the tropics*. Manila: ICLARM.

Eger, H. 1989. *Ecological land management and resource conservation on the Central Plateau of Burkina Faso*. GTZ Seminar on Low-External-Input and Sustainable Agriculture, 5 – 7 March 1989, Tanga/Lushoto, Tanzania.

Egger, K. 1987. Ein Weg aus der Krise: Möglichkeiten des ökologischen Landbaus in den Tropen. In: Heske, H. (ed.), *Landwirtschaft zwischen Agrobusiness, Gentechnik und traditionellem Landbau* (Giessen: Focus), pp. 72 – 97.

El Amami, S. 1983. *Les aménagements hydrauliques traditionnels en Tunisie*. Tunis: Centre de Recherche du Génie Rural.

Eldin, M. and Milleville, P. 1989. *Le risque en agriculture*. Paris: ORSTOM.

ENDA. 1987. *Pour une recherche-formation action sur la fertilité des sols: une étude de cas en milieu sahélian*. Essais Documents de Base 270. Dakar: ENDA.

Essers, S. 1988. Bitter cassava as a drought-resistant crop. *ILEIA Newsletter* 4 (4): 10 – 11.

Ethiopian Red Cross Society. 1988. *Rapid rural appraisal: a closer look at rural life in Wollo*. Addis Ababa: Ethiopian Red Cross Society; London: IIED.

FAO. 1977. *Organic materials and soil productivity*. Soils Bulletin 35. Rome: FAO.

FAO. 1984a. *Traditional (indigenous) systems of veterinary medicine for small farmers in Nepal*. Bangkok: FAO-RAPA.

FAO. 1984b. *Traditional (indigenous) systems of veterinary medicine for small farmers in Thailand*. Bangkok: FAO-RAPA.

FAO. 1985. *Learning from rural women: a manual for village-level training to promote women's activities in marketing*. Rome: FAO.

FAO. 1988a. *Traditional food plants: a resource book for promoting the exploitation and consumption of food plants in arid, semi-arid and sub-humid lands of Eastern Africa.* Food and Nutrition Paper 42. Rome: FAO.

FAO. 1988b. *Participatory monitoring and evaluation: handbook for training field workers.* Bangkok: FAO-RAPA.

FAO. 1990. *Integrated Plant Nutrient Systems: state of the art.* Commission on Fertilizers, 11th Session, 4–6 April 1990. Rome: FAO.

Farrington, J. and Amanor, K. 1990. *NGOs, the state and agricultural technology: preliminary evidence from a global review.* Paper presented at Asian Farming Systems Research and Extension Symposium, 19–22 November 1990, AIT, Bangkok, Thailand

Farrington, J. and Martin, A. 1988. *Farmer participation in agricultural research: a review of concepts and practices.* Agricultural Administration Unit Occasional Paper 9. London: ODI.

Feder, G. and Slade, R. 1985. The role of public policy in the diffusion of improved agricultural technology. *American Journal of Agricultural Economics* 67 (2): 423–8.

Fernandes, F.R.M., Oktingatie, A. and Maghembe, J. 1984. The Chagga home-gardens: a multistoried agroforestry cropping system on Mount Kilimanjaro (Northern Tanzania). *Agroforestry Systems* 2 (2): 73–86.

Fernandez, M. 1988. Towards a participatory system approach: new demands on researchers and research methodologies. *ILEIA Newsletter* 4 (3): 15–17.

Fernandez, M. (in press). Women's Agricultural Production Committees and the participative-research-action approach. In: Feldstein, H. and Jiggins, J. (eds), *Methodologies handbook: gender analysis in agriculture.* (West Hartford: Kumarian).

Flanagan, J.C. 1954. The critical incident technique. *Psychological Bulletin* 51: 327–58.

Francis, C.A. (ed.) 1986. *Multiple cropping systems.* New York: Macmillan.

Fresco, L.O. 1984. *Techniques agricoles ameliorées pour le Kwangu-Kwilu.* Kinshasa: INADES.

Fresco, L.O. 1986. *Cassava in shifting cultivation: a systems approach to agricultural technology development in Africa.* Amsterdam: KIT.

Friis-Hansen, E. 1989. Village-based seed production. *ILEIA Newsletter* 5 (4): 26–7.

Frisch, J.E. and Vercoe, J.E. 1978. Utilizing breed differences in growth of cattle in the tropics. *World Animal Review* 25: 8–12.

Fujisaka, S. 1989. *Participation by farmers, researchers and extension workers in soil conservation.* Gatekeeper Series 16. London: IIED.

Garcia-Padilla, V. 1990. *Working towards LEISA: a register of organizations and experiences on Low-External-Input and Sustainable Agriculture in the Philippines.* Pangasinen: AGTALON/ILEIA.

van Gelder, B. and Voskuil, B. 1990. Fodder hedges within dairy projects. *ILEIA Newsletter* 6 (2): 16–17.

Gilbert, E. 1990. Non-governmental organisations and agricultural research: the experience of The Gambia. *Agricultural Administration (Research and Extension) Network Paper* 12. London: ODI.

Gips, T. 1986. What is sustainable agriculture? In: Allen P. and van Dusen, D. (eds), *Global perspectives on agroecology and sustainable agricultural systems: proceedings of the 6th International Scientific Conference of the International Federation of Organic Agriculture Movements* (Santa Cruz: Agroecology Program, University of California), Vol. 1, pp. 63–74.

Gips, T. 1987. *Breaking the pesticide habit: alternatives to 12 hazardous pesticides.* Minneapolis: IASA.

Glaeser, B. (ed.) 1987. *The green revolution revisited.* London: Allen and Unwin.

Glass, E.H. and Thurston, H.D. 1978. Traditional and modern crop protection in perspective. *Bioscience* 28: 109–15.

Gonsalves, J.F. 1989. Bio-intensive gardening: alternatives to conventional external-input based gardening. *ILEIA Newsletter* 5 (3): 21–3.

Gonsalves, J.F. 1990. Saving seeds. *ILEIA Newsletter* 6 (1): 27–8.

GRAAP. 1987. *Pour une pédagogie de l'autopromotion.* 5th ed. Bobo Dialasso, Burkina Faso: GRAAP.

Grandstaff, T.B. and Grandstaff, S.W. 1986. Choice of rice technology: a farmer perspective. In: Korten, D.C. (ed.), *Community management: Asian experience and perspectives* (West Hartford: Kumarian), pp. 51–61.

Greeley, M. and Farrington, J. 1989. Potential implications of agricultural biotechnology for the Third World. In: Farrington, J. (ed.), *Agricultural biotechnology: prospects for the Third World* (London: ODI), pp. 49–65.

Greenfield, J.C. 1988. Moisture conservation: fundamental to rainfed agriculture. *ILEIA Newsletter* 4 (4): 15–17.

Greenland, D.J. 1986. Nitrogen and food production in the tropics: contributions from fertilizer nitrogen and biological nitrogen fixation. In: Kang, B.T. and van der Heide, J. (eds), *Nitrogen management in farming systems in the humid and sub-humid tropics*

(Haren: Institute for Soil Fertility), pp. 9–38.

Greenwood, P.J., Cleaver, T.J., Turner, P.K., Niendarf, K.B. and Loquens, S.M.H. 1980. Comparison of the effects of nitrogen fertiliser on the yield, nitrogen content and quality of 21 different vegetable and agricultural crops. *Journal of Agricultural Science* (Cambridge) 95: 471–85.

Grigg, D.B. 1974. *The agricultural systems of the world: an evolutionary approach.* Cambridge University Press.

Grillo Fernandez, E. and Rengifo Vasquez, G. 1988. Agricultura y cultura en el Peru. In: PRATEC, *Agricultura andina y saber campesino.* Lima: PRATEC.

Gubbels, P. 1988. Peasant farmer agricultural self-development: the World Neighbors experience in West Africa. *ILEIA Newsletter* 4 (3): 11–14.

Gupta, A.K. and IDS Workshop. 1989. Maps drawn by farmers and extensionists. In: Chambers, R., Pacey, A. and Thrupp, L.A. (eds), *Farmer first: farmer innovation and agricultural research* (London: ITP), pp. 86–92.

Hart, R.D. 1980. A natural ecosystem analog approach to the design of a successional crop system for tropical forest environments. *Tropical Succession* 12 (2): 73–95.

Haverkort, B., Hiemstra, W., Reijntjes, C. and Essers, S. 1988. Strengthening farmers' capacity for technology development. *ILEIA Newsletter* 4 (3): 3–7.

Haverkort, B., van der Kamp, J. and Waters-Bayer, A. 1991. *Joining farmers' experiments: experiences in Participatory Technology Development.* London: ITP.

Hegde, N. 1990. Markets for tree products needed. *ILEIA Newsletter* 6 (2): 18–19.

van der Heide, J. and Hairiah, K. 1989. The role of green manures in rainfed farming systems in the humid tropics. *ILEIA Newsletter* 5 (2): 11–13.

Hildebrand, P.E. 1981. Combining disciplines in rapid appraisal: the sondeo approach. *Agricultural Administration* 8: 423–32.

Hoeksema, J. 1989. *Woman and social forestry: how women can play an active role in programming and implementing forestry projects.* BOS Document 10. Wageningen: BOS Foundation.

Hoffmann-Kuehnel, M. 1989. African women farmers utilize local knowledge. *ILEIA Newsletter* 5 (4): 8–9.

van Hoof, W.C.H. 1987. *Mixed cropping of groundnuts and maize in East Java.* Wageningen: Dept of Tropical Crop Science, Agricultural University.

van den Houdt, F. 1988. We moeten beter luisteren naar de Afrikaanse boeren. *Internationale Samenwerking,* March 1988, pp. 34–5.

Hudson, N. 1989. Soil conservation strategies for the future. In: *Land conservation for future generations: proceedings of the 5th International Soil Conservation Conference* (Bangkok: Dept of Land Development), Vol. 1, pp. 117–30.

Hulugalle, N.R. 1989a. Properties of tied ridges in the Sudan savannah of the West African semi-arid tropics. In: *Land conservation for future generations* (Bangkok: Dept of Land Development), Vol. 2. pp. 693–709.

Hulugalle, N.R. 1989b. Effect of tied ridges and undersown *Stylosanthes hamata* on soil properties and growth of maize in the Sudan savannah of Burkina Faso. *Agriculture, Ecosystems and Environment* 25 (1): 39–51.

IASA. 1990. *Planting the future: a resource guide to sustainable agriculture in the Third World.* Minneapolis: IASA.

IBSRAM. 1987. *Management of vertisols under semi-arid conditions: proceedings of the 1st Regional Seminar on Management of Vertisols under Semi-arid Conditions, 1–6 December 1986, Nairobi, Kenya.* Bangkok: IBSRAM.

ICDA News. 1985. Traditional varieties: the testimony of a farmer. In: *Special report on seeds,* July 1985, p. 5.

ICRISAT. 1986. *Annual report.* Patancheru: ICRISAT.

IIED. 1988. *Glossary of selected terms in sustainable agriculture* (compiled by J.A. McCracken and J.N. Pretty). Gatekeeper Series 6. London: IIED.

IITA. 1985. *Research highlights 1984.* Ibadan: IITA.

ILCA. 1988. *Annual report 1987.* Addis Ababa: ILCA.

ILEIA. 1988. *Proceedings of the ILEIA Workshop on Operational Approaches for Participatory Technology Development in Sustainable Agriculture.* Leusden: ILEIA.

ILEIA. 1989. Farmers' hands on alternatives to chemical pesticides. *ILEIA Newsletter* 4 (3): 1–36.

IRRI. 1988. *Green manure in rice farming: proceedings of a symposium on Sustainable Agriculture: The Role of Green Manure Crops in Rice Farming Systems, 25–29 May 1987, Los Baños, Philippines.* Los Baños: IRRI.

Ison, R.L. 1990. *Teaching threatens sustainable agriculture.* Gatekeeper Series 21. London: IIED.

ITDG/GRET 1991. *Tools for agriculture.* 4th ed. London/Paris, ITDG/GRET.

Janssens, M.J.J., Neumann, I.F. and Froidevaux, L. 1990. Low-input ideotypes. In: Gliessman, S.R. (ed.), *Agroecology: researching the ecological basis for sustainable agriculture* (New York: Springer), pp. 130–45.

Jiggins, J. 1990. *Crop variety mixtures in marginal environments.* Gatekeeper Series 19. London: IIED.

Jodha, N.S. 1990. Mountain agriculture: the search for sustainability. *Journal of Farming Systems Research-Extension* 1 (1): 55–76.

Johnson, A.W. 1972. Individuality and experimentation in traditional agriculture. *Human Ecology* 1 (2): 149–59.

Juma, C. 1987. *Ecological complexity and agricultural innovation: the use of indigenous genetic resources in Bungoma, Kenya.* Paper presented at IDS Workshop on Farmers and Agricultural Research: Complementary Methods, 26-31 July 1987, University of Sussex, Brighton, UK.

Juma, C. 1989. *The gene hunters: biotechnology and the scramble for seeds.* London: Zed Books/Princeton University Press.

Juo, A.S. and Lal, R. 1977. The effect of fallow and continuous cultivation on the chemical and physical properties of an alfisol in Western Nigeria. *Plant and Soil* 63: 567–84.

Jutzi, S., Anderson, F.M. and Abiye Astatke. 1987. Low-cost modifications of the traditional Ethiopian tine plough for land shaping and surface drainage of heavy clay soils: preliminary results from on-farm verification trials. *ILCA Bulletin* 27: 28–31.

Kenmore, P.E., Litsinger, J.A., Bandong, J.P., Santiago, A.C. and Salac, M.M. 1987. Philippine rice farmers and insecticides: thirty years of growing dependency and new options for change. In: Tait, J. & Napompeth, B. (eds), *Management of pests and pesticides* (Boulder: Westview), pp. 98–109.

Kerkhof, P. 1990. *Agroforestry in Africa: a survey of project experience.* London: Panos.

Kessler, J.J. and Breman, H. 1991. Agroforestry in the Sahel and the savannah zone of West Africa. *Agroforestry Systems* 13: 42–62.

van Keulen, H. and Breman, H. 1990. Agricultural development in the West African Sahelian region: a cure against land hunger? *Agriculture, Ecosystems and Environment* 32: 177–97.

King, J.M. 1983. *Livestock water needs in pastoral Africa in relation to climate and forage.* ILCA Research Report 7. Addis Ababa: ILCA.

Kirschenmann, F. 1988. *Switching to a sustainable system: strategies for converting from conventional/chemical to sustainable/organic farming systems.* Windsor: NPSAP.

Klee, G.A. (ed.) 1980. *World systems of traditional resource management.* London: Arnold.

Kotschi, J. 1990. Agroforestry for soil fertility maintenance in the semi-arid areas of Zimbabwe. In: Kotschi, J. (ed.) *Ecofarming practices for tropical smallholdings* (Weikersheim: Margraf/GTZ), pp. 7–28.

Kotschi, J., Waters-Bayer, A., Adelhelm, R. and Hoesle, U. 1989. *Ecofarming in agricultural development.* Weikersheim: Margraf/GTZ.

Kumar, R. 1984. *Insect pest control with special reference to African agriculture.* London: Arnold.

Kwatpata, M.B. 1984. Shifting cultivation: problems and solutions in Malawi. In: *The future of shifting cultivation in Africa and the task of universities* (Rome: FAO), pp. 77–85.

Lacey, C.J., Walker, J. and Noble, I.R. 1982. Fire in Australian tropical savannas. In: Huntley, B.J. and Walker, B.H. (eds), *Ecology of tropical savannas* (Berlin: Springer), pp. 246–72.

Lagemann, J. 1977. *Traditional African farming systems in Eastern Nigeria.* Munich: Weltforum.

Lal, R. 1975. *Role of mulching techniques in tropical soil and water management.* Ibadan: IITA.

Lal, R. 1987. *Tropical ecology and physical edaphology.* Chichester: Wiley and Sons.

Lal, R. and Stewart, B.A. 1990. *Soil degradation.* New York: Springer.

Lawton, H.W. and Wilke, P.J. 1979. Ancient agricultural systems in dry regions. In: Hall, A.E., Cannell, G.H. and Lawton, H.W. (eds), *Agriculture in semi-arid environments*, pp. 1–44 (New York: Springer).

de Leener, P.P.M. and Perier, J.P. 1989. *Evaluation of the trial actions concerning the participation of the population in afforestation in rural areas.* Final Report, Vol 1: Synthesis. Dakar: ENDA.

Leggett, J. (ed.) 1990. *Global warming: the Greenpeace report.* Oxford University Press.

Lightfoot, C. 1987. Indigenous research and on-farm trials. *Agricultural Administration and Extension* 24: 79–89.

Lightfoot, C. 1990. Integration of aquaculture and agriculture: route to sustainable farming systems. *Naga (The ICLARM Quarterly)* January 1990: 9–12.

Lightfoot, C. and Minnick, D.R. 1990. *Farmer-first qualitative methods: farmers' diagrams for improving methods of experimental design in integrated farming systems.* AFSRE Symposium on the Role of Farmers in FSRE and Sustainable Agriculture, 14–17 October 1990, Michigan State University, East Lansing, USA.

Lightfoot, C. and Ocado, F. 1988. A Philippine case on participative technology development. *ILEIA Newsletter* 4 (3): 18–19.

Lightfoot, C., de Guia Jr., O. and Ocado, F. 1988. A participatory method for systems-problem research rehabilitating marginal uplands in the Philippines. *Experimental Agriculture* 24: 301 – 9.

Little, D. and Muir, J. 1987. *A guide to integrated warm water aquaculture.* Stirling, Scotland: Institute of Aquaculture, University of Stirling.

Loevinsohn, M. 1989. *A participatory approach to valley-bottom rice development.* Letter of 13 January 1989 to Dr J. Farrington, ODI, London.

Long, N. 1990. *From paradigm lost to paradigm regained: the case of actor-oriented sociology of development.* Wageningen Agricultural University.

Lundqvist, J. (In press). Right food, right way and right people: aspects of resource management in semi-arid Third World regions. In: *Combating famine in a changing world: famine between subsistence and market economy* (University of Bayreuth).

MacKay, K.T. 1990. *Farming Systems Research, sustainability and entry points.* Paper presented at the Asian Farming Systems Research and Extension Symposium, 19 – 22 November 1990, AIT, Bangkok, Thailand.

Marten, G.G. (ed.) 1986. *Traditional agriculture in Southeast Asia: a human ecology perspective.* Boulder: Westview.

Maseko, P., Scoones, I. and Wilson, K. 1988. Farmer-based research and extension. *ILEIA Newsletter* 4 (4): 18 – 19.

Matthewman, R.W. 1980. Small ruminant production in the humid tropical zone of southern Nigeria. *Tropical Animal Health and Production* 12: 234 – 42.

Mathias-Mundy, E. and McCorkle, C.M. 1989. *Ethnoveterinary medicine: an annotated bibliography.* Ames: Technology and Social Change Program, Iowa State University.

Matzigkeit, U. 1990. *Natural veterinary medicine: ectoparasites in the tropics and subtropics.* Weikersheim: Margraf/AGRECOL.

Mazo, C.I. 1986. *Informe anual programa de microproyectos.* Quibdo: Proyecto DIAR.

McCall, M. 1987. *Indigenous knowledge systems as the basis for participation: East African potentials.* Working Paper 36. Enschede: Technology and Development Group, University of Twente.

McCorkle, C.M., Brandstetter, R.H. and McClure, G.D. 1988. *A case study on farmer innovations and communication in Niger.* Washington DC: Academy for Educational Development.

McCracken, J., Pretty, J.N. and Conway, G.R. 1988. *An introduction to Rapid Rural Appraisal fo. ricultural development.* London: IIED.

McGuahey, M. 1986. *Impact of forestry initiatives in the Sahel.* Washington DC: Chemonics International.

Mearns, R. 1988. Direct matrix ranking in highland Papua New Guinea. *RRA Notes* 3: 11 – 14.

Medina, J.R. 1988. *MASIPAG: farmer-scientist partnership in rice project.* Paper prepared for ILEIA Workshop on Operational Approaches for Participative Technology Development in Sustainable Agriculture, 11 – 12 April 1988, Leusden, Netherlands.

Meerman, J., Oudejans, J. and Takken, W. 1989. *Gewasbescherming, vectorbestrijding en bestrijdingsmiddelen gebruik in de derde wereld.* Wageningen, Dept of Entomology, Agricultural University.

Michon, G., Bompard, J., Hecketsweiler, P. and Ducatillon, C. 1983. Tropical forest architectural analysis as applied to agroforestry in the humid tropics: the example of traditional village agroforests in West Java. *Agroforestry Systems* 1 (2): 117-30.

Millington, A. 1987. Soil conservation and shifting cultivators. *ILEIA Newsletter* 3 (1): 15 – 16 and 3 (2): 18 – 19.

Milton, F.B. 1989. Velvetbeans: an alternative to improve small farmers' agriculture. *ILEIA Newsletter* 5 (2): 8 – 9.

Mishra, M.M. and Bangar, K.C. 1986. Rock phosphate composting: transformation of phosphate forms and mechanisms of solubilization. *Biological Agriculture and Horticulture* 3: 331 – 40.

Mkandawire, A.B. 1988. Productivity of Malawian landrace dry beans under intercropping and drought conditions. *Michigan State University Pulse Beat* (Winter I), p. 5.

Mooney, P. 1983. The law of the seed: another development and plant genetic resources. *Development Dialogue* 1 – 2: 1 – 173.

Mountjoy, D.C. and Gliessman S.R. 1988. Traditional management of a hillside agro-ecosystem in Tlaxcala, Mexico: an ecologically based maintenance system. *American Journal of Alternative Agriculture* 3 (1): 3 – 10.

Nair, P.K.R. 1984. *Soil productivity aspects of agroforestry.* Nairobi: ICRAF.

Nair, P.K.R., Fernandes, E.C.M. and Wambugu, P.N. 1984. Multipurpose leguminous trees and shrubs for agroforestry. *Agroforestry Systems* 2 (3): 145 – 65.

NCSU. 1980. *Agronomic – economic research on soils in the tropics: annual report 1978/79.* Raleigh: Soil Science Dept, North Carolina State University.

Netting, R.M., Stone, P. and Stone, G. 1990. Development by farmers: seizing the opportunities. *ILEIA Newsletter* 6 (3): 12 – 14.

Niamir, M. 1990. *Herders' decision-making in natural resources management in arid and semi-arid Africa*. Community Forestry Note 4. Rome: FAO.

Norman, D., Baker, D., Heinrich, G. and Worman, F. 1989. Technology development and farmer groups: experiences from Botswana. *Experimental Agriculture* 24: 321 – 31.

O'Sullivan-Ryan, J. and Kaplun, M. 1979. *Communication methods to promote grassroots participation for an endogenous development process*. Paris: UNESCO.

Obi, J.K. and Tuley, P. 1973. *The bush fallow and ley farming in the oilpalm belt of South-Eastern Nigeria*. Surbiton: Land Resources Division, Overseas Development Administration.

Okubo, T., Hirakawa, M., Okajima, T. and Kayama, R. 1983. Comparison of radiant energy distribution and conversion efficiency in grazed pastures between temperate grass sward and subtropical grass sward. *Journal of Japanese Grassland Science* 29: 73 – 4.

OTA. 1988. *Enhancing agriculture in Africa: a role for US development assistance*. Washington DC: US Government Printing Office.

Otchere, E.O. 1986. Traditional cattle production in the subhumid zone of Nigeria. In: von Kaufmann, R., Chater, S. and Blench, R. (eds), *Livestock Systems Research in Nigeria's subhumid zone* (Addis Ababa: ILCA), pp. 110 – 40.

Otsyina, R.M., von Kaufmann, R.R., Mohamed Saleem, M.A. and Suleiman, H. 1987. *Manual on fodder bank establishment and management*. Kaduna: ILCA Subhumid Zone Programme.

Owusu, D.Y. 1990. Experiences with agroforestry. *ILEIA Newsletter* 6 (2): 9 – 10.

Pacey, A. and Cullis, A. 1986. *Rainwater harvesting: the collection of rainfall and runoff in rural areas*. London: ITP.

Palte, J.G.L. 1989. *Upland farming on Java, Indonesia: a socioeconomic study of upland agriculture and subsistence under population pressure*. Utrecht: Geographical Institute, University of Utrecht.

Panel of Experts in Integrated Pest Control. 1967. Rome. Quoted in: van der Weel, J.J. and van Huis, A. (1989), *Integrated pest management in the tropics: needs and constraints of information and documentation, a feasibility study*. Leidschendam: Ministry of Housing, Physical Planning and Environment.

Papendick, R.I., Sanchez, P.A. and Triplett, G.B. (eds) 1976. *Multiple cropping*. Madison: American Society of Agronomy.

Pedro, R.C.de and Mercado, C.B. 1989. Soil erosion control and maximum utilization of limited area in a contour farm. *Farm and Resource Management Institute Information Service Newsletter* 3 (1): 9 – 10.

Peters, K.J. 1987. Unconventional livestock: classification and potential uses. *ILCA Bulletin* 27: 36 – 42.

Pieri, C. 1985. Food crop fertilization and soil fertility: the IRAT experience. In: Ohm, H. and Nagy, J. (eds), *Appropriate technologies for farmers in semi-arid West Africa* (West Lafayette: Purdue University), pp. 74 – 110.

van der Ploeg, J.D. 1990. Farmers' knowledge as line of defence. *ILEIA Newsletter* 6 (3): 27.

Plucknett, D.L. 1987. *Gene banks and the world's food*. Princeton University Press.

Porter, E. 1978. *Water management in England and Wales*. 2nd ed. Cambridge University Press.

Poats, S.V., Schmink, M. and Spring, A. (eds) 1988. *Gender issues in Farming Systems Research and Extension*. Boulder: Westview.

Prasad, R. and De Datta, S.K. 1979. Increasing fertiliser nitrogen efficiency in wetland rice. In: *Nitrogen and rice* (Los Baños: IRRI).

Primavesi, A. 1990. Soil life and chemical fertilizers. *ILEIA Newsletter* 6 (3): 8 – 9.

Prinz, D. 1986. Increasing the productivity of smallholder farming systems by introduction of planted fallows. *Plant Research and Development* 24: 31 – 56.

Prior, C. 1989. *Biological pesticides for low-external-input agriculture*. Wallingford: CAB International.

Project Consult. 1986. *Rural production and use of plant preparations for crop and post-harvest protection*. Eschborn: GATE.

Radwanski, S.A. and Wickens, G.E. 1981. Vegetative fallows and potential value of the neem tree *(Azadirachta indica)* in the tropics. *Economic Botany* 35: 398 – 414.

Rangarajan, M. 1988. The living soil. In: Essers, S. (ed.) *Proceedings of the Seminar on Ecological Agriculture for Researchers* (Leusden: ETC Foundation).

Rao, P. 1991. Protecting fruit trees from stray animals. *ILEIA Newsletter* 7 (1/2): 67.

Reij, C. 1987. *Soil and water conservation in sub-Saharan Africa: the need for a bottom-up approach*. Paper prepared for IDS Workshop on Farmers and Agricultural Research: Complementary Methods, 26 – 31 July 1987, University of Sussex, Brighton, UK.

Reij, C. 1990. *Indigenous soil and water conservation in Africa: an assessment of current knowledge*. Workshop on Conservation in Africa: Indigenous Knowledge and Conservation Strategies, 2 – 7 December 1990, Harare, Zimbabwe. Amsterdam: CDCS.

Reij, C., Turner, S. and Kuhlman, T. 1986. *Soil and water conservation in sub-Saharan Africa: issues and options*. Amsterdam: CDCS/IFAD.

Reijntjes, C. 1986a. Old techniques for new concepts. *ILEIA Newsletter* 5: 12 – 13.

Reijntjes, C. 1986b. Water harvesting: a review of different techniques. *ILEIA Newsletter* 5: 7 – 8.

Reynolds, L. 1989. Legume trees for integrated crop/livestock farming in southern Nigeria. *ILEIA Newsletter* 5 (2): 14 – 15.

Rhoades, R.E. 1984. *Breaking new ground: agricultural anthropology.* Lima: CIP.

Rhoades, R.E. 1988. Thinking like a mountain. *ILEIA Newsletter* 4 (1): 3 – 5.

Rhoades, R.E. and Bebbington, A. 1988. *Farmers who experiment: an untapped resource for agricultural research and development?* Paper presented at the International Congress on Plant Physiology, 15 – 20 February 1988, New Delhi, India.

Richards, B.N. 1974. *Introduction to the soil ecosystem.* London: Longman.

Richards, P. 1979. Community environmental knowledge in African rural development. *IDS Bulletin* 10 (2): 28 – 36.

Richards, P. 1985. *Indigenous agricultural revolution: ecology and food production in West Africa.* London: Hutchinson.

Richards, P. 1986. *Coping with hunger: hazard and experiment in an African rice farming system.* Boston: Allen and Unwin.

Richards, P. 1988. *Experimenting farmers and agricultural research.* Paper prepared for ILEIA Workshop on Operational Approaches for Participative Technology Development in Sustainable Agriculture, 11 – 12 April 1988, Leusden, Netherlands.

Rocheleau, D.E. 1987. The user perspective and the agroforestry research and action agenda. In: Gholz, H.L. (ed.), *Agroforestry: realities, possibilities and potentials* (Dordrecht: Nijhoff/Junk), pp. 59 – 87.

Rocheleau, D.E. 1988. Gender, resource management and the rural landscape: implications for agroforestry and Farming Systems Research. In: Poats, S.V., Schmink, M. and Spring, A. (eds), *Gender issues in Farming Systems Research and Extension* (Boulder: Westview), pp. 149 – 70.

Roder, W. 1990. Traditional use of nutrient inputs. *ILEIA Newsletter* 6 (3): 3 – 4.

Rojas, M. 1989. *Women and community forestry: a field guide for project design and implementation.* Rome: FAO/SIDA.

Röling, N. 1988. *Extension science: information systems in agricultural development.* Cambridge University Press.

Röling, N. 1990. The agricultural research-technology transfer interface: a knowledge systems perspective. In: Kaimowitz, D. (ed.), *Making the link: agricultural research and technology transfer in developing countries* (Boulder: Westview/ISNAR), pp. 1 – 42.

Ruddle, K. and Zhong, G. 1988. *Integrated agriculture-aquaculture in South China.* Cambridge University Press.

Rupper, G. 1987. Sunnhemp: experiences in Tanzania. *ILEIA Newsletter* 3 (1): 11 – 12.

Russo, R.O. and Budowski G. 1986. Effect of pollarding frequency on biomass of *Erythrina poeppigiana* as a coffee shade tree. *Agroforestry Systems* 4: 145 – 62.

Ruthenberg, H. 1980. *Farming systems in the tropics.* Oxford: Clarendon.

Sachs, I. 1987. Towards a second green revolution. In: Glaeser, B. (ed.) *The green revolution revisited* (London: Allen & Unwin), pp. 193 – 8.

Sanwal, M. 1989. What we know about mountain development: common property, investment priorities and institutional arrangements. *Mountain Research and Development* 9 (1): 3 – 14.

Sashi, S. and D'Silva, B. 1989. Catch a moth and you'll kill a caterpillar. *ILEIA Newsletter* 5 (3): 29 – 30.

Sautier, D. and Amaral, H. 1989. Integrated pest management or integrated system management? *ILEIA Newsletter* 5 (3): 6 – 8.

Savory, A. 1988. *Holistic resource management.* Washington DC: Island Press.

Scheuermeier, U. 1988. *Approach Development: a contribution to participatory development of techniques based on practical experience in Tinau Watershed Project, Nepal.* Lindau: LBL.

de Schlippe, P. 1956. *Shifting cultivation in Africa: the Zande system of agriculture.* London: Routledge and Kegan Paul.

van Schoubroeck, F.H.J., Herens, M., de Louw, W., Louwen, J.M. and Overtoom T. 1990. *Managing pests and pesticides in small-scale agriculture.* Wageningen: CON.

Schrimpf, B. and Dziekan, I. 1989. Working with farmers on natural crop protection. *ILEIA Newsletter* 5 (3): 23 – 4.

Scoones, I. 1989. Preference and direct matrix rankings, Sudan. *RRA Notes* 7: 28 – 30.

Shaner, W.W., Philipp, P.F. and Schmehl, W.R. 1982. *Farming Systems Research and Development: guidelines for developing countries.* Boulder: Westview.

Sharland, R. 1990. A trap, a fish poison and culturally significant pest control. *ILEIA Newsletter* 6 (1): 12 – 13.

Sharma, R. 1985. Nutrient drain. In: Agarwal, A. and Narain, S. (eds), *The state of India's environment 1984 – 85: the 2nd Citizens' Report* (New Delhi: CSE), p. 20.

Shiva, V. 1988. *Staying alive: women, ecology and development.* London: Zed Books.

Simaraks, S., Khammaeng, T. and Uriyapongson, S. 1986. *Farmer-to-farmer workshops on small farmer dairy cow raising in three villages, northeast Thailand.* 3rd Thailand National Farming Systems Seminar, 2 – 4 April 1986, Chiang Mai University, Chiang Mai, Thailand.

Singh Hara, J. 1989. Punjab's problems of plenty. *The Hindu Survey of Indian Agriculture* (Madras), p. 15.

Smaling, E.M.A. 1990. Two scenarios for the sub-Sahara: one leads to disaster. *Ceres* 126: 20 – 4.

Soegito and Siemonsma, J.S. 1985. On-farm programme Wonorejo. Soybeans improvement Paper 1. *Quarterly Technical Progress Reports* 13. Malang: MARIF.

Soegito, Siemonsma, J.S., Sutrisno and Kuntyastuti, H. 1986. Soybean on-farm yield trials in Pasuruan. *Penelitian Palawija* 1 (1): 16 – 25.

Srivastava, P.B.L. 1986. *Shifting cultivation: problems and alternatives.* Field Document 10. Bangkok: FAO.

van Steekelenburg, P.N.G. 1988. *Irrigated agriculture in Africa: Proceedings of CTA Seminar, 25 – 29 April 1988, Harare, Zimbabwe.* Wageningen: CTA/ILRI.

Steinberg, D. 1988. Tree planting for soil conservation: the need for a holistic and flexible approach. *ILEIA Newsletter* 4 (4): 20 – 1.

Steiner, K.G. 1984. *Intercropping in tropical smallholder agriculture with special reference to West Africa.* Eschborn: GTZ.

Stigter, C.J. 1984. Examples of mulch use in microclimate management by traditional farmers in Tanzania. *Agriculture, Ecosystems and Environment* 11: 173 – 6.

Stigter, C.J. 1987a. Mulching with organic materials: knowledge is power. *ILEIA Newsletter* 3 (3): 10 – 11.

Stigter, C.J. 1987b. Traditional manipulation of microclimate factors: knowledge to be used. *ILEIA Newsletter* 3 (3): 5 – 6.

Stinner, B.R. and House, G.J. 1987. Role of ecology in lower-input, sustainable agriculture: an introduction. *American Journal of Alternative Agriculture* 2 (4): 146 – 7.

Stocking, M. 1986. *The cost of soil erosion in Zimbabwe in terms of the loss of three major nutrients.* Consultants' Working Paper 3, Soil Conservation Programme, Land and Water Development Division. Rome: FAO/Overseas Development Group.

Stoll, G. 1986. *Natural crop protection based on local farm resources in the tropics and subtropics.* Langen: Margraf.

Stoop, W.A. and Bingen, J.R. 1989. *Towards sustainable institutes for small NARS: organizational and technical issues in the integration of on-station and on-farm research – the Rwanda case.* The Hague: ISNAR.

Stotz, D. 1983. *Production techniques and economics of smallholder livestock production systems in Kenya.* Farm Management Handbook of Kenya, Vol. IV. Nairobi: Ministry of Livestock Development.

Stout, B.A. 1990. *Handbook of energy for world agriculture.* London: Elsevier.

Subba Rao, N.S. 1977. *Soil microorganisms and plant growth.* New Delhi: Oxford/IBH Publishing Co.

Subramanya, S. and Sastry, K.N.R. 1990. Indian peasants have long used vetiver grass. *ILEIA Newsletter* 6 (1): 26.

Swift, M.J. (ed.) 1984. Soil biological processes and tropical soil fertility: a proposal for a collaborative programme of research. *Biology International,* Special Issue 5. Paris: International Union of Biological Sciences.

TAC/CGIAR 1988. *Sustainable agricultural production: implications for international agricultural research.* Rome: FAO.

Tacio, H.D. 1988. SALT: Sloping Agricultural Land Technology. *ILEIA Newsletter* 4 (1): 8 – 9.

Tandon, H.L.S. 1990. Where rice devours the land. *Ceres* 126: 25 – 9.

Taylor, S.R. 1987. Diversified alley cropping. *ILEIA Newsletter* 3 (2): 13.

Tekelenburg, T. 1988. The multi-purpose use of the tuna plant: agricultural development in the Bolivian highlands. *ILEIA Newsletter* 4 (4): 22 – 3.

Thiam, A. 1987. Composting: soil improvement in a semi-arid Sahelian environment. *ILEIA Newsletter* 4 (4): 12 – 13.

Thrupp, L.A. 1987. *Building legitimacy of indigenous knowledge: empowerment for Third World people or 'scientized packages' to be sold by development agencies?* Paper presented at IDS Workshop on Farmers and Agricultural Research: Complementary Methods, 26 – 31 July 1987, University of Sussex, Brighton, UK.

Thurston, H.D. 1990. Plant disease management practices of traditional farmers. *Plant Diseases* 74 (2): 96 – 102.

Tilakaratna, S. 1981. *Participatory rural development: two case studies.* Colombo: Konrad Adenauer Foundation.

Ubels, J. (ed.) 1990. *Design for sustainable farmer-managed irrigation schemes in sub-Saharan Africa.* Wageningen: Dept of Irrigation and Soil and Water Conservation, Agricultural University.

UNESCAP. 1979. *Transfer of knowledge and skills among peer groups: a manual on methodology*. Bangkok: FAO.

Unger, P.W. 1987. Possibilities of zero-tillage for small-scale farmers in the tropics. *ILEIA Newsletter* 3 (3): 12 – 13.

Upawansa, G.K. 1989. Ancient methods for modern dilemmas. *ILEIA Newsletter* 5 (3): 9 – 11.

Vaughan, M. 1987. The story of an African famine. In: *Gender and famine in twentieth-century Malawi* (Cambridge University Press).

Vel, J., van Veldhuizen, L. and Petch, B. 1989. Beyond the PTD approach. *ILEIA Newsletter* 5 (1): 10 – 12.

Veneracion, C.C. 1987. *Developing farming systems in the rainfed areas: the Bicol experience*. Manila: Institute of Philippine Culture.

Warren, D.M. and Cashman, K. 1989. *Indigenous knowledge for sustainable agricultural and rural development*. Gatekeeper Series 10. London: IIED.

Warren, D.M., Slikkerveer, L.J. and Titilola, S.O. (eds) 1989. *Indigenous knowledge systems: implications for agriculture and international development*. Ames: Technology and Social Change Program, Iowa State University.

Waters-Bayer, A. 1988. Soybean daddawa: an innovation by Nigerian women. *ILEIA Newsletter* 4 (3): 8 – 9.

Waters-Bayer, A. 1990. Trials by scientists and farmers: opportunities for cooperation in ecofarming research. In: Kotschi, J. (ed.), *Ecofarming practices for tropical smallholdings* (Weikersheim: Margraf/GTZ), pp. 161 – 85.

Waters-Bayer, A. and Bayer, W. 1987. Building on traditional resource use by cattle-keepers in central Nigeria. *ILEIA Newsletter* 3 (4): 7 – 9.

WCED. 1987. *Our common future*. Oxford University Press.

Weiskel, T.C. 1989. The ecological lessons of the past: an anthropology of environmental decline. *The Ecologist* 19 (3): 98 – 103.

van der Werf, E. 1985. *Pest management in ecological agriculture*. Leusden: AME Programme, ETC Foundation.

van der Werf, E. 1989. *Ecological farming principles*. Leusden: AME Programme, ETC Foundation.

Werner, D. 1989. Fitting iguanas and forests into Central American farms. *ILEIA Newsletter* 5 (4): 16 – 17.

Werter, F. 1987. *The implementation of the ASE intervention model: manual for field staff*. Addis Ababa: PPE Services, Agri-Service Ethiopia.

West, O. 1965. *Fire in vegetation and its use in pasture management with special reference to tropical and sub-tropical Africa*. Hurley: Commonwealth Bureau of Pastures and Field Crops.

Whitington, D., MacDicken, K.G., Sastry, C.B. and Adams, N.R. 1988. *Multipurpose tree species for small-farm use: Proceedings of an international workshop, 2 – 5 November 1987, Pattaya, Thailand*. Bangkok: Winrock/IDRC.

Wilken, G.C. 1987. *Good farmers: traditional agricultural resource management in Mexico and Central America*. Berkeley: University of California Press.

Winter, W.H. 1985. Pasture and animal nutrition inter-relations. In: Adams, G.T. (ed.), *CSIRO Division of Tropical Crops and Pastures Annual Report 1984 – 85* (Brisbane: CSIRO), pp. 115 – 17.

Wolf, E.C. 1986. *Beyond the green revolution: new approaches for Third World agriculture*. Washington DC: Worldwatch Institute.

Wolf, E.C. 1987. Mimicking nature: the sustainability of many traditional farming practices lies in the ecological models they follow. *Ceres* 20: 20 – 4.

World Bank. 1984. *World development report*. Oxford University Press.

World Neighbors. 1988. Saving labor, saving soil with in-row tillage. *World Neighbors in Action* 19 (14E): 1 – 4.

Worman, F.D., Heinrich, G.M. and Masikara, S. 1988. *Strengthening the link among farmers: extension and research for agricultural development in Botswana*. Gaborone: Dept of Agricultural Research, Ministry of Agriculture.

Woudmansee, R.G. 1984. Comparative nutrient cycles of natural and agricultural ecosystems: a step towards principles. In: Lowrance, R., Stinner, B.R. and House, G.J. (eds), *Agricultural ecosystems: unifying concepts* (New York: John Wiley and Sons), pp. 145 – 56.

Young, A. 1990. *Agroforestry for soil conservation*. Wallingford: CAB International.

Index